Springer Geology

The book series Springer Geology comprises a broad portfolio of scientific books, aiming at researchers, students, and everyone interested in geology. The series includes peer-reviewed monographs, edited volumes, textbooks, and conference proceedings. It covers the entire research area of geology including, but not limited to, economic geology, mineral resources, historical geology, quantitative geology, structural geology, geomorphology, paleontology, and sedimentology.

More information about this series at http://www.springer.com/series/10172

Oleg Petrov
Editor

Isotope Geology of the Norilsk Deposits

Springer

Editor
Oleg Petrov
A. P. Karpinsky Russian Geological
Research Institute (VSEGEI)
St. Petersburg, Russia

ISSN 2197-9545 ISSN 2197-9553 (electronic)
Springer Geology
ISBN 978-3-030-05215-7 ISBN 978-3-030-05216-4 (eBook)
https://doi.org/10.1007/978-3-030-05216-4

Library of Congress Control Number: 2019935843

This Springer imprint is published by the registered company Springer Nature Switzerland AG.
The registered company address is: Gewerbestrasse 11, 6330 Cham, Switzerland

Contents

Helium and Argon Isotopes 1
Vladimir Khalenev, Edward Prasolov, Konstantin Gruzdov,
Dmitry Zavilejsky, Kirill Lokhov, Edward Prilepsky and Vera Badinova

Sulphur Isotopes 49
Edward Prasolov, Vladimir Khalenev, Boris Belyatsky, Edward Prilepsky
and Tatiana Nazarova

Copper and Nickel Isotopes 73
Sergey Sergeev, Igor Kapitonov, Robert Krymsky, Dmitriy Sergeev,
Elena Adamskaya and Nikolay Goltsin

Strontium and Neodymium Isotopes 89
Yevgeny Bogomolov, Boris Belyatsky, Robert Krymsky
and Yury Pushkarev

Lead Isotopes 133
Boris Belyatsky, Yury Pushkarev, Edward Prasolov, Igor Kapitonov,
Robert Krymsky and Sergey Sergeev

Lutetium and Hafnium Isotopes in Zircons 189
Igor Kapitonov, Kirill Lokhov, Dmitriy Sergeev, Elena Adamskaya,
Nikolay Goltsin and Sergey Sergeev

**Isotope Correlations in Rocks and Ores of Major Intrusions
in the Norilsk District** 207
Oleg Petrov, Edward Prasolov, Sergey Sergeev and Yury Pushkarev

Isotope Chronology of Geological Processes 215
O. Petrov, S. Sergeev, R. Krymsky, S. Presnyakov, N. Rodionov,
A. Larionov, E. Lepekhina and D. Sergeev

Conclusion 303

Introduction

The methodology of isotope and geochemical studies involves the most complicated analytical procedures to obtain high-quality results and application to the analytical data of the genetic criteria developed by isotope geochemistry. Our results and recommendations can be regarded as innovative; they are for the first time based on such a broad and complete (11 systems) set of studies of independent isotope systems with nano- and micro-amounts of matter.

None of the known isotope geochemical and geochronological methods is universal and self-sufficient in solving various geological problems. A successful application of such methods is achieved through an integrated approach. It is possible to ensure, technically and organizationally, the necessary efficiency of the studies using a wide range of methods including the finest ones only within a single internationally certified analytical centre. In this regard, the importance of using the facilities of the Centre of Isotopic Research (CIR) at VSEGEI for the creation, setting-up, adaptation, development, approval, certification and practical application of the innovative methods of isotopic studies of geological objects cannot be overestimated.

In samples of ores, rocks and minerals, isotopic composition of the components were studied including those in the fluids of the mineral formation environment in general and ore formation environment, in particular. On the basis of the analytical results, the genetic isotope criteria were formulated making it possible to diagnose the sources of the matter of rocks and ores, to reveal the formation conditions of mineral deposits. Especially important is the diagnosis of the contribution of matter of different geospheres—crust and mantle.

Many researchers now regard the crust–mantle interaction as a determining factor in the ore generation processes. An effective use of the obtained empirical isotope data allows not only to create new, more correct models of deposit formation, but also to find the isotopic search criteria.

Isotopic composition of helium ($^3He/^4He$ isotope ratio) is used to diagnose the contribution of the mantle helium. The above ratio in the upper mantle is about 600-fold versus that in helium forming in crustal rocks. This is the strongest and most unambiguous criterion of the mantle origin of fluids.

$^{40}Ar/^{36}Ar$ isotope ratio makes it possible to unambiguously and with a high degree of precision detect the atmospheric component of argon and other gases as well as the association of mineral-forming fluids with near-surface sedimentation and infiltration water. The study of noble gas isotopes offers extensive possibilities for determining the formation conditions of the sources of mineral matter.

We studied the isotopic composition of noble gases (helium and argon) in gas–liquid microinclusions (but not in the crystal lattice of minerals) to avoid distortions caused by radioactive processes in the "hard" part of the rocks. Isotopic composition of helium giving a possibility of a numerical determination of the contribution of the crustal and mantle components in fluids as well as the composition of argon was studied in rock samples from a number of intrusions in the Norilsk district.

Isotope helium–argon systematics is most reliable for determining the source of the fluid and a quantitative assessment of the contribution of different sources to the formation of ore-transporting fluids. It allows a quantitative evaluation of the main reservoirs of volatiles—the upper mantle, the continental crust, surface water. The systematics allows a reliable control of the evolution of fluid systems.

Isotopic composition of sulphur ($\delta^{34}S$ value) which is in the ore minerals—sulphides, in some cases—allows diagnosing their genesis. In particular, it is possible to assess the presence of mantle sulphur as well as sulphur of evaporites.

Results of rubidium–strontium (Rb–Sr) and samarium–neodymium (Sm–Nd) analysis point to the source of the matter of rock-forming and metasomatic minerals. To establish the possible sources of ore matter, isotopic composition of lead in sulphides is determined using local sampling (LA–ICP) as well as sulphur in the same samples.

A lot of data were obtained on the age of rocks and occurrence time of geological processes on the basis of rhenium–osmium (Re–Os) and in situ uranium–lead dating of accessory zircons and ore minerals from a broad range of rocks.

Multisystem isotopic studies of a series of intrusions were carried out at CIR VSEGEI in two stages. In 2003–2008 (I stage) and 2012–2014 (II stage), they were aimed at determining the sources of matter, time and generation pattern of ore bodies which are of great importance. A large collection of new samples (more than 250) taken from the exploratory wells in 22 intrusions in the Norilsk district was studied. New quantitative genetic indicators of 11 different isotopic systems were obtained which, in general, allow refining the generation models of unique Cu–Ni–PGE-deposits.

For example, the origin of silicate matter of the Norilsk intrusions was traditionally regarded as mantle. However, the first study of isotopic composition of fluid components of He and Ar carried out by S. S. Neruchev and E. M. Prasolov and of sulphur performed by L. N. Grinenko revealed an extensive crust–mantle interaction.

Under governmental contracts with the Federal Agency on Mineral Resources and contracts with the Norilsk Nickel, VSEGEI Centre of Isotopic Research made over 5000 isotopic analyses including measurements of helium and argon of fluid microinclusions in rocks and ores, as well as sulphur, copper and nickel from sulphides. The absolute (isotopic) age of sulphide ores and zircons from rocks in intrusions characterized by different ore potential was determined. Eleven isotopic systems were studied in different substances—in silicate matter, in ore (in sulphide), in paleofluids. Data of Russian and foreign researchers of the Norilsk region were also used to obtain an integrated picture.

Team of authors is grateful to the PJSC MMS Norilsk Nickel for financial support of the publication and fruitful cooperation in studying geology and metallogeny of the Norilsk ore district.

Helium and Argon Isotopes

Vladimir Khalenev, Edward Prasolov, Konstantin Gruzdov, Dmitry Zavilejsky, Kirill Lokhov, Edward Prilepsky and Vera Badinova

Abstract The results of helium and argon isotope composition measurements ($^3He/^4He$, $^{40}Ar/^{36}Ar$ ratios and others) are presented in the chapter. In paleofluids from the Norilsk intrusions, the crustal helium is dominant, and the fraction of mantle helium is in the range from 0.1 to 22%. The contribution of crustal helium (0.1–4%) in rich and medium intrusions is especially low. In this parameter, poor intrusions (4–22%) are significantly different, with much more mantle helium. In fluid inclusions of the studied targets, the share of air argon is high, 60–100%. It is especially high in rich intrusions, from 88 to 100%. Consequently, air-saturated waters from enclosing sedimentary rocks actively participated in the formation of rocks and ores of the intrusions in the region. Average (by reserves) intrusions differ significantly from the rich ones; they contain only 60–85% of atmospheric argon.

Geochemistry of isotopes of noble gases offers to the researchers unique possibilities of determining the origin of fluids. Helium isotopic composition ($^3He/^4He$ isotope ratio) is considered to be, virtually, the only reliable and strong criterion of the association of mineral-forming fluids with mantle. It was found that helium isotope ratio in the Earth's upper mantle (1.2×10^{-5}) is approximately 600 times higher than in helium forming in crustal rocks (about 2×10^{-8}) [1–3]. This makes it possible to reveal and calculate the share of mantle helium, if it exceeds 1% of total helium in the sample.

Isotopic composition of argon ($^{40}Ar/^{36}Ar$ ratio) enables to precisely calculate the share of argon of atmospheric origin in mineral-forming fluids. The only path for air argon penetration into the subsurface is migration with infiltration and sedimentation water. Therefore, isotope ratio indicates the extent of participation of near-surface water in mineral formation, or the degree of openness of geosystems for near-surface fluids.

V. Khalenev (✉) · E. Prasolov · K. Gruzdov · D. Zavilejsky · K. Lokhov · E. Prilepsky · V. Badinova
Russian Geological Research Institute (VSEGEI), St. Petersburg, Russia
e-mail: Vladimir_Khalenev@vseei.ru

O. Petrov (ed.), *Isotope Geology of the Norilsk Deposits*, Springer Geology,
https://doi.org/10.1007/978-3-030-05216-4_1

1 Methodology, Samples

According to some methodological considerations, gas-liquid microinclusions were used as a direct study object.

Extraction of gases from inclusions was performed by mechanical destruction of samples (weighing 2 g) in vacuum applying the methodologies developed in [4, 5]. The released noble gases were cleaned from active components using getters in an inflow system. Light gases (He + Ne) were separated from heavy gases (Ar + Kr + Xe) using activated carbon at −196 °C. Light and heavy gases were successively transferred to the analyzer cell of mass spectrometer Micromass NG5400 (England, Manchester), where under the static pumping mode $^{3}He/^{4}He$, $^{4}He/^{20}Ne$, $^{40}Ar/^{36}Ar$, $^{38}Ar/^{36}Ar$ ratios were determined as well as the number of ^{4}He and ^{40}Ar isotopes.

Helium and argon blank of the entire equipment set: ^{4}He of the inlet systems 4×10^{-9} cm^3, full 2.2×10^{-8} cm^3 or 1.1×10^{-8} cm^3/g; Ar^{40} of the inlet system 6×10^{-9} cm^3, full 1.0×10^{-8} cm^3 or 5×10^{-9} cm^3/g; isotope ratio close to that of air.

Random errors (1σ) based on the results of repeated isotope measurements in samples at 2–5 V signal constituted $^{3}He/^{4}He$ ratio of 4% with at value $\sim 6 \times 10^{-8}$ and (1.2)% at (2–8) $\times 10^{-7}$. Measurement errors of $^{4}He/^{20}Ne$ ratio are $\sim$10%; those of $^{40}Ar/^{36}Ar$, 0.05–0.18%; and $^{38}Ar/^{36}Ar$, 0.05–0.15%.

Measurement accuracy appeared to be quite sufficient for a correct data interpretation. Analytical data on helium and argon isotopes obtained at the I stage [6] using MI9303 and MS10 mass spectrometers had approximately the same metrological characteristics. They were slightly better when studying gases from the crystal lattice; gases were extracted by melting samples at 1700 °C in a high-vacuum jacketed resistance furnace.

Presumably, the attained metrological characteristics are sufficient for obtaining correct results on the spread of noble gas isotopes in rocks of the Norilsk-Taimyr district intrusions.

For the convenience of geological interpretation of data, isotope ratios were transformed to the values showing the share of mantle helium (**m**), atmospheric (**a**) and radiogenic (**r**) argon. Calculation of the first of these values was carried out proceeding from the values of $^{3}He/^{4}He$ ratio in the Earth's mantle and crust 1.2×10^{-5} and 2×10^{-8} from the following formula:

$$\mathbf{m}(\%) = He_{\text{mantle}}/\text{He} = \frac{\left(^{3}\text{He}/^{4}\text{He}\right)_{\text{sample}} - \left(^{3}\text{He}/^{4}\text{He}\right)_{\text{crust}}}{\left(^{3}\text{He}/^{4}\text{He}\right)_{\text{mantle}}} \times 100.$$

Share of atmospheric helium in the vast majority of samples was very small (the first thousandths), and the corrections made [2] practically did not change the measured value of $^{3}He/^{4}He$ ratio in samples.

Calculation of the shares of air and radiogenic argon was performed taking into account the atmospheric ratio of $^{40}Ar/^{36}Ar_a$ 296:

$$\mathbf{a}(\%) = Ar_a/Ar = \frac{\left(^{40}Ar/^{36}Ar\right)_a}{\left(^{40}Ar/^{36}Ar\right)_{sample}} \times 100.$$

$$\mathbf{r}(\%) = Ar_r/Ar = 100 - \mathbf{a}.$$

The first studies of helium and argon isotopes in 32 samples from the Talnakh and some other intrusions in the Norilsk district [6] showed that the inclusions from mafic rocks and ores were dominated by crustal helium and atmospheric argon. At the II stage (2004–2005), the distribution of isotopes of noble gases in gas-liquid inclusions from rocks and crystal lattice of minerals in Norilsk-1 intrusion was investigated (8 samples). Later, at the III stage (2006–2008), at CIR VSEGEI, over 70 samples of noble gas isotopes were studied in 16 intrusions of the Norilsk-Taimyr district with varying degrees of ore content (inclusions were dominated by crustal helium and air argon).

It also appeared that the isotopic composition of noble gases was different in intrusions depending on their ore content. This made it possible to propose a He–Ar isotopic criterion of ore presence in intrusions. At the last, IV stage (2012–2014), isotopic composition of helium and argon was studied to improve the reliability of

Table 1 Intrusives studied at the I, II and III stages (1995, 2005, 2008)

Intrusive	Borehole	Types of intrusives
Kharaelakh	KZ-963	Commercial ore-bearing (rich)
Talnakh	OUG-2, KZ-1710, KZ-1739	
Norilsk-1	MN-2 (MS-33)	
Chernogorsk	MP-2bis	Non-commercial ore-bearing (average)
Vologochan	OV-29	
South Pyasina	OV-25	
Zub-Marksheidersky	MP-27	
Zelenaya Griva	F-233	Weakly ore-bearing (poor)
Tulaj-Kiryak	2.18	
Bootankaga	8A	
Lower Fokino	NF	
Maslovsky	OM-31	Satellites of commercial ore-bearing intrusives
Lower Talnakh	TG-31	
Lower Norilsk	NP-37	
Mikchangda	MD-48	Potentially ore-bearing
Binyuda	S-1	

Table 2 Objects studied at the IV stage (2014)

Object	Borehole	Number of samples
Kharaelakh intrusive, Oktyabrskoe deposit		
Western flank	ZF-13, ZF-30, KZ-931, KZ-952, KZ1319	20, including 6 (ore)
Central part	PT-2, KZ-1089, KZ-1112, KZ-1535	17, including 9 (ore)
Southern flank	KZ-361, KZ-1084	5, including 5 (ore)
Total	11 boreholes	42, including 20 (ore)
Deposits		
Talnakh	KZ-774	1
Zub-Marksheidersky	MP-25, 13074	2
Chernogorsk	MP-2	1
Maslovsky	OM-32, OM-123, OM-10	12, rocks
Areas		
Vologochan	OV-28, OV-36	10, rocks
Koevo	PK-11	7, basalts
Tangnarylakh	13090, 13096, 13097	8, gabbro
Mokulaj	13005, 13016, 13020, 13033, 13049	5, basalts
Oganer		2
Listvyanka		2
Krasny Kamen		2

the proposed criteria. In particular, isotopic measurements were performed on individual intrusions in a group of wells, but not in a single well. Isotopic heterogeneity of intrusions was estimated, and the amount of the material was obtained (wells, samples) required for a correct evaluation of ore presence. In addition, isotopic studies of ores proper were significantly expanded—of massive (rich) and disseminated ones. The list of the studied objects is presented in Tables 1 and 2.

At all the four stages, noble gases from gas-liquid inclusions were investigated. This was due to the fact that the contents of microinclusions disseminated in minerals and inter-mineral space of rocks is the only reliable evidence of the origin of mineral-forming fluids. The study of the gaseous phase of inclusions is particularly important when investiga-ting noble gases due to the hazard of radiogenic isotope formation in the crystal lattice of minerals.

In this work, we investigate the occurrence of noble gas isotopes in gas-liquid inclusions. Isotopic composition of helium and argon was studied in relic fluids preserved in gas-liquid microinclusions in minerals and rock pores. Rock and ore samples were taken from different drill core depth intervals. Each intrusion was characterized, on the average, by 5–6 bulk samples. At the I, II and III stages, the information about helium and argon isotopes in more than 100 rock and mineral samples was obtained.

Fig. 1 Area of work

The investigated intrusions represented three main geological economic types: commercially mineralized containing solid (massive), disseminated and vein-disseminated ores (**rich**); mineralized (non-commercial) containing disseminated and vein-disseminated ores (**average**); weakly mineralized, not containing commercial mineralization (**poor**). **Satellites** of commercially mineralized intrusions are distinguished as an independent type. Also investigated were poorly studied intrusions—objects of a prognostic assessment. Names of intrusions, well numbers and belonging to the above types are presented in Tables 1 and 2, and their distribution within the Norilsk-Taimyr district is shown in Figs. 1 and 2.

One part of the samples was provided by the researchers of IGEM RAS and KNIIGiMS, another one was collected in the field by the geologists of VSEGEI and VNIIOkeangeologia. Published data on the Talnakh, Norilsk-1 and certain Taimyr intrusions were used [6–12].

The use of bulk rock samples significantly simplified the procedure of sample preparation for the analysis. No significant differences were revealed in isotopic composition of bulk samples and monomineral fractions (olivine, pyroxene). The identified isotopic variations along the section of the intrusions appeared to be relatively low, and the averaged isotopic characteristics were acceptable for determining the difference (or similarity) between the intrusions.

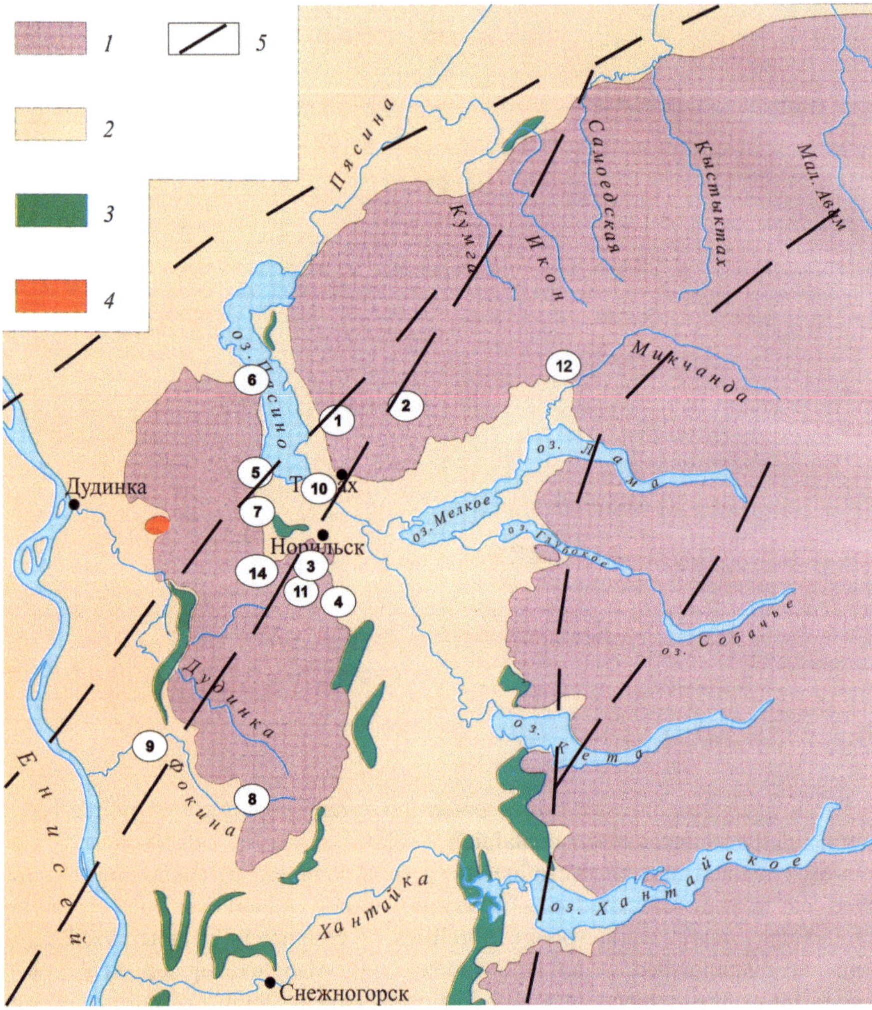

Fig. 2 Location of the studied intrusions in the Norilsk district. 1—effusive traps; 2—sedimentary deposits; 3—trappean intrusions; 4—granitoid intrusions; 5—regional faults. Numbers of intrusives (in circles): 1—Kharaelakh; 2—Talnakh; 3—Norilsk-1; 4—Chernogorsk; 5—Vologochan; 6—South Pyasina; 7—Zub-Marksheidersky; 8—Zelenaya Griva; 9—Lower Fokino; 10—Lower Talnakh; 11—Lower Norilsk; 12—Mikchangda

A comparison of measurement results showed the lack of significant differences between the rocks and ores. This is important not only for approval of the notions about the unity of the processes of formation/alteration of rocks and ores. It is, undoubtedly, convenient for the practical diagnosis of ore presence prospects, since it allows using as test samples not only ores (not yet discovered!), but also rocks from the section of the intrusion.

Table 3 Helium and neon in rocks of Norilsk-1 intrusion

Fluid inclusions (II stage)[a]

Sample number	Q, $\times 10^{-3}$ cm^3/g	He, $\times 10^{-6}$ cm^3/g	$^3He/^4He$ (meas.) $\times 10^{-6}$	$^4He/^{20}Ne$ (meas.)	$^3He/^4He$ (corr.) $\times 10^{-6}$	R_A (corr.)	He_m/He, %	$^4He/^{40}Ar_r$
441,1	3.1	0.091	0.51	5.7	0.46	0.33	3.6	0.27
441,2	4.9	0.15	0.56	0.91	0.11	0.078	0.8	5.6
441,7	3.9	0.13	0.35	4.7	0.27	0.19	2.2	–
442,6	3.1	0.20	0.38	5.1	0.31	0.22	2.4	1.1
444,1	4.0	0.11	0.50	4.7	0.43	0.31	3.4	–
448,2	2.3	0.051	0.51	2.8	0.40	0.29	3.1	3.1
449,1	3.5	0.20	0.78	0.088	<0.70	<0.50	<6	6.2
449,1п	2.8	0.37	0.50	0.18	<0.50	<0.40	<4	5.1

Crystalline lattice *(gases were released by melting in vacuum according to the procedure of VSEGEI)*

Sample number	Sample type	He, $\times 10^{-6}$ cm^3/g	$^3He/^4He$ (meas.) $\times 10^{-6}$	$^4He/^{20}Ne$ (meas.)	$^3He/^4He$ (corr.) $\times 10^{-6}$	R_A (corr.)	He_m/He, %	$^4He/^{40}Ar_r$
441,2	Px	12.8	0.030	1660	0.029	0.021	0.1	71.0
441,2	bulk	15.6	0.031	3980	0.030	0.021	0.1	8.2
441,7	Px	15.2	0.026	3000	0.026	0.019	0.1	25.0
441,7	bulk	14.3	0.016	2840	0.015	0.011	0	1.5
442,6	Px	4.9	0.027	–	0.027	0.019	0.1	27.0
442,6	bulk	8.8	0.008	3200	0.008	0.006	0	2.3
444,1	bulk	5.5	0.007	2080	0.007	0.005	0	2.9
449,1	heavy fract.	8.1	0.015	3180	0.015	0.011	0	8.9

[a]Upper platinum-bearing horizons. Gases were released by sample crushing in vacuum according to the procedure of VNIGRI-SPbSU. Mass spectral measurements were taken on Micromass NG 5400 at CIR VSEGEI; Q—amount of gases non-condensed in vacuum at −196 °C, i.e., except CO_2, HH, H_2S, H_2O vapours etc.; $^3He/^4He$(corr.) $\times 10^{-6}$—value of isotopic ratio corrected for air component of helium calculated from $^4He/^{20}Ne$ ratio, i.e. for isotopic composition of "deep" helium; R_A(corr.)—corrected ratio of helium isotopes divided by ratio in atmospheric helium (1.4×10^{-6}); He_m/He—share of mantle helium calculated for $^3He/^4He$ values in the upper mantle (1.2×10^{-5}) and the Earth's crust (2×10^{-8})

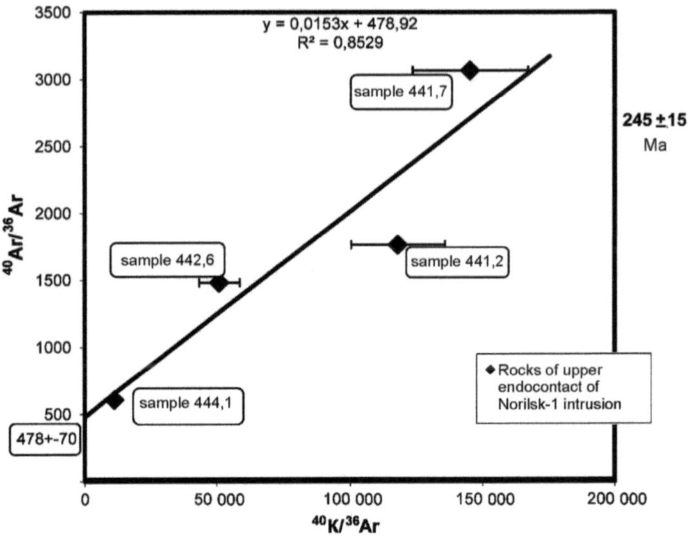

Fig. 3 Potassium-argon isochron on rocks of low-sulphide platinum-bearing -horizon of Norilsk-1 intrusion

As is known, gas-liquid inclusions can occur at different stages of existence of the investigated objects and can, accordingly, be subdivided into primary, secondary etc. The concept of "secondary" ones is relative. There may be situations when the processes of rock and ore formation are separated in time, and the coeval inclusions might appear to be secondary with respect to rocks and primary in the ores that formed later. Perhaps, it is more convenient to typify the inclusions on the basis of their formation temperature. The extensive papers by Aplonov [13, 14]) showed that, according to decreptometric analysis, rocks and ores of the Talnakh deposit are dominated by low-temperature gas-liquid inclusions (200–400 °C).

One of the important methodological problems is a possibility of contamination of the inclusions by helium migrating from the crystal lattice of minerals during rock existence. This difficult question cannot be, apparently, solved in the general case. Indeed, radiogenic helium forms in the lattice of minerals and can differ in isotopic composition from that captured into inclusions. Helium can leave the lattice of accessory minerals in which it appeared, the faster, the higher rock temperature throughout its existence time. Thermal history of rocks here can be decisive. Discharge channel can be, most likely, the system of open fractures and pores, where partial pressure of helium is low (at least, in rocks of the Norilsk intrusions). For assessing a possible impact of helium contamination in a specific situation in the Norilsk district intrusions, objective (rational) criteria are required:

1. Rocks and ores were not subjected to heating after the cooling of intrusions. No signs of metamorphism are recorded in the enclosing sedimentary rocks. Most researchers believe that the Norilsk intrusions formed at a depth of 1.5–2 km [14, p. 49] without further plunging. Estimates of pressure during formation of

Table 4 Argon in rocks of Norilsk-1 intrusion

Fluid inclusions (II stage)[a]

Sample number	Ar, $\times 10^{-6}$ cm^3/g	$^{40}Ar/^{36}Ar$ (meas.)	$^{38}Ar/^{36}Ar$ (meas.)	$^{40}Ar/^{36}Ar$ (corr.)	Ar_a/Ar, %	Ar_r/Ar, %	$^{40}Ar_r$, $\times 10^{-6}$ cm^3/g	$^{40}Ar_a$, vol. %
441,1	16.0	294.1	0.1856	302	97.9	2.1	0.330	0.51
441,2	1.7	304.5	0.1893	300	98.4	1.6	0.027	0.034
441,7	13.0	297.0	0.1894	294	100.0	0	–	0.34
442,6	12.0	295.0	0.1864	300	98.4	1.6	0.190	0.38
444,1	9.3	297.0	0.1894	293	100.0	0	–	0.22
448,2	2.7	298.2	0.1883	297	99.4	0.6	0.016	0.10
449,1	4.9	313.6	0.1930	297	99.3	0.7	0.032	0.11
449,1п	3.0	299.6	0.1870	303	97.6	2.4	0.072	0.11

Crystalline lattice (gases were released by melting in vacuum according to the procedure of VSEGEI)

Sample number	Sample type	Ar, $\times 10^{-6}$, cm^3/g	$^{40}Ar/^{36}Ar$, (meas.)	$^{38}Ar/^{36}Ar$, (meas.)	$^{40}Ar/^{36}Ar$, (corr.)	Ar_a/Ar, %	Ar_r/Ar, %	$^{40}Ar_r$, $\times 10^{-6}$ cm^3/g	$^{40}Ar_a$, vol. %
441,2	Px	0.7	–	–	392	75.4	24.6	0.18	0.52
441,2	bulk	2.3	1805	0.1902	1762	16.8	83.2	1.9	1.4
441,7	Px	2.1	–	–	413	71.6	28.4	0.61	1.5
441,7	bulk	10.8	3131	0.1900	3064	9.6	90.4	9.8	1.0
442,6	Px	0.6	–	–	421	70.3	29.7	0.18	0.42
442,6	bulk	4.9	–	–	1358	21.8	78.2	3.8	1.1
444,1	bulk	4.1	–	–	557	53.0	47.0	1.9	2.2
449,1	heavy fract.	1.4	–	–	793	37.3	62.7	0.91	0.49

[a]Upper platinum-bearing horizons. Gases were released by sample crushing in vacuum according to the procedure of VNIGRI-SPbSU. Mass spectral measurements were taken on Micromass NG 5400. $^{40}Ar/^{36}Ar$ (corr.)—value of isotopic ratio corrected for fractionation of argon isotopes detected by deviation of measured $^{40}Ar/^{36}Ar$ ratio from air ratio. Ar_a—argon of atmospheric origin, Ar_r—radiogenic argon

Table 5 Calculation of $^{40}K/^{36}Ar$ ratio in rocks

Sample number	K, $\times 10^{-6}$ cm^3/g	^{40}K, $\times 10^{-6}$ cm^3/g	Ar, $\times 10^{-6}$ cm^3/g	$^{36}Ar_r$, $\times 10^{-6}$ cm^3/g	$^{40}Ar/^{36}Ar$	$^{40}K/^{36}Ar$
442,6	2800	0.33	4.86	6.4×10^{-6}	1481.7	51,068
444,1	1300	0.15	4.09	1.3×10^{-5}	609.5	11,576
441,2	2340	0.27	2.28	2.3×10^{-6}	1762.7	118,100
441,7	7860	0.92	10.8	6.3×10^{-6}	3064.8	145,360

intrusions given in the paper by V. S. Aplonov (9–190 atm) [14, p. 50] point to even shallower depths. Consequently, the temperature of intrusion occurrence during 250 Ma (after cooling) was unlikely to exceed 50 °C; substantial helium losses from the crystal lattice of minerals are not likely.

It is interesting to note that in the crystal lattice, as expected, helium and argon isotopic composition differs due to neogeneration of radiogenic isotopes (Table 3). From these data, an isochrone was plotted for rocks of Norilsk-1 intrusion in coordinates $^{40}Ar/^{36}Ar$ and $^{40}K/^{36}Ar$ (Fig. 3, Table 4). The resulting age value (245 ± 15 Ma) virtually coincided with that determined using other methods. This points to a lack of argon losses throughout the time of rock existence and to soft thermal conditions in this interval (250 Ma).

Table 6 Th, U and He contents in Norilsk rocks

Sample	Th, ppm	U, ppm	0,28Th + 1,2U, ppm	He, $\cdot 10^{-6}$ cm^3/g
T-3	2.67	1.24	2.24	0.51
T-15	0.55	0.18	0.37	0.33
F-2	1.23	0.35	0.76	0.04
F-13	0.67	0.20	0.43	0.14
25-17	0.89	0.29	0.60	1.14
25-38	0.69	0.27	0.52	1.60
CH-6	0.69	0.23	0.47	0.07
CH-11	0.55	0.20	0.39	0.06
31-7	0.75	0.21	0.46	0.34
31-16	1.19	0.49	0.92	1.50
963-25	0.05	0.02	0.04	0.43
963-35	0.39	0.13	0.27	0.09
1514.5	8.09	2.26	4.98	0.76
1620.3–1622.2	0.50	0.19	0.37	0.41
29-22	0.65	0.29	0.53	0.73
29-24	0.42	0.25	0.42	2.10
27-5	1.01	0.39	0.75	0.18
27-10	0.30	0.12	0.23	0.13
Sl-5.5	0.25	0.07	0.15	0.04
Sl-58	0.37	0.09	0.22	0.03
48-9	0.47	0.18	0.35	0.35
48-32	1.10	0.34	0.72	1.55

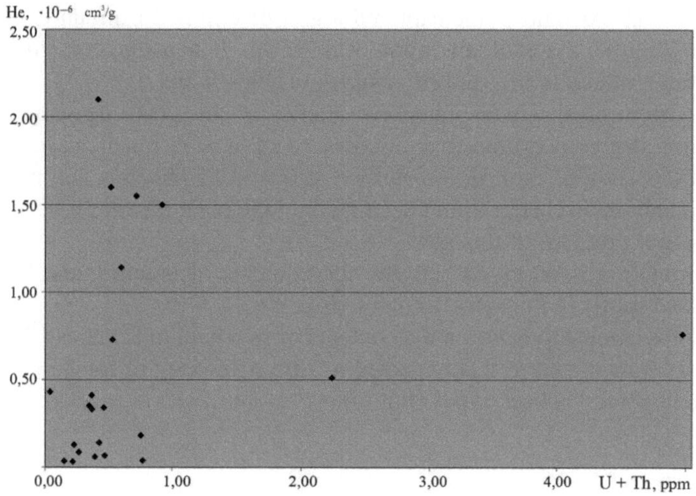

Fig. 4 Interrelation of Th, U and He contents in Norilsk rocks

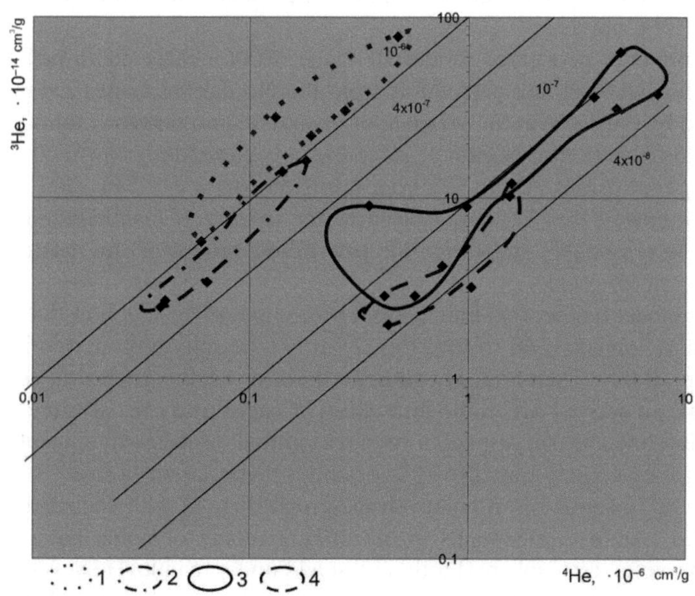

Fig. 5 Interrelation of the contents of ^{3}He and ^{4}He isotopes in inclusions of intrusives. Intrusives: 1—Tulaj-Kiryak; 2—Zelenaya Griva; 3—Mikchangda; 4—Lower Norilsk. Values of ^{3}He/^{4}He isotope ratios are given

Initial $^{40}Ar/^{36}Ar$ ratio in the captured argon determined from an isochrone was 480 ± 70 (the share of air argon was 62%). It actually coincides with the maximum values in the studied samples (Tables 5 and 6).
2. Rocks of intrusions were intensely altered at the post-magmatic stage. The alteration degree is estimated at 30–60% (data of S. F. Sluzhenikin). Secondary alterations should result in a disturbed structure of minerals and rocks—channels of helium discharge from the lattice into the open fracture-pore space with a low partial pressure of this gas.
3. No correlation is observed between the contents of uranium and helium from microinclusions in the same samples (Fig. 4).
4. Type of relationship between the contents of two helium isotopes (^3He and ^4He) does not correspond to that expected in case of mixing of the "microincluded" and radiogenic "lattice" types (Fig. 5).

2 Results and Discussion

Measurement results of isotopic composition of helium and argon ($^3He/^4He$, $^{40}Ar/^{36}Ar$ ratios etc.) in certain intrusions are presented in Tables 3, 4, 7, 8, 9, 10, 11, 12, 13, 14 and 15.

Measurements performed at the III stage (2006–2008) confirmed the earlier results. They showed that paleofluids from the Norilsk intrusions are ubiquitously dominated by crustal helium, the share of mantle helium ranges within 0.1 and 22% (Fig. 6). Contribution of helium of such origin is particularly low (0.1–4%) in rich and average intrusions (Table 13). Poor intrusions (**m** = 4–22%) differ significantly by this parameter; they contain a much larger amount of mantle helium.

It can be reasonably stated that the **prevailing source of the matter of paleofluids** are crustal rocks.

Measurement results also pointed to a high share of air argon in fluid inclusions in the studied objects—60–100% (Fig. 7). It is especially high in rich intrusions—from 88 to 100%. Therefore, **air-saturated water of the enclosing sedimentary rocks took an active part in the formation of rocks and ores of intrusions in the district**. Intermediate (in respect of reserves) intrusions differ significantly from the rich ones; they contain only 60–85% of atmospheric argon (Table 13).

In general, location of all points on the graph (Fig. 8) with coordinates $^3He/^4He$ (**m**)—$^{40}Ar/^{36}Ar$ (**a**) corresponds to the mixing model of noble gases from three global reservoirs: atmosphere, Earth's crust and upper mantle. However, direct fluid sources can be various mixtures of gases from the mentioned global reservoirs. Their number can exceed three. The contribution of atmospheric helium determined from $^4He/^{20}Ne$ ratio is extremely low in most samples.

There is a strikingly close correlation between helium and argon isotopic composition in three intrusions richest in ore (Fig. 9). As can be seen, the greatest

Table 7 Helium in fluid inclusions of rocks in intrusions of Norilsk District (I stage) (collection of S. S. Neruchev, 1994 [6])

Borehole number/Sampling depth, m	Object, rock	He, $\times 10^{-6}$ cm^3/g	$^3He/^4He$ (meas.) $\times 10^{-6}$	R_A	He_m/He, %	$^4He/^{40}Ar_r$	Gas Q 10^{-3}, cm^3/g	P_1/P_2
1799/1318	Tal	0.052	–	–	–	1.7	0.35	31
1799/1320	Tal	0.17	0.59	0.42	4.7	13	0.2	40
1739/1646	Tal, 3	0.19	0.34	0.24	2.7	1.2	0.81	1.3
1739/1660	Tal, 4	0.26	0.50	0.36	4.0	–	1.3	1.1
1739/1679	Tal, 1	0.59	0.14	0.10	1.0	20	0.79	3.9
1739/1683	Tal, 1	0.31	0.24	0.17	1.8	–	0.32	3.2
1739/1698	Tal, 7	0.95	$\leq$ 1.0#	$\leq$0.7	$\leq$8	7.1	0.37	3.0
1739/1703	Tal, 7	1.60	0.077	0.055	0.5	5.5	17	1.8
1739/1708	Tal, 1, 7	0.48	0.13	0.093	0.9	–	1.2	5.0
1739/1716	Tal, 1, 7	1.00	0.072	0.051	0.4	4.0	3.7	3.3
1739/1748	Tal, 2	0.40	0.23	0.16	1.8	10	0.3	1.3
1739/1771	Tal, 6	0.085	0.59	0.042	4.7	7.1	0.81	1.7
1739/1801,5	Tal, 6	0.32	0.36	0.26	2.8	–	13	1
1739/1817	Tal, 3	0.48	0.32	0.23	2.5	2.8	2.9	1.1
1739/1828,5	Tal, 9	0.13	0.77	0.55	6.2	–	0.37	3.8
1739/1839	Tal, 9	0.17	0.70	0.50	5.7	–	0.35	7.0
1739/1843,3	Tal, 9	0.16	0.97	0.69	7.9	–	0.28	9.8
1739/1848	Tal, 5	0.52	0.21	0.15	1.6	2.1	4.37	1.1
1739/1857	Tal, 8	0.60	0.21	0.15	1.6	2.5	0.45	1.2
1710/862	Tal, 1	0.27	0.25	0.18	1.9	1.1	1.5	1.4
1710/1027,5	Tal, 3	0.15	0.42	0.30	3.3	0.56	1.7	1.6
1710/1033	Tal, 4	0.14	0.76	0.43	6.2	1.7	1.1	1.1
1710/1050	Tal, 5	0.40	0.23	0.16	1.8	2.6	1.6	1.3

(continued)

Table 7 (continued)

Borehole number/Sampling depth, m	Object, rock	He, $\times 10^{-6}$ cm^3/g	$^3He/^4He$ (meas.) $\times 10^{-6}$	R_A	He_m/He, %	$^4He/^{40}Ar_r$	Gas Q 10^{-3}, cm^3/g	P_1/P_2
1710/1063,5	Tal, 9[a]	2.20	0.12	0.086	0.8	7.9	2.2	6.0
1710/1066,5	Tal, 9[b]	0.26	0.34	0.24	2.7	–	0.23	7.0
PE-2/975	Cher, 4	0.59	0.23	0.16	1.8	–	3.6	1.1
8A	Boo, 3	0.20	2.52	1.80	21	0.19	6.0	1.1
103-1	Boo, 4	0.15	2.65	1.90	22	–	1.9	1.2
2.16/19	Tul, 7	0.06	0.95	0.68	7.8	0.71	2.0	1.4
2.18/16	Tul, 3?	0.13	2.16	1.54	18	–	3.8	1.0
2.18/7	Tul, 4	0.47	1.67	1.19	14	0.42	12.6	1.0
2446-1	Rut, 5	0.19	1.17	0.83	9.6	–	4.3	1.0
2446-2	Rut, 5	0.27	1.13	0.81	9.2	–	12.3	1.0

Notes Tal—Talnakh, Cher—Chernogorsk, Boo—Bootankaga, Tul—Tulaj-Kiryak, Rut—Ruchej Travyanoj

Rocks: 1—gabbrodiorites; 2–6—gabbro-dolerites (2—olivine-free and olivine-bearing, 3—olivinic, 4—picritic, 5—taxitic, 6—picrite-troctolitic); 7—leucogabbro and upper taxitic gabbro with platinoids; 8—hornfels; 9—solid sulphide ores ([a]—chalcopyrite, [b]—pyrrhotite)

Gases were released by samples crushing in vacuum according to the procedure of VNIGRI–SPbSU. Q—measured amount of gases, non-condensed in vacuum at −196 °C, i.e., without CO_2, H_2S, HH, H_2O vapours. P_1/P_2—ratio of total amount of gases including that condensed in vacuum, to Q, is indicated by pressure change (P)

Value of $^3He/^4He$ isotope ratio for air component of helium was not corrected. $^4He/^{20}Ne$ ratio > 6 in all samples, except for Bh. 1739 (depth 1698 m), in which it is ≈1. For this sample, an assessment of the corrected value of $^3He/^4He$ (#) ratio is given. Isotopic analysis of helium was performed on magneto resonance mass spectrometer at the PTI of the RAS by L. V. Khabarin

R_A—ratio of helium isotopes divided by such ratio in atmospheric helium (1.4×10^{-6}), He_m/He—share of mantle helium calculated for $^3He/^4He$ values in the upper mantle (1.2×10^{-5}) and the Earth's crust (2×10^{-8})

Table 8 Argon in fluid inclusions in rocks of intrusions in the Norilsk District (I stage) (collection of S. S. Neruchev, 1994 [6])

Borehole number Sampling depth, m	Object, rock	Ar, $\times 10^{-6}$ cm^3/g	$^{40}Ar/^{36}Ar$ (meas.)	Ar_a/Ar, %	Ar_r/Ar, %	$^{40}Ar_r$, $\times 10^{-6}$ cm^3/g	Ar_a, vol. %
1799/1318	Tal	0.5	314	94.1	5.9	0.030	0.13
1799/1320	Tal	0.3	309	95.7	4.3	0.013	0.14
1739/1646	Tal, 3	1.3	336	88.0	12.0	0.16	0.14
1739/1660	Tal, 4	0.3	296	100.0	0.0	0	0.023
1739/1679	Tal, 1	1.6	301	98.2	1.8	0.029	0.20
1739/1683	Tal, 1	3.1	294	100.0	0.0	0	0.97
1739/1698	Tal, 7	4.3	305	96.9	3.1	0.13	1.2
1739/1703	Tal, 7	4.3	317	93.2	6.8	0.29	0.023
1739/1708	Tal, 1,7	4.2	295	100.0	0.0	0	0.35
1739/1716	Tal, 1,7	4.2	314	94.1	5.9	0.25	0.10
1739/1748	Tal, 2	1.6	303	97.6	2.4	0.038	0.51
1739/1771	Tal, 6	1.5	298	99.2	0.8	0.012	0.19
1739/1801,5	Tal, 6	4.1	294	100.0	0.0	0	0.031
1739/1817	Tal, 3	2.6	317	93.2	6.8	0.18	0.084
1739/1828,5	Tal, 9	1.3	292	100.0	0.0	0	0.35
1739/1839	Tal, 9	1.0	296	100.0	0.0	0	0.29
1739/1843,3	Tal, 9	0.9	296	100.0	0.0	0	0.32
1739/1848	Tal, 5	1.7	345	85.7	14.3	0.24	0.033
1739/1857	Tal, 8	1.0	389	76.0	24.0	0.24	0.17
1710/862	Tal, 1	4.7	319	92.7	7.3	0.34	0.30
1710/1027,5	Tal, 3	4.1	316	93.5	6.5	0.27	0.22
1710/1033	Tal, 4	2.6	305	96.9	3.1	0.081	0.22
1710/1050	Tal, 5	1.9	322	91.8	8.2	0.16	0.11
1710/1063,5	Tal, 9[a]	8.4	309	96.7	3.3	0.28	0.37

(continued)

Table 8 (continued)

Borehole number Sampling depth, m	Object, rock	Ar, $\times 10^{-6}$ cm^3/g	$^{40}Ar/^{36}Ar$ (meas.)	Ar_a/Ar, %	Ar_r/Ar, %	$^{40}Ar_r$, $\times 10^{-6}$ cm^3/g	Ar_a, vol. %
1710/1066,5	Tal, 9[b]	3.4	291	100.0	0.0	0	1.50
PE-2/975	Cher, 4	3.9	–	–	–	–	0.10
8A	Boo, 3	4.0	404	73.2	26.8	1.07	0.048
103-1	Boo, 4	2.4	–	–	–	–	0.012
2.16/19	Tul, 7	5.6	300	98.5	1.5	0.084	0.27
2.18/16	Tul, 3?	1.7	275	100.0	0.0	0	0.044
2.18/7	Tul, 4	3.3	450	65.7	34.3	1.13	0.017
2446-1	Rut, 5	2.8	–	–	–	–	0.064
2446-2	Rut, 5	3.4	–	–	–	–	0.027

Notes Tal—Talnakh, Cher—Chernogorsk, Boo—Bootankaga, Tul—Tulaj-Kiryak, Rut—Ruchej Travyanoj

Rocks: 1—gabbrodiorites, 2–6—gabbro-dolerites (2—olivine-free and olivine-bearing, 3—olivinic, 4—picritic, 5—taxitic, 6—picrite-troctolitic), 7—leucogabbro and upper taxitic gabbro with platinoids, 8—hornfels, 9—solid sulphide ores ([a]—chalcopyrite, [b]—pyrrhotite ore)

Gases were released by sample crushing in vacuum according to the procedure of VNIGRI-SPbSU. Q—measured amount of gases, non-condensed in vacuum at −196 °C, i.e. without CO_2, H_2S, HH, H_2O vapours. Content of gas component in vol. % in non-condensed gas. $^{40}Ar/^{36}Ar$ ratio was measured on mass spectrometer MS10. Ar_a—argon of atmospheric origin, Ar_r—radiogenic argon

Table 9 Isotopic composition of argon and helium in fluid inclusions of rocks of commercial ore-bearing intrusives (III stage)

Sample number	Ar, $\times 10^{-6}$ cm^3/g	$^{40}Ar/^{36}Ar$ (meas.)	$^{38}Ar/^{36}Ar$ (meas.)	$^{40}Ar/^{36}Ar$ (corr.)	Ar_a/Ar, %	Ar_r/Ar, %	$^{40}Ar_r$, $\times 10^{6}$ cm^3/g	He, $\times 10^{-6}$ cm^3/g	$^3He/^4He$ (meas.) $\times 10^{-6}$	$^4He/^{20}Ne$ (meas.)	$^3He/^4He$ (corr.) $\times 10^{-6}$	R_A (corr.)	He_m/He, %	$^4He/^{40}Ar_r$
Talnakh intrusive, Bh. OUG-2														
T-3	1.64	304.2	0.1854	313	94.6	5.4	0.1	0.51	0.0570	088	0.05	0.0390	0.3	5.7
T-6	0.67	342.7	0.1875	345	85.8	14.2	0.1	1.11	0.1670	631	0.17	0.1200	1.2	12.0
T-10	0.52	356.7	0.1878	358	82.7	17.3	0.1	0.76	0.1330	538	0.13	0.0930	0.9	8.4
T-13	1.72	314.0	0.1870	317	93.1	6.9	0.1	0.54	0.1860	217	0.18	0.1300	1.4	4.6
T-15	3.34	304.6	0.1866	309	95.6	4.4	0.2	0.33	0.1910	144	0.19	0.1400	1.4	2.2
T-16	0.90	359.3	0.1883	358	82.5	17.5	0.2	0.40	0.2600	224	0.26	0.1900	2.0	2.5
T-21-2M	0.21	301.9	0.1870	305	96.9	3.1	0.0	0.17	0.4300	127	0.43	0.3100	3.4	26.0
T-22-2M	0.24	309.4	0.1875	311	95.0	5.0	0.0	0.21	0.2350	171	0.23	0.1660	1.8	17.5
Kharaelakh intrusive, KZ-963														
963-21	1.26	364.4	0.1878	365	80.9	19.1	0.2	1.55	0.1500	733	0.15	0.1100	1.1	6.4
963-25	0.68	301.0	0.1875	303	97.6	2.4	0.0	0.43	0.0720	165	0.07	0.0490	0.4	27.0
963-29	0.47	348.5	0.1869	353	83.8	16.2	0.1	0.24	0.2200	297	0.21	0.1500	1.6	3.2
963-30	0.56	326.3	0.1872	329	89.8	10.2	0.1	0.20	0.1490	243	0.15	0.1100	1.1	3.6
963-35	0.45	348.3	0.1878	349	84.7	15.3	0.1	0.09	0.3800	093	0.37	0.2600	3.0	1.3
963-65	1.71	328.3	0.1860	335	88.2	11.8	0.2	0.37	0.1370	171	0.13	0.0930	1.0	1.8
963-86	0.43	313.7	0.1874	316	93.6	6.4	0.0	0.35	0.1430	239	0.14	0.1000	1.0	13.0

Table 10 Isotopic composition of argon and helium in fluid inclusions of rocks of ore-bearing intrusives (III stage)

Sample number	Ar, $\times 10^{-6}$ cm^3/g	$^{40}Ar/^{36}Ar$ (meas.)	$^{38}Ar/^{36}Ar$ (meas.)	$^{40}Ar/^{36}Ar$ (corr.)	Ar_a/Ar. %	Ar_r/Ar, %	$^{40}Ar_r$, $\times 10^{-6}$ cm^3/g	He, $\times 10^{-6}$ cm^3/g	$^3He/^4He$ (meas.) $\times 10^{-6}$	$^4He/^{20}Ne$ (meas.)	$^3He/^4He$ (corr.) $\times 10^{-6}$	R_A (corr.)	He_m/He, %	$^4He/^{40}Ar_r$
Chernogorsk intrusive. Bh. MP-2bis														
CH-6	0.58	342.8	0.1887	340	86.9	13.1	0.08	0.0670	0.46	46.5	0.45	0.32	3.6	0.9
CH-9	0.74	350.1	0.1880	350	84.4	15.6	0.11	0.0820	0.26	51.9	0.26	0.19	2.0	0.7
CH-11	0.69	411.3	0.1877	413	71.7	28.3	0.20	0.0590	1.36	54.7	1.36	0.97	11.0	0.3
Repeated	0.71	440.0	0.1886	437	67.6	32.4	0.23	0.0640	0.89	–	0.89	0.64	7.3	0.3
CHb-15	1.07	363.0	0.1889	360	82.2	17.8	0.19	0.0510	0.65	30.2	0.65	0.46	5.2	0.3
Repeated	1.25	396.1	0.1873	399	74.1	25.9	0.32	0.0600	0.49	34.6	0.48	0.34	3.8	0.2
Zub-Marksheidersky intrusive. Bh. MP-27														
27 5	1.45	463.3	0.1875	466	63.5	36.5	0.53	0.1820	0.11	90.9	0.10	0.07	0.7	0.3
27 6	0.81	345.7	0.1878	346	85.3	14.7	0.12	0.1140	0.31	63.9	0.30	0.21	2.3	1.0
27 10	1.65	327.2	0.1883	326	90.6	9.4	0.16	0.1280	0.60	61.7	0.60	0.43	4.8	0.8
27-14	0.85	351.9	0.1883	351	84.3	15.7	0.13	0.1380	0.04	61.8	0.03	0.02	0.1	1.0
Vologochan intrusive. Bh. OV-29														
29-2	1.15	486.6	0.1885	484	61.1	38.9	0.45	0.7300	0.09	337.0	0.09	0.06	0.6	1.6
29 11	6.60	506.4	0.1872	510	57.9	42.1	2.80	3.1000	0.05	414.0	0.05	0.04	0.3	1.1
29-22	1.30	369.1	0.1881	369	80.2	19.8	0.25	0.9300	0.08	363.0	0.08	0.06	0.5	3.7
29-24	3.20	459.2	0.1881	459	64.4	35.6	1.10	2.1000	0.13	453.0	0.13	0.09	0.9	1.9
South Pyasina intrusive. Bh. OV 25														
25-17	0.42	544.3	0.1835	571	51.8	48.2	0.22	1.1400	0.17	527.0	0.17	0.12	1.2	5.7
25-20	1.16	354.6	0.1881	354	83.5	16.5	0.19	0.6900	0.11	350.0	0.11	0.08	0.8	3.8
25-22	0.32	431.7	0.1875	434	68.1	31.9	0.10	1.0600	0.16	510.0	0.16	0.11	1.2	10.5
25-29	0.49	379.9	0.1857	389	76.1	23.9	0.12	0.8800	0.13	656.0	0.13	0.09	0.9	7.8
25-38	0.59	398.2	0.1863	405	72.9	27.1	0.16	1.6000	0.12	803.0	0.12	0.09	0.9	10.5

Table 11 Isotopic composition of argon and helium in fluid inclusions of rocks of potentially ore-bearing intrusives (III stage)

Sample	Ar. $\times 10^{-6}$ cm^3/g	$^{40}Ar/^{36}Ar$ (meas.)	$^{38}Ar/^{36}Ar$ (meas.)	$^{40}Ar/^{36}Ar$ (corr.)	Ar_a/ Ar. %	Ar_r/ Ar. %	$^{40}Ar_r$. $\times 10^{-6}$ cm^3/g	He. $\times 10^{-6}$ cm^3/g	$^3He/^4He$ (meas.) $\times 10^{-6}$	$^4He/^{20}Ne$ (meas.)	$^3He/^4He$ (corr.) $\times 10^{-6}$	R_A (corr.)	He_m/ He. %	$^4He/^{40}Ar_r$
Mikchangda intrusive. Bh. MD-48														
48-9	1.13	333.6	0.1873	336	87.9	12.1	0.14	0.3500	0.2580	197	0.2560	0.1830	2.0	2.5
48-16	0.53	489.2	0.1880	489	60.4	39.6	0.21	4.9800	0.1290	1380	0.1280	0.0910	0.9	24.0
48-18	0.51	416.6	0.1860	425	69.5	30.5	0.16	3.7700	0.0960	1200	0.0950	0.0680	0.6	24.0
48-23	0.69	553.3	0.1870	559	52.9	47.1	0.33	4.7800	0.0650	1250	0.0650	0.0460	0.4	15.0
48-25	0.67	422.3	0.1877	424	69.7	30.3	0.20	0.5700	0.0500	352	0.0490	0.0350	0.2	2.8
48-27	0.97	412.0	0.1881	412	71.8	28.2	0.27	0.9800	0.0910	467	0.0900	0.0640	0.6	3.6
48-30	1.15	620.1	0.1875	624	47.4	52.6	0.61	7.4200	0.0510	1220	0.0510	0.0360	0.3	12.0
48-32	0.95	665.4	0.1878	667	44.3	55.7	0.53	1.5500	0.0660	659	0.0660	0.0470	0.4	2.9
Binyuda intrusive. Bh. S-1														
S 1-5	1.21	314.0	0.1878	315	93.9	6.1	0.07	0.0680	0.4890	33	0.4800	0.3430	3.8	0.9
S 1-5.5	0.62	320.2	0.1879	321	92.2	7.8	0.05	0.0350	0.4080	31	0.4000	0.2860	3.2	0.7
S 1-38	0.41	324.2	0.1877	325	90.9	9.1	0.04	0.0280	0.6690	19	0.6600	0.4710	5.3	0.8
S 1-58	0.36	360.1	0.1876	362	81.7	18.3	0.07	0.0320	0.2320	23	0.2200	0.1570	1.6	0.5

Table 12 Isotopic composition of argon and helium in fluid inclusions of rocks of weakly ore-bearing and non-ore-bearing intrusives (III stage)

Sample number	Ar. $\times 10^{-6}$ cm^3/g	$^{40}Ar/^{36}Ar$ (meas.)	$^{38}Ar/^{36}Ar$ (meas.)	$^{40}Ar/^{36}Ar$ (corr.)	Ar_a/Ar. %	Ar_r/Ar. %	$^{40}Ar_r$. $\times 10^{-6}$ cm^3/g	He. $\times 10^{-6}$ cm^3/g	$^3He/^4He$ (meas.) $\times 10^{-6}$	$^4He/^{20}Ne$ (meas.)	$^3He/^4He$ (corr.) $\times 10^{-6}$	R_A (corr.)	He_m/He. %	$^4He/^{40}Ar_r$
Lower Talnakh intrusive. Bh. TG-31														
311	1.40	380.8	0.1876	383	77.3	22.7	0.33	0.6	0.1490	267	0.1500	0.1100	1.1	1.8
31 3	1.00	373.3	0.1881	373	79.3	20.1	0.20	0.8	0.1080	434	0.1100	0.0790	0.7	3.7
31 7	0.64	323.7	0.1886	322	91.9	8.1	0.05	0.3	0.0860	211	0.0840	0.0600	0.5	6.5
31-11	1.20	342.6	0.1878	341	86.1	13.9	0.16	0.8	0.0620	372	0.0610	0.0440	0.3	5.0
31-13	1.00	323.3	0.1885	322	91.9	8.1	0.08	0.6	0.0780	468	0.0770	0.0550	0.5	6.7
31-16	1.20	380.9	0.1883	380	77.8	22.2	0.26	1.5	0.0410	625	0.0400	0.0290	0.2	6.0
Lower Norilsk intrusive. Bh. NP-37														
1514.5	2.46	331.3	0.1867	336	88.0	12.0	0.30	0.8	0.0550	100	0.0510	0.0360	0.2	2.6
1514.5	0.67	373.5	0.1872	377	78.5	21.5	0.14	1.0	0.0310	462	0.0300	0.0210	0.1	7.2
1584	1.23	336.0	0.1886	334	88.5	11.5	0.14	1.6	0.0760	464	0.0750	0.0530	0.5	11.0
1614	1.72	346.4	0.1875	348	84.9	15.1	0.26	0.4	0.0460	134	0.0430	0.0310	0.2	1.7
1620.3-1622.2	2.31	312.3	0.1878	313	94.5	5.5	0.13	0.4	0.0700	97	0.0650	0.0460	0.38	3.2
Zelenaya Griva intrusive. Bh. F-233														
F2	1.05	324.3	0.1882	324	91.4	8.6	0.09	0.0	0.6340	23	0.6200	0.4400	5.0	0.5
F4	1.56	336.5	0.1851	347	85.2	14.8	0.23	0.0	0.6530	15	0.6400	0.4600	5.1	0.2
F7	1.19	344.5	0.1866	349	84.6	15.4	0.18	0.1	0.5350	33	0.5300	0.3800	4.2	0.4
F-10	0.94	479.2	0.1884	477	61.9	38.1	0.36	0.2	0.8900	95	0.8900	0.6400	7.3	0.5
F-13	0.80	526.0	0.1886	522	56.6	43.4	0.35	0.1	1.0000	101	0.9900	0.7100	8.1	0.4
Oganer intrusive. Bh. MD-48														
48-7	1.06	356.7	0.1864	363	81.5	18.5	0.20	0.5	0.0940	177	0.0920	0.0660	0.6	2.3
Agatsky intrusive. Bh. OV-25														
25-26	0.45	443.1	0.1887	440	67.2	32.8	0.15	0.3	0.1630	133	0.1600	0.1100	1.2	1.7
Daldykan intrusive. Bh. NP-37														
37-44	0.54	426.5	0.1885	424	69.7	30.3	0.16	0.5	0.1240	259	0.1200	0.0900	0.8	2.8

Table 13 Isotopic composition of argon and helium in fluid inclusions of rocks and ores (IV stage)

Sample number	Rock	Ar, $\times 10^{-6}$ cm^3/g	$^{40}Ar/^{36}Ar$ (meas.)	Ar_a/Ar, %	Ar_r/Ar, %	$^{40}Ar_r$, $\times 10^{-6}$ cm^3/g	He, $\cdot 10^{-6}$ cm^3/g	$^3He/^4He$ (meas.) $\times 10^{-6}$	$^4He/^{20}Ne$ (meas.)	$^3He/^4He$ (corr.) $\times 10^{-6}$	R_A (corr.)	He_m/He, %	$^4He/^{40}Ar_r$
Oktyabrskoe deposit													
Western flank													
ZF-13-429.9	Metasomatite epidote-chlorite-carbonate. mottled. with carbonate segregations to 1 cm	0.68	313	94.4	5.6	0.038	0.99	0.031	401	0.030	0.021	0.08	26
ZF-13-441.9	Gabbro. coarse-grained. olivine. melanocratic Secondary: serpentine. Ore: segregations 3–8 mm	1.28	348	85	15	0.192	0.56	0.061	386	0.060	0.043	0.33	4.7
ZF-13-452.3	Gabbro. coarse-grained. olivinic. melanocratic. Ore: impregnations 8–10% 0.3–2 mm in size	0.91	351	84.3	15.7	0.143	0.89	0.077	409	0.076	0.054	0.5	6.4
ZF-13-462.3	Metasomatite. epidote-pyroxenic. albite and carbonate segregations. Ore: impregnations to 3%	0.95	309	95.6	4.4	0.042	3.77	0.029	712	0.028	0.02	0.07	89
ZF-13-483.2	Inequigranular pyroxenite. Secondary: serpentine. Ore: impregnations to 2%	0.93	333	88.8	11.2	0.104	0.3	0.078	243	0.076	0.054	0.50	2.9
ZF-13-521.8	Dolerite. medium-coarse grained. poikiloophitic. intensely altered. Secondary: amphibole. serpentine. epidote. biotite. Ore: to 3%	1.32	347	85.1	14.9	0.196	0.85	0.061	415	0.06	0.043	0.3	4.3
ZF-30-346.1	Pyroxenite. inequigranular. from fine- to coarse-grained. Ore: accumulations to 3 mm in size	0.76	352	84	16	0.122	0.58	0.084	367	0.082	0.058	0.5	4.8
ZF-30-357.1	Metasomatite. mottled. anhydrite-chlorite-feldspar-epidotic. Ore: rare impregnations to 1 mm	1.49	307	96.2	3.8	0.057	0.39	0.09	254	0.088	0.063	0.6	6.9

(continued)

Table 13 (continued)

Sample number	Rock	Ar, $\times 10^{-6}$ cm^3/g	$^{40}Ar/^{36}Ar$ (meas.)	Ar_a/Ar, %	Ar_r/Ar, %	$^{40}Ar_r$, $\times 10^{-6}$ cm^3/g	He, $\cdot 10^{-6}$ cm^3/g	$^3He/^4He$ (meas.) $\times 10^{-6}$	$^4He/^{20}Ne$ (meas.)	$^3He/^4He$ (corr.) $\times 10^{-6}$	R_A (corr.)	He_m/He, %	$^4He/^{40}Ar_r$
ZF-30-375.9	Leucogabbro medium-grained. intensely altered. Secondary: saussurite. carbonate. anhydrite. epidote. Ore: rare impregnations to 1 mm in size	2.45	328	90.2	9.8	0.239	0.62	0.16	120	0.16	0.114	1.2	2.6
ZF-30-387.7	Leucogabbro. coarse-grained. olivine-bearing mesocratic. intensely altered. Secondary: epidote. carbonate. chlorite. Ore: segregations to 5 mm with chlorite	1.89	368	80.3	19.7	0.373	1.27	0.06	328	0.058	0.041	0.3	3.4
ZF-30-458.7	Metasomatite. amphibole-epidote-albitic	0.60	345	85.6	14.4	0.086	0.83	0.16	468	0.15	0.107	1.2	9.6
KZ931-643.9	Disseminated ore	0.52	402	73.5	26.5	0.138	0.18	0.148	199	0.146	0.104	1.05	1.3
KZ931-645	Rich ore	0.31	330	89.4	10.6	0.032	0.18	0.123	93	0.118	0.084	0.8	5.4
KZ952-971.1	Disseminated ore	0.33	307	96.3	3.7	0.012	0.33	0.118	254	0.116	0.083	0.8	27
KZ952-1010.4	»	0.96	306	96.8	3.2	0.031	0.87	0.082	453	0.081	0.058	0.5	28
KZ952-1013.5	»	0.31	305	97	3	0.009	0.38	0.115	560	0.114	0.081	0.8	41
KZ1319-598.4	Rich ore	0.24	308	96	4	0.01	0.58	0.133	813	0.132	0.094	0.9	59
KZ1319-626.8	Dolerite. coarse-grained. poikilitic. olivine-bearing. Ore: rare impregnations to 0.5 mm in size	0.68	355	83.3	16.7	0.114	0.84	0.118	491	0.117	0.084	0.8	7.4
KZ1319-638.4	Plagiowehrlite. medium-grained. intensely serpentinized. Ore: 15–20% in interstices to 3 mm	0.92	306	96.5	3.5	0.032	0.18	0.166	259	0.164	0.117	1.2	5.7
KZ1319-641.4	Plagiowehrlite. medium-grained. analogue of KZ1319-638.4 totally serpentinized. Ore: 15–20% in interstices to 3 mm	8.17	299	99	1	0.079	0.12	0.588	36	0.581	0.415	4.7	1.5

(continued)

Table 13 (continued)

Sample number	Rock	Ar, $\times 10^{-6}$ cm^3/g	$^{40}Ar/^{36}Ar$ (meas.)	Ar_a/Ar, %	Ar_r/Ar, %	$^{40}Ar_r$, $\times 10^{-6}$ cm^3/g	He, $\cdot 10^{-6}$ cm^3/g	$^3He/^4He$ (meas.) $\times 10^{-6}$	$^4He/^{20}Ne$ (meas.)	$^3He/^4He$ (corr.) $\times 10^{-6}$	R_A (corr.)	He_m/He, %	$^4He/^{40}Ar_r$
Central part													
PT-2-1364.1	Hornfels	1.27	1621	18.2	81.8	1.04	2.36	0.035	1260	0.035	0.025	0.1	2.3
PT-2-1371.2	Dolerite. medium-grained. poikiloophitic. mesocratic. Secondary: amphibole. biotite. Ore: rare segregations to 2 mm in size	0.71	485	61	39	0.278	0.5	0.066	522	0.065	0.046	0.4	1.8
PT-2-1425.0	Plagiowehrlite. fine-medium-grained. Ore: 5–8%. accumulations to 2 mm	0.89	349	84.7	15.3	0.136	4.41	0.108	1593	0.107	0.076	0.7	32
PT-2-1436.2	Plagiowehrlite. fine-medium-grained. pyroxene grains to 5 mm with olivine inclusions. Ore: to 15%. accumulations to 10 mm	0.57	409	72.2	27.8	0.158	1.58	0.193	1098	0.192	0.137	1.4	9.9
PT-2-1453.7	Metasomatite. albite-chlorite-epidotic. Ore: rare grains to 0.4 mm	1.19	665	44.5	55.5	0.66	7.65	0.035	2006	0.035	0.025	0.1	12
KZ1089-1154.4	Rich ore	0.49	306	96.7	3.3	0.016	0.13	0.145	144	0.142	0.101	1	8.2
KZ1089-1155.6	»	0.16	317	93.2	6.8	0.011	0.11	0.201	134	0.198	0.141	1.5	10
KZ1089-1156.7	»	2.38	312	94.7	5.3	0.125	0.68	0.112	217	0.11	0.078	0.8	5.5
KZ1112-1092.4	Disseminated ore	0.43	467	63.3	36.7	0.157	0.3	0.152	433	0.151	0.108	1.1	1.9
KZ1112-1094.8	»	0.54	327	90.3	9.7	0.053	0.27	0.243	145	0.24	0.171	1.8	5
KZ1112-1098.4	»	0.41	354	83.4	16.6	0.069	0.28	0.102	249	0.1	0.071	0.7	4.1
KZ1112-1100.4	»	0.23	330	89.6	10.4	0.024	0.18	0.332	186	0.33	0.236	2.6	7.4
KZ1112-1102.4	»	0.7	332	89.2	10.8	0.075	0.17	0.163	123	0.16	0.114	1.2	2.2
KZ1112-1103.8	»	0.18	318	92.9	7.1	0.012	0.08	0.548	129	0.546	0.39	4.4	6.1
		0.16	325	91.1	8.9	0.014							
KZ1535-1489.2	Gabbro. coarse-grained. olivinic. Ore: 10% impregnations to 1.5 mm in size	0.45	377	78.5	21.5	0.097	0.53	0.07	453	0.069	0.049	0.4	5.4

(continued)

Table 13 (continued)

Sample number	Rock	Ar, $\times 10^{-6}$ cm^3/g	$^{40}Ar/^{36}Ar$ (meas.)	Ar_a/Ar, %	Ar_r/Ar, %	$^{40}Ar_r$, $\times 10^{-6}$ cm^3/g	He, $\cdot 10^{-6}$ cm^3/g	$^3He/^4He$ (meas.) $\times 10^{-6}$	$^4He/^{20}Ne$ (meas.)	$^3He/^4He$ (corr.) $\times 10^{-6}$	R_A (corr.)	He_m/He, %	$^4He/^{40}Ar_r$
KZ1535-1491.0	Plagiowehrlite. medium-grained. analogue of KZ1319-638.4. totally serpentinized. Ore: to 15% in interstices to 3 mm	0.29	378	78.1	21.9	0.063	0.33	0.122	190	0.12	0.086	0.8	5.2
KZ1535-1495.9	Plagiowehrlite. medium-coarse grained. intensely serpentinized. Ore: to 15% in interstices to 3 mm	0.97	328	90.2	9.8	0.095	0.44	0.154	409	0.153	0.109	1.1	4.7
	Southern flank												
KZ361bis-1062.8	Rich ore	0.7	385	76.8	23.2	0.162	1.06	0.045	515	0.044	0.031	0.2	6.5
KZ361bis-1063.4	The same	0.28	332	89	11	0.031	0.17	0.31	193	0.308	0.22	2.4	5.5
KZ361bis-1074.4	»	0.33	366	80.8	19.2	0.063	0.1	0.14	84	0.135	0.096	1	1.5
KZ361bis-1075.6	»	0.35	306	96.5	3.5	0.012	0.13	0.283	139	0.28	0.2	2.2	10.5
KZ1084-1146.9	Disseminated ore	0.64	467	63.3	36.7	0.235	0.48	0.07	375	0.069	0.049	0.4	2
Maslovsky deposit													
OM-32-1057.3	Plagiowehrlite. coarse-grained. intensely serpentinized. Secondary: chlorite. serpentine. saussurite. Ore: 3% segregations to 3 mm	1.7	318	92.9	7.1	0.121	0.21	0.177	91	0.173	0.124	1.3	1.7
OM-32-1090.0	Gabbro. medium-coarse-grained. olivinic. poikiloophitic. Secondary: serpentine. chlorite. amphibole. talc. Ore: rare impregnations to 1 mm in size	1.64	335	88.1	11.9	0.195	3.76	0.023	697	0.022	0.016	0.02	19
OM-32-1114.4	Gabbro. medium-coarse-grained. olivinic. poikiloophitic. Secondary: serpentine. chlorite. amphibole. talc. Ore: rare impregnations to 1 mm in size	0.53	350	84.5	15.5	0.083	0.27	0.041	226	0.039	0.28	0.2	6.4

(continued)

Table 13 (continued)

Sample number	Rock	Ar, $\times 10^{-6}$ cm^3/g	$^{40}Ar/^{36}Ar$ (meas.)	Ar_a/Ar, %	Ar_r/Ar, %	$^{40}Ar_r$, $\times 10^{-6}$ cm^3/g	He, $\cdot 10^{-6}$ cm^3/g	$^3He/^4He$ (meas.) $\times 10^{-6}$	$^4He/^{20}Ne$ (meas.)	$^3He/^4He$ (corr.) $\times 10^{-6}$	R_A (corr.)	He_m/He, %	$^4He/^{40}Ar_r$
OM-32-1149.8	Mudstone (silt-rich mudstone). black with carbonate. plagioclase and quartz grains from 0.1 to 2 in size and discontinuous veinlets of a similar composition	1.37	309	95.7	4.3	0.059	0.95	0.04	117	0.036	0.26	0.1	16
OM-123-952.7	Gabbro. coarse-grained. olivine-bearing. intensely altered. Secondary: serpentine. chlorite. amphibole. talc. Ore: rare impregnations to 1 mm in size	0.73	333	88.6	11.4	0.083	0.54	0.082	66	0.076	0.054	0.5	6.4
OM-123-980.0	Gabbro. medium-coarse-grained. mesocratic poikiloophitic. Secondary: saussurite. serpentine. chlorite. Ore: rare impregnations to 1 mm in size	1.53	332	89.2	10.8	0.166	2.99	0.032	766	0.032	0.023	0.1	18
OM-123-1009.6	Gabbro. coarse-grained. poikiloophitic. olivinic. Secondary: saussurite. serpentine. chlorite. Ore: rare impregnations to 1 mm in size	1.7	325	91	9.0	0.153	0.81	0.022	261	0.021	0.015	0.01	5.3
OM-123-1089.1	Basalt. fine-grained. rare porphyric. with itersertal groundmass. Secondary: saussurite. epidote. chlorite	1.25	307	96.3	3.7	0.046	0.91	0.064	178	0.061	0.044	0.3	19.4
OM-123-1112.0	Basalt. pyroxene-porphyric. impregnations to 5 mm with inclusions of plagioclase laths. groundmass intensely chloritized. single apatite crystals. Ore: 3–5% to 0.5 mm	1.31	394	75	25	0.327	0.32	0.172	236	0.17	0.121	1.2	0.94

(continued)

Table 13 (continued)

Sample number	Rock	Ar, $\times 10^{-6}$ cm^3/g	$^{40}Ar/^{36}Ar$ (meas.)	Ar_a/Ar, %	Ar_r/Ar, %	$^{40}Ar_r$, $\times 10^{-6}$ cm^3/g	He, $\cdot 10^{-6}$ cm^3/g	$^3He/^4He$ (meas.) $\times 10^{-6}$	$^4He/^{20}Ne$ (meas.)	$^3He/^4He$ (corr.) $\times 10^{-6}$	R_A (corr.)	He_m/He, %	$^4He/^{40}Ar_r$
Vologochan area													
OV-28-690.1	Metasomatite. fine-grained. anhydrite-carbonate-epidote-pyroxenic	0.53	410	72.1	27.9	0.148	0.59	0.085	260	0.083	0.059	0.5	4
OV-28-711.1	Gabbro. coarse-grained. mesocratic. amphibolitized. analogue of OV28-703.2. Ore: to 3%. accumulations to 10 mm in size	0.65	346	85.4	14.6	0.095	0.16	0.157	94	0.153	0.109	1.1	1.7
OV-28-745.4	Gabbro. medium-grained. olivine-bearing amphibolitized. Ore: rare impregnations to 1 mm in size	0.67	350	84.3	15.7	0.105	0.44	0.117	239	0.117	0.083	0.8	4.2
OV-28-813.8	Gabbro-dolerite. picrite-like	0.3	330	87	13		0.3	0.172	161	0.17	0.121	1.2	3
OV-28-840.1	Hornfels. fine-grained. mottled. chlorite-biotite-pyroxene-feldspar	1.96	532	55.6	44.4	0.87	0.52	0.126	186	0.124	0.089	0.9	2.2
South Pyasina intrusive													
Vologochan area													
OV-36-1403.1	Siltstone. fine-grained. Secondary: carbonate-epidote. mottled. Ore: rare grains (pyrite) to 0.3 mm	0.55	340	87	13	0.071	0.72	0.093	274	0.092	0.066	0.6	10
OV-36-1424.0	Hornfels after siltstone. carbonate-epidote-pyroxene. feldspar	0.93	305	96.9	3.1	0.029	0.74	0.086	258	0.084	0.06	0.5	26
OV-36-1438.0	Metasomatite. fine-medium-grained. mottled. carbonate-epidote-amphibole-pyroxenic. Ore: rare impregnations in carbonate veinlets	1.2	316	93.8	6.2	0.076	1.53	0.087	452	0.086	0.061	0.6	20

(continued)

Table 13 (continued)

Sample number	Rock	Ar, $\times 10^{-6}$ cm^3/g	$^{40}Ar/^{36}Ar$ (meas.)	Ar_a/Ar, %	Ar_r/Ar, %	$^{40}Ar_r$, $\times 10^{-6}$ cm^3/g	He, $\cdot 10^{-6}$ cm^3/g	$^3He/^4He$ (meas.) $\times 10^{-6}$	$^4He/^{20}Ne$ (meas.)	$^3He/^4He$ (corr.) $\times 10^{-6}$	R_A (corr.)	He_m/He, %	$^4He/^{40}Ar_r$
OV-36-1463.7	Dolerite. fine-medium-grained. olivine-bearing. ophitic. intensely altered. Secondary: saussurite. serpentine. epidote. Ore: 3–5% to 1 mm in size	0.66	425	69.5	30.5	0.201	1.88	0.083	868	0.082	0.059	0.5	9.3
OV-36-1539.2	Gabbro. medium-grained. olivine-bearing. Ore: 3–5% impregnations to 3 mm in size	0.83	427	69.2	30.8	0.255	1.84	0.093	1173	0.092	0.066	0.6	7.2
Koevo area													
PK-11-320.1	Basalt. medium-grained. with doleritic structure. olivine-bearing. glass decrystallized. Ore: rare impregnations to 0.3 mm	0.23	303	97.5	2.5	0.006	0.03	0.917	59	0.70[a]	0.5	5.7	5.2
PK-11-353.3	Basalt. medium-grained. amygdaloidal. with doleritic structure. intensely altered. Secondary: amphibole. chlorite. carbonate. Ore: rare impregnations to 0.1 mm	0.64	306	96.5	3.5	0.023	0.08	0.363	21	0.22[a]	0.157	1.7	3.6
PK-11-416.0	Dolerite. medium-coarse grained. olivinic. Ore: 3% impregnations to 2 mm in size	2.16	656	45.1	54.9	1.189	0.19	0.277	145	0.275	0.196	2.1	0.16
PK-11-449.8	Basalt. plagioporphyric. groundmass fine-grained. ophitic. Ore: rare impregnations to 0.3 mm in size	0.32	767	38.5	61.5	0.197	0.06	0.501	88	0.45[a]	0.32	3.6	0.29
PK-11-477.3	Basalt. medium-grained. amygdaloidal (quartz-carbonate). with doleritic structure. glass decrystallized. Ore: rare impregnations to 0.2 mm	0.39	376	78.6	21.4	0.084	0.2	0.247	306	0.246	0.176	1.9	2.3

(continued)

Table 13 (continued)

Sample number	Rock	Ar, $\times 10^{-6}$ cm^3/g	$^{40}Ar/^{36}Ar$ (meas.)	Ar_a/Ar, %	Ar_r/Ar, %	$^{40}Ar_r$, $\times 10^{-6}$ cm^3/g	He, $\cdot 10^{-6}$ cm^3/g	$^3He/^4He$ (meas.) $\times 10^{-6}$	$^4He/^{20}Ne$ (meas.)	$^3He/^4He$ (corr.) $\times 10^{-6}$	R_A (corr.)	He_m/He, %	$^4He/^{40}Ar_r$
PK-11-503.5	Basalt. medium-grained. amygdaloidal (quartz-carbonate). with doleritic structure. glass decrystallized. Ore: rare impregnations to 0.2 mm	0.41	306	96.7	3.3	0.014	0.19	0.133	174	0.131	0.094	0.9	14
PK-11-529.3	Basalt. fine-grained. amygdaloidal (quartz-carbonate). with doleritic and hyalopelitic structure. to 40% decrystallized glass. Ore: rare impregnations to 0.2 mm	0.24	309	95.6	4.4	0.011	0.36	0.127	366	0.126	0.09	0.9	34
Tangnarylakh area													
13096a	Dolerite. coarse-grained. poikiloophitic. olivine-bearing. Secondary: iddingsite. chlorite. talc. Ore: rare impregnations to 1 mm in size	0.67	553	53.5	46.5	0.313	0.17	0.609	162	0.607	0.433	4.9	0.55
13096b	Gabbro. medium-coarse-grained. Secondary: iddingsite. amphibole. chlorite. Ore: rare accumulations to 1.5 mm in size	0.4	473	62.5	37.5	0.149	0.1	0.79	146	0.788	0.563	6.4	0.7
13097a	Gabbro. coarse-grained. Secondary: amphibole. chlorite. Ore: rare accumulations of decay structure to 3 mm in size	0.14	586	50.5	49.5	0.068	0.12	0.562	226	0.561	0.401	4.5	1.7
13097b	Gabbro. coarse-grained. Secondary: iddingsite. amphibole. chlorite. Ore: rare accumulations of decay structure to 3 mm in size	0.51	1249	23.7	76.3	0.387	0.22	0.359	314	0.358	0.256	2.8	0.58

(continued)

Table 13 (continued)

Sample number	Rock	Ar, $\times 10^{-6}$ cm^3/g	$^{40}Ar/^{36}Ar$ (meas.)	Ar_a/Ar, %	Ar_r/Ar, %	$^{40}Ar_r$, $\times 10^{-6}$ cm^3/g	He, $\cdot 10^{-6}$ cm^3/g	$^3He/^4He$ (meas.) $\times 10^{-6}$	$^4He/^{20}Ne$ (meas.)	$^3He/^4He$ (corr.) $\times 10^{-6}$	R_A (corr.)	He_m/He, %	$^4He/^{40}Ar_r$
Area of the Mokulaj Creek													
13005	Basalt. fine-grained. ophitic. olivine-bearing. Secondary: saussurite. iddingsite. chlorite. Ore: to 1% disseminated impregnations 0.1 mm in size	0.17	368	80.4	19.6	0.032	0.03	0.741	52	0.46[a]	0.33	3.7	0.96
13016	Basalt. poikilitic. pyroxene crystals with frequent plagioclase ingrowths in recrystallized glass. Ore: rare impregnations to 0.1 mm in size	0.07	555	53.3	46.7	0.034	0.02	0.832	58	0.36[a]	0.26	2.8	0.68
13020	Basalt. rare plagioporphyric. groundmass 70% of pyroxene and plagioclase crystals in dark brown glass	0.88	305	97.1	2.9	0.026	0.06	0.562	89	0.41[a]	0.29	3.2	2.4
13033	Basalt. fine-medium-grained. poikilitic. 80% of pyroxene crystals with plagioclase inclusions in brown glass	0.07	439	67.3	32.7	0.023	0.15	0.84	124	0.84	0.6	6.8	6.6
13049	Basalt. fine-grained. ophitic. with directive texture. Secondary: iddingsite. chlorite. epidote. Ore: to 1% disseminated impregnations 0.01 mm in size	0.06	483	61.2	38.8	0.024	0.02	0.569	68	~0[a]	~0	~0	0.94

[a] - part of the atmospheric argon, % (see formula a(%)=...)

Table 14 Helium and neon in fluid inclusions of rocks of intrusions in the Norilsk District (IV stage)

Object	Sample number	He, $\times 10^{-6}$ cm^3/g	$^3He/^4He$, $\times 10^{-6}$ meas.	$^4He/^{20}Ne$ (meas.)	$^3He/^4He$ (corr.) $\times 10^{-6}$	R_A (corr.)	He_m/He, %	$^4He/^{40}Ar_r$
Tal	KZ 774 1049.2–1049.7	0.12	0.28	91	0.28	0.20	2.1	0.7
Mas	OM 10 1082.4–1083.0	0.47	0.26	157	0.26	0.19	2.0	2.5
Mas	OM 32 1084.8	1.44	0.12	626	0.12	0.09	0.8	19
Mas	OM 123 952.7	0.26	0.16	160	0.15	0.11	1.1	6.7
Cher	MP-2 206 M3	0.06	0.44	91	0.44	0.32	3.5	0.68
ZMa	MP-25–23	1.44	0.19	276	0.19	0.14	1.4	4.0
Oga	12N 26a	0.31	0.23	142	0.23	0.16	1.7	0.77
Oga	12N 26b	0.93	0.17	686	0.17	0.12	1.2	3.2
Lis	12N07	0.09	0.62	73	0.62	0.44	5.0	2.4
Lis	12N05	0.58	0.40	231	0.40	0.29	3.2	8.0
KrK	12N33	0.14	0.17	286	0.17	0.12	1.3	1.1
KrK	12N34	0.17	0.56	379	0.55	0.40	4.4	4.4
ZMa	13074	0.16	0.16	181	0.16	0.12	1.1	0.92
Tan	13097A	0.18	0.72	153	0.72	0.52	5.8	6.5
Tan	13097B	0.06	1.05	72	1.05	0.75	8.6	0.77
Tan	13096B	0.18	0.14	210	0.14	0.10	1.0	2.9
Tan	13090	0.19	2.89	104	2.89	2.06	24	0.86
Lis repeat.	12H05	0.08	0.28	66	0.28	0.20	2.2	0.84
ZfO	ZF 13 441.9–442.5	0.26	0.250	129	0.25	0.18	1.9	1.9

Notes Tal—Talnakh. Mas—Maslovsky. ZMa—Zub-Marksheidersky. Cher—Chernogorsk. areas: Oga—Oganer. Lis—Listvyanka. KrK—Krasnye Kamni. Tan—Tangnarylakh. ZfO—western flank of Oktyabrskoe deposit. Gases were released by samples crushing in vacuum. $^3He/^4He$ $\times 10^{-6}$ corr.—value of isotopic ratio corrected for air component of helium calculated from $^4He/^{20}Ne$ ratio. i.e. for isotopic composition of "deep" helium. R_A—corrected ratio of helium isotopes divided by such ratio in atmospheric helium (1.4×10^{-6}). He_m/He—share of mantle helium calculated for $^3He/^4He$ values in the upper mantle (1.2×10^{-5}) and the Earth's crust (2×10^{-8})

Table 15 Argon in fluid inclusions of rocks of intrusions in the Norilsk District (IV stage)

Object	Sample number	Ar, $\times 10^{-6}$ cm^3/g	$^{40}Ar/^{36}Ar$ (meas.)	Ar_a/Ar, %	Ar_r/Ar, %	$^{40}Ar_r$, $\times 10^{-6}$ cm^3/g
Tal	KZ 774 1049.2–1049.7	0.69	389.7	75.8	24.2	0.170
Mas	OM 10 1082.4–1083.0	1.43	339.6	87.0	13.0	0.180
Mas	OM 32 1084.8	0.56	342.9	86.2	13.8	0.077
Mas	OM 123 952.7	0.66	314.2	94.1	5.9	0.039
Cher	MP-2 206 M3	0.55	354.9	83.3	16.7	0.091
ZMa	MP-25–23	0.96	471.1	62.7	37.3	0.358
Oga	12 N 26a	0.67	729.6	40.5	59.5	0.396
Oga	12 N 26b	0.57	588.8	50.2	49.8	0.284
Lis	12N07	0.34	334.4	88.4	11.6	0.039
Lis	12N05	0.15	573.0	51.6	48.4	0.073
KrK	12N33	0.32	503.3	58.7	41.3	0.130
KrK	12N34	0.16	388.1	76.2	23.8	0.039
ZMa	13074	0.38	541.1	54.6	45.4	0.172
Tan	13097A	0.17	350.5	84.3	15.7	0.027
Tan	13097B	0.14	652.6	45.3	54.7	0.076
Tan	13096B	0.39	350.1	84.4	15.6	0.061
Tan	13090	0.48	554.7	53.3	46.7	0.224
Lis repeat.	12N05	0.23	478.6	61.8	38.2	0.090
ZfO	ZF 13 441.9–442.5	0.91	347.0	85.2	14.8	0.135

Notes Tal—Talnakh. Mas—Maslovsky. ZMa—Zub-Marksheidersky. Cher—Chernogorsk. areas: Oga—Oganer. Lis—Listvyanka. KrK—Krasnye Kamni. Tan—Tangnarylakh. ZfO—western flank of Oktyabrskoe deposit. Gases were released by samples crushing in vacuum

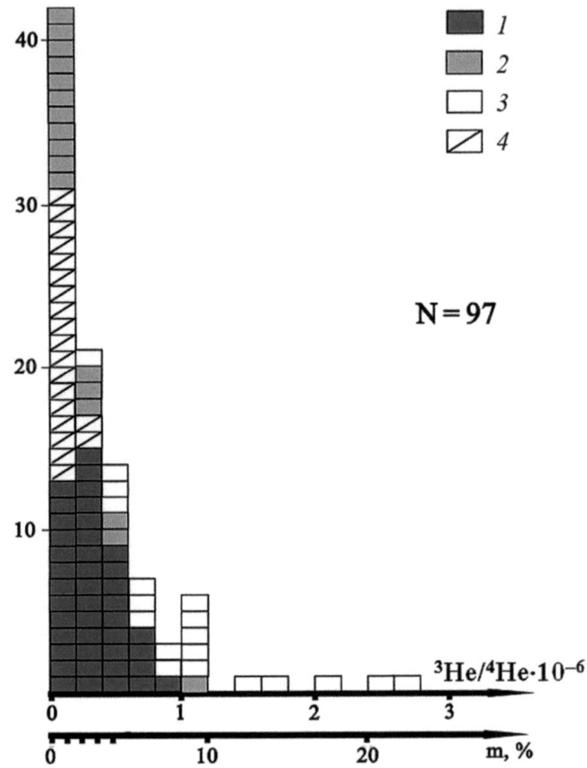

Fig. 6 He isotopes in fluid inclusions from intrusives of Norilsk-Taimyr District. Intrusives: 1—commercial ore-bearing (rich); 2—ore-bearing (intermediate); 3—weakly ore-bearing (poor); 4—commercial ore-bearing satellites

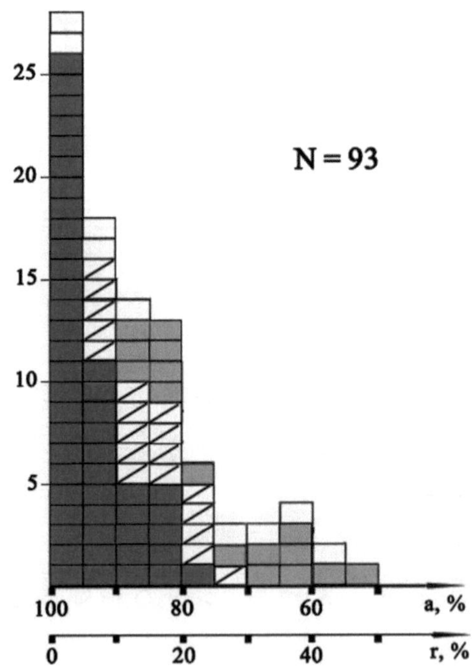

Fig. 7 Ar isotopes in fluid inclusions from -intrusives of Norilsk-Taimyr District. See legend in Fig. 6

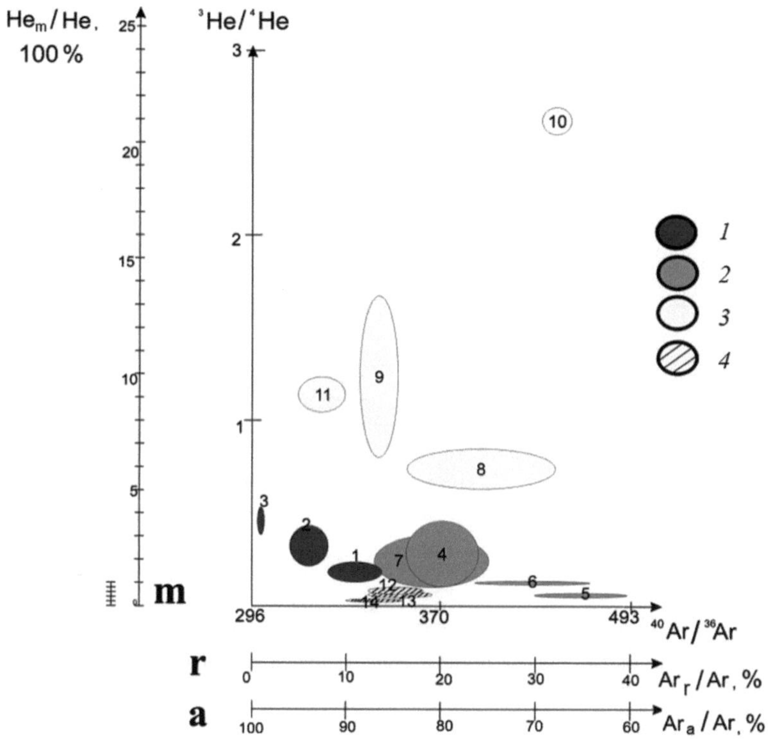

Fig. 8 Relationship of He and Ar isotopic composition (mean values) in Norilsk-1 intrusives (stages I-III). Intrusives: 1—rich (1—Kharaelakh; 2—Talnakh; 3—Norilsk-1); 2—intermediate (4—Chernogorsk; 5—Vologochan; 6—South Pyasina; 7—Zub-Marksheidersky); 3—poor (8—Zelenaya Griva; 9—Tulaj-Kiryak; 10—Bootankaga; 11—Lower Fokino); 4—satellites (12—Maslovsky; 13—Lower Talnakh; 14—Lower Norilsk). Ellipses—variability limits of mean values (σ_{mean})

share of mantle helium (**m** = 3.7%, Norilsk-1) corresponds to the maximum contribution of air argon (**a** = 99%).

For the statistical validity of the conclusions it was necessary to expand the sample of isotope data on certain objects, in particular, to expand the scope of ore studies. At the IV stage, helium and argon isotopic composition was investigated in 42 samples from 11 wells at the Oktyabrsky deposit (Kharaelakh intrusion), three wells at the Maslovskoye deposit etc. (Table 2). Overall, approximately 95 samples were studied including 20 samples of massive and disseminated ores (Tables 13, 16, 17 and 18).

It should be noted that the results of isotopic analysis did not differ significantly from those obtained earlier on a larger object—Kharaelakh intrusion (Table 13). It is important to stress that similar values of parameters (mean values and RMSE) were obtained for both 7 samples at the II stage, and 42 samples at the IV stage. This means, that seven samples are enough to assess ore presence in intrusions. Let us discuss the generalized data.

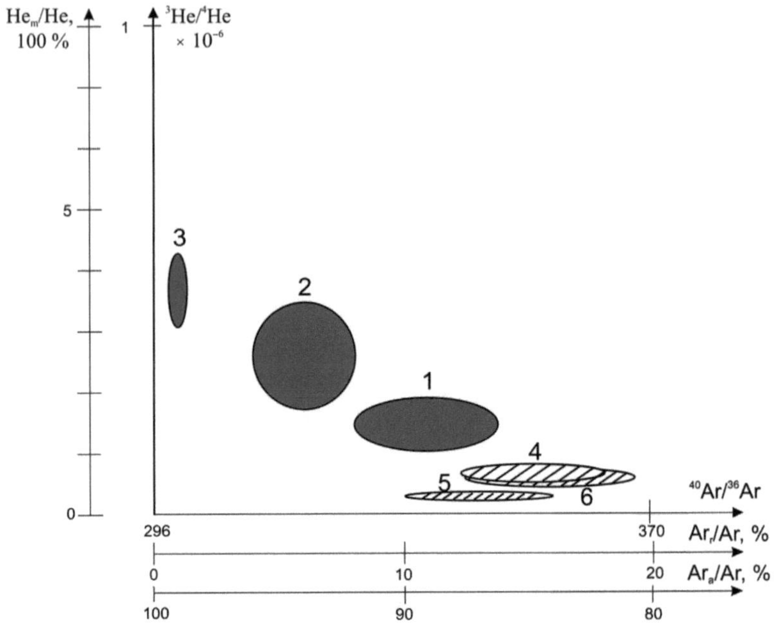

Fig. 9 Interrelation of He and Ar isotopic composition in rich intrusives and their satellites. Intrusives: rich (1—Kharaelakh; 2—Talnakh; 3—Norilsk-1); satellites (4—Maslovsky; 5—Lower Norilsk; 6—Lower Talnakh)

Isotopic composition of helium presented as **m** units—the share of mantle helium—in samples of intrusions of the IV stage in 90% of cases lies in the range of 0.1–1.5% (Figs. 23, 24), reaching 4.7% in one sample. Newly received data confirm the statement on the predominance of crustal helium in the mineral-forming environment. The range of variations almost fully coincides with that obtained at the I–III stages, despite the fact that then the data for a dozen and a half intrusions were available (Table 1); and at the IV stage, for three intrusions. Variations of **m** value in 42 samples from 11 wells from the whole area of the Oktyabrsky deposit (Kharaelakh intrusion) in the vast majority of samples are in the range of 0.1–1.5% (Figs. 10, 11). Such a heterogeneity is acceptable, especially as the data not on one, but on several samples are used to evaluate ore presence.

Mean **m** value at the Maslovskoye deposit is low (0.4%); it is close to the earlier received 0.7% (Table 16). On the Vologochan area, the mean **m** value is 0,7; earlier, 0.6%. In basalts, dolerites and gabbro of the Koevsky, Tangnarylakh and Mokulay areas, the share of mantle helium is 4–5 times higher along with major variations (**m** = 1–6.5%).

No significant differences are recorded in helium isotopic composition between massive (rich) and disseminated ores, ores and rocks of intrusions (Fig. 11). In other words, the contribution of mantle components in the mineral-forming environment of both massive and disseminated ores is equally low. Apparently, fluids of

Table 16 Isotopes of noble gases in intrusives of the Norilsk-Taimyr District (I–III stages)

Intrusion	Number of samples	Helium				Argon			
		$^{3}He/^{4}He \times 10^{-6}$	**m**	σ_i, %	σ_{cp}, %	$^{40}Ar/^{36}Ar$	**a**	σ_i, %	σ_{cp}, %
Commercial ore-bearing intrusions									
Kharaelakh	7	0.17	1.3	0.82	0.31	335	88.3	5.8	2.2
Talnakh	29	0.32	2.7	1.8	0.38	315	94.0	4.0	0.8
Norilsk-1	7	0.45	3.7	1.2	0.43	299	98.9	0.83	0.32
Ore-bearing intrusions									
Chernogorsk	4	0.52	4.3	2.9	1.3	371	79.8	7.7	3.8
Vologochan	4	0.07	0.6	0.25	0.12	449	65.9	9.9	5.0
South Pyasina	5	0.12	1.0	0.19	0.08	420	70.5	12	5.3
Zub-Marksheidersky	4	0.26	2.0	2.24	1.12	365	81.0	12	6
Weakly ore-bearing intrusions									
Zelenaya Griva	5	0.73	5.9	1.7	0.74	390	75.9	15.6	7
Tulaj-Kiryak	4	1.20	9.8	7	3.6	340	87.0	4	2
Bootankaga	2	2.60	22	1.2	0.8	435	68.0	3	2
Lower Fokino	7	1.10	9.2	1.5	0.61	320	92.5	5	2
Satellites of commercial ore-bearing intrusions									
Maslovsky	8	0.10	0.7	0.3	0.1	349	84.8	5.8	2.1
Lower Talnakh	6	0.07	0.6	0.32	0.13	351	84.2	6.8	2.8
Lower Norilsk	5	0.06	0.3	0.15	0.07	340	87.0	5.8	2.6
Potentially ore-bearing intrusions									
Mikchangda	8	0.10	0.7	0.54	0.19	455	65.0	15	5.3
Binyuda	4	0.42	3.5	1.5	0.75	329	89.7	5.4	2.7

Notes **m**—share of mantle helium (%). **a**—share of air argon (%), σ_{cp}—mean standard deviation

Table 17 Isotope-argon characteristics of intrusives and Oktyabrskoe deposit

Parameter	Oktyabrskoe deposit			Intrusives	
	In general	Western flank	Central part	Maslovsky	Vologochan
a, %	84/88	89/0	77/0	89/85	80/66
σ, %	16/6	7/0	20/0	5/6	15/10
n	42/7	20/0	17/0	9/8	10/4

Notes In the numerator—work at the IV stage; in denominator—at the I–III. **a**—share of air argon calculated from its isotopic composition; σ—root-mean-square deviation of a single sample; **n**—number of samples

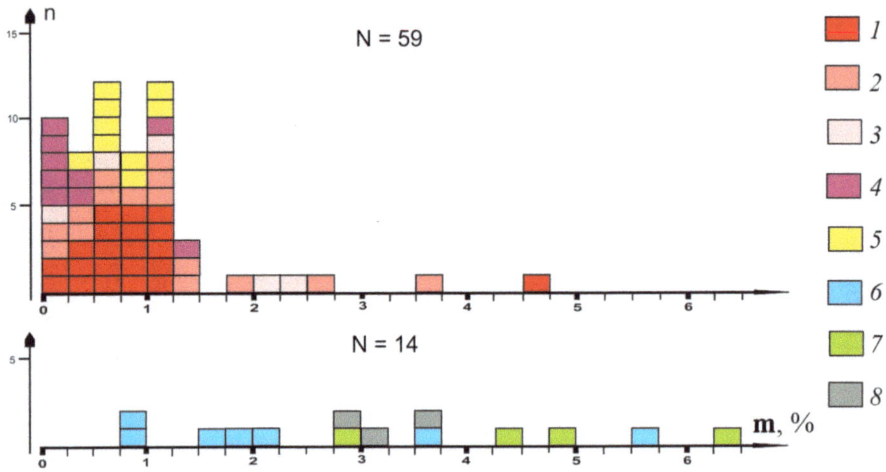

Fig. 10 Helium isotopes in rocks of productive and weakly productive areas in the Norilsk District (stage IV). 1–3, Oktyabrskoe deposit: 1—western flank; 2—central part; 3—southern flank; 4—Maslovsky deposit; 5–8—areas: 5—Vologochan; 6—Koevo; 7—Tangnarylakh; 8—Mokulaj Creek

the same type took part in the formation of ores of the two above varieties. From Fig. 11, no significant differences are seen in helium isotopic composition in different rock types of layered intrusions. Such isotopic similarity of helium in rocks and ores of the rich and intermediate intrusions may indicate a general and strong contamination of deep fluids by crustal ones at the stage of ore mineral formation in the Norilsk intrusions. Fluid contamination showed up in the formation and filling of gas-liquid inclusions, secondary in respect of rocks of the intrusions and primary in ore minerals. The contribution of mantle helium in intrusions of the Norilsk-Taimyr District is within 0.3–10% (except for the Bootankag, poor in ore, where **m** = 22%). Thus, **the crustal component clearly dominates the material balance of helium**. One should discard other notions as contradicting virtually the only reliable genetic criterion—helium isotopic composition.

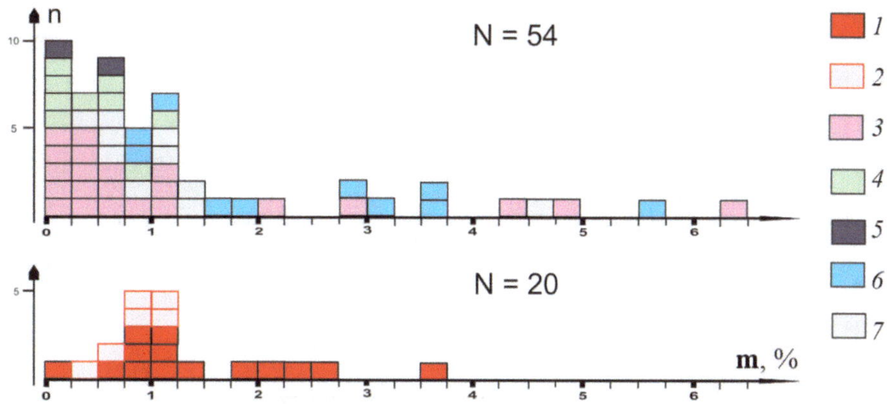

Fig. 11 Helium isotopes in rocks and ores of the studied areas in the Norilsk District. 1—rich ores; 2—disseminated ore; 3—gabbro-dolerites including leucogabbro; 4—metasomatites, hornfels; 5—mudstone, siltstone; 6—basalt; 7—plagiowehrlite, pyroxenite

Similarity of isotope pattern of helium from gas-liquid inclusions in rocks and ores of the rich and intermediate intrusions, apparently, points to the genetic unity of fluids and allows using helium isotopic composition as a criterion for distinguishing rich, intermediate and poor intrusions.

Argon isotopic composition presented as **a** units—the share of atmospheric argon (%), is in most samples from 96 to 80 reaching 20% in some samples (Fig. 12). This confirms the earlier statement on the dominance of air argon in rocks of intrusions rich in ore. Intrusions with average ore presence are noted by a much lower contribution of air argon. Since air gases, as mentioned above, penetrate into the subsurface with near-surface water (infiltration and sedimentation), such a high

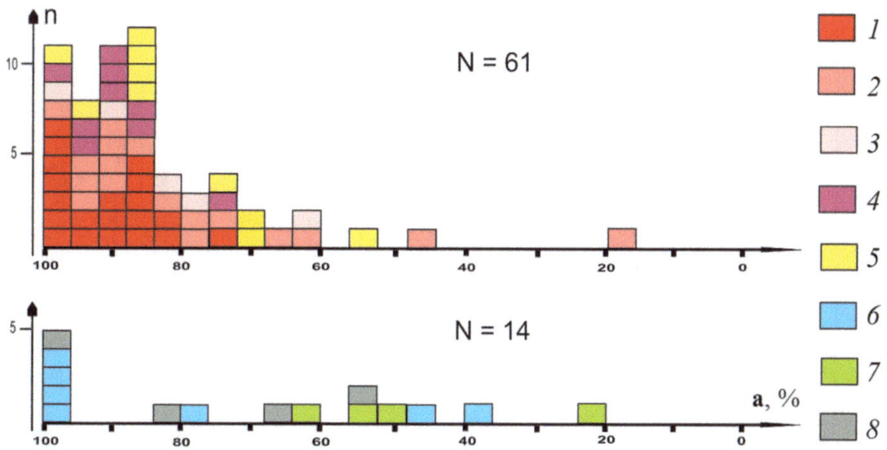

Fig. 12 Argon isotopes in rocks of productive and weakly productive areas in the Norilsk -district (stage IV). See legend in Fig. 10

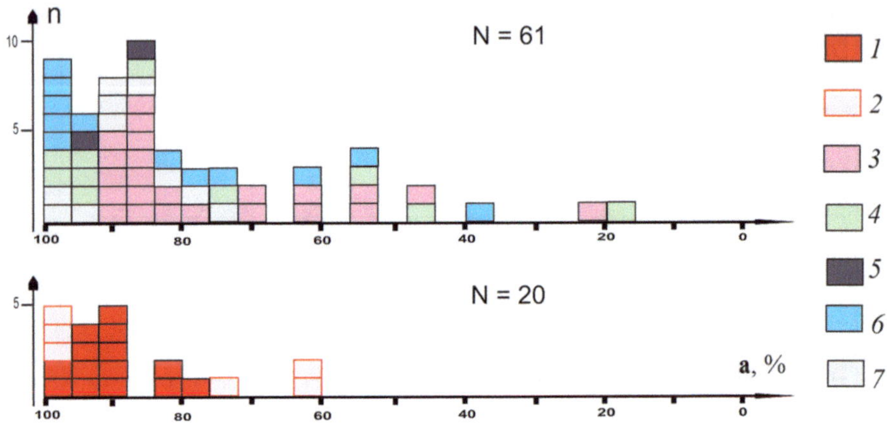

Fig. 13 Argon isotopes in rocks and ores of the studied areas in the Norilsk district. See legend in Fig. 11

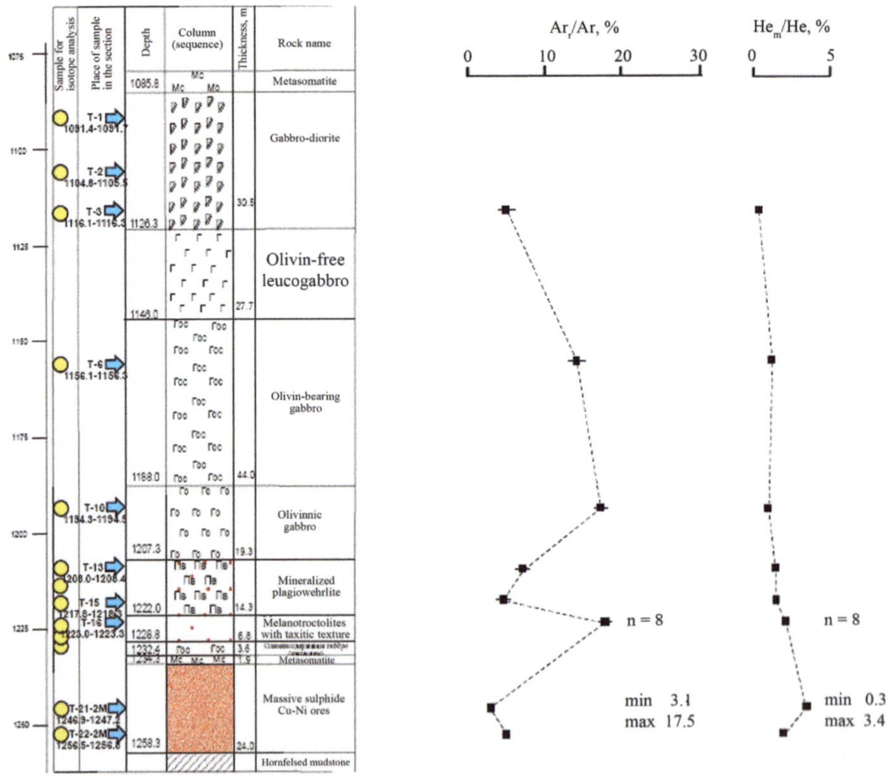

Fig. 14 Variations in He and Ar isotopic composition in rocks of Talnakh intrusive, Bh. OUG-2

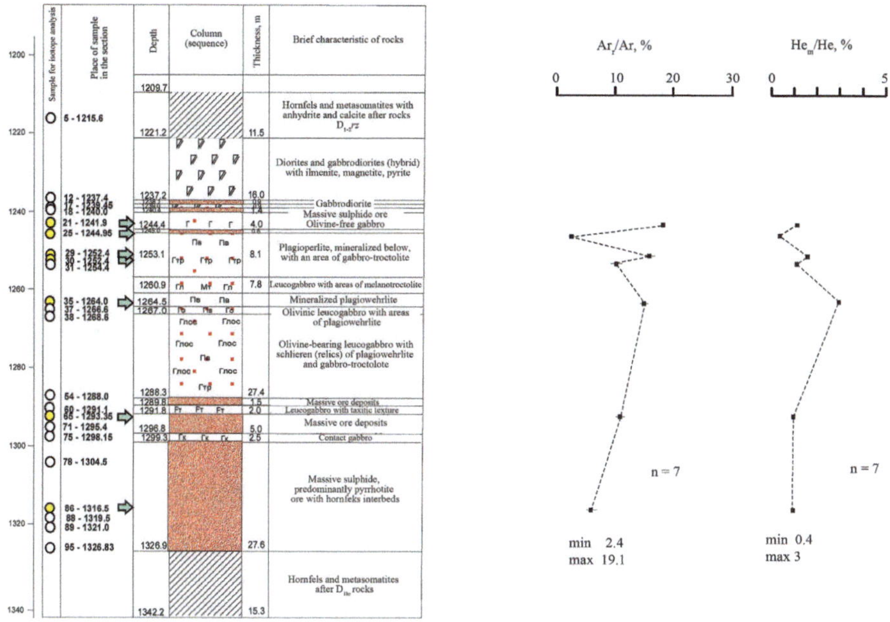

Fig. 15 Variations of He and Ar isotopic composition in rocks of Kharaelakh intrusive, Bh. KZ-963

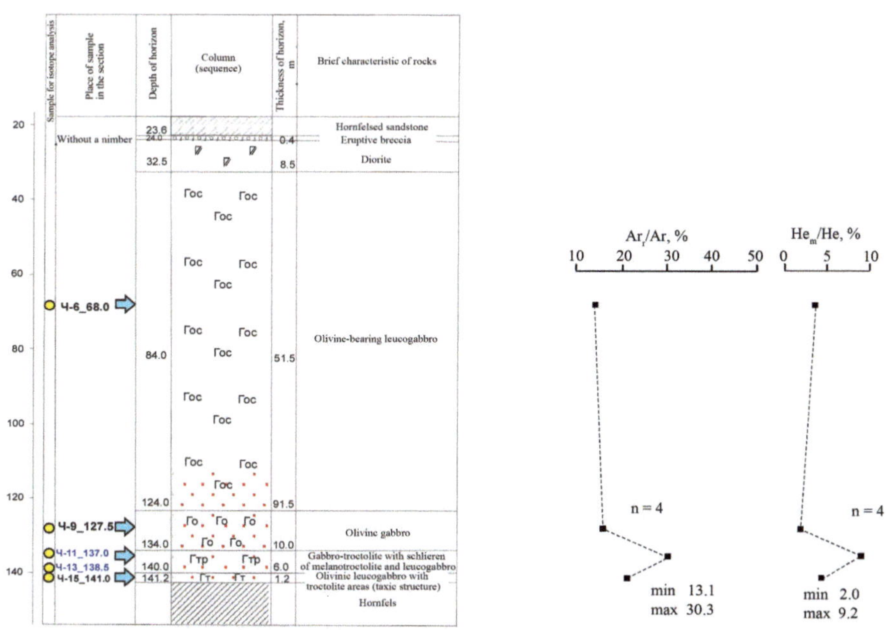

Fig. 16 Variations of He and Ar isotopic composition in rocks of Chernogorsk intrusive, Bh. MP-2bis

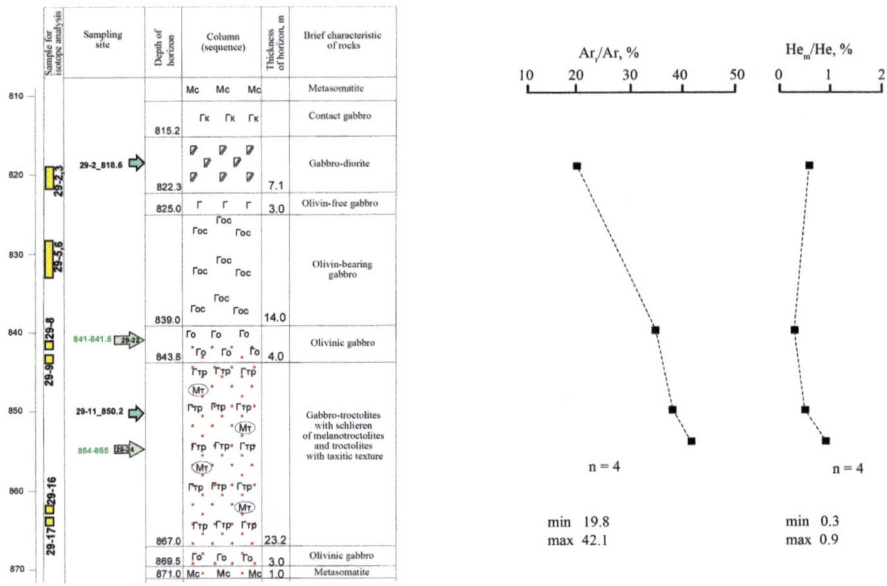

Fig. 17 Variations of He and Ar isotopic composition in rocks of Vologochan intrusive, Bh. OV-29

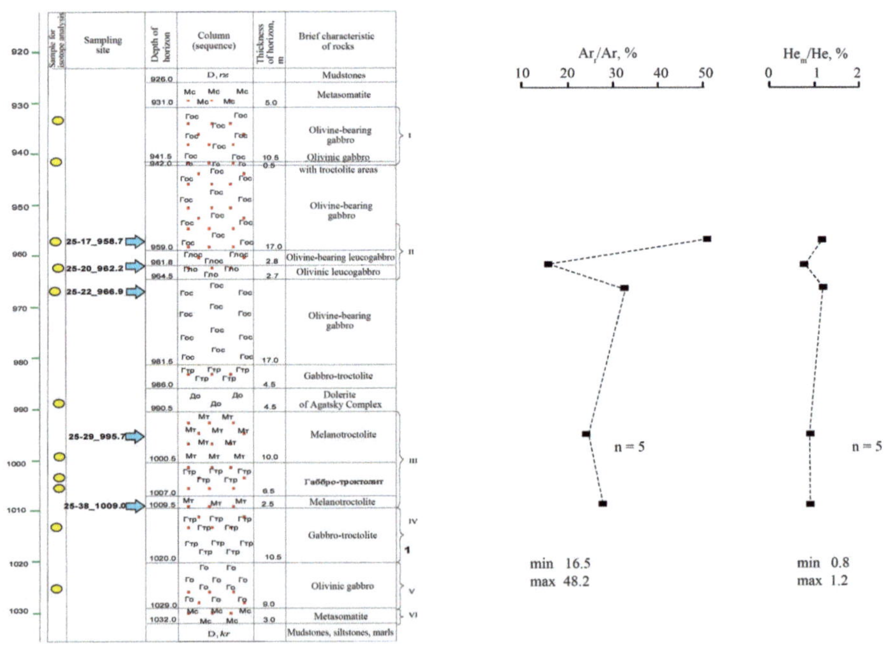

Fig. 18 Variations of He and Ar isotopic composition in rocks of South Pyasina intrusive, Bh. OV-25

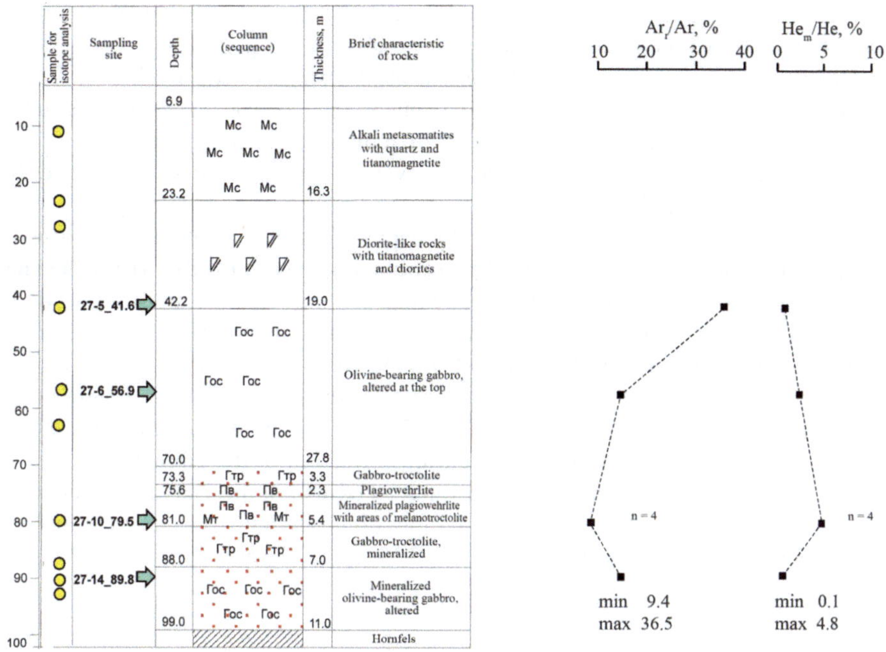

Fig. 19 Variations of He and Ar isotopic composition in rocks of Zub-Marksheidersky intrusive, Bh. MP-27

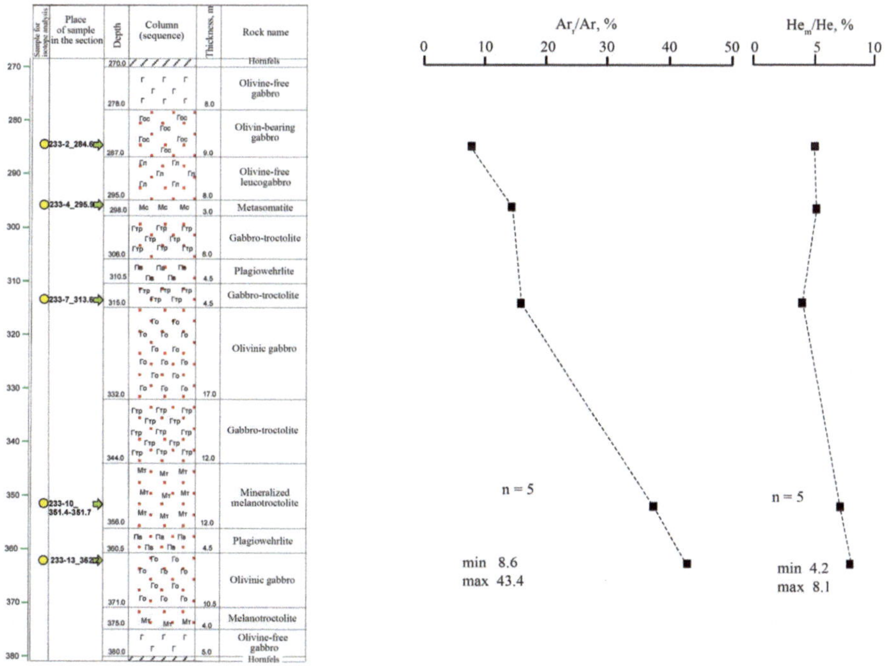

Fig. 20 Variations of He and Ar isotopic composition in rocks of Zelenaya Griva intrusive, Bh. F-233

contribution of argon of air origin points to active participation of this water in the formation of minerals, including that in the course of ore genesis.

Distribution of **a** values in intrusions at the IV stage of the work (Fig. 13) is similar to that earlier obtained at the I–III stages (Fig. 7). The share of air argon is slightly higher in plagiowehrlites and pyroxenites as compared to gabbro and dolerites. Metasomatites and hornfels, as well as basalts, are characterized by extremely broad variations.

On the western flank of the Oktyabrsky deposit (Fig. 13), **disseminated ores (4 samples) do not differ significantly by argon isotopic composition from the rich ones (massive—two samples)**. It is very important that the amount of air argon in ores is much larger than in rocks. This means that air contamination, i.e. the **impact of near-surface water was the highest during formation of ores**. In the central and southern parts of the deposit, two samples of disseminated ores contain much less air argon (63%) than rich ores (77–97% in 12 samples), indicating a less effective impact of near-surface water during formation of disseminated ores.

The central and southern parts are distinguished within the Oktyabrsky deposit, where the rocks have a much lower share of air argon (to 63%), as can be seen from Table 17 and Figs. 14, 15, 16, 17, 18, 19, 20, 21, 22, 23 and 24.

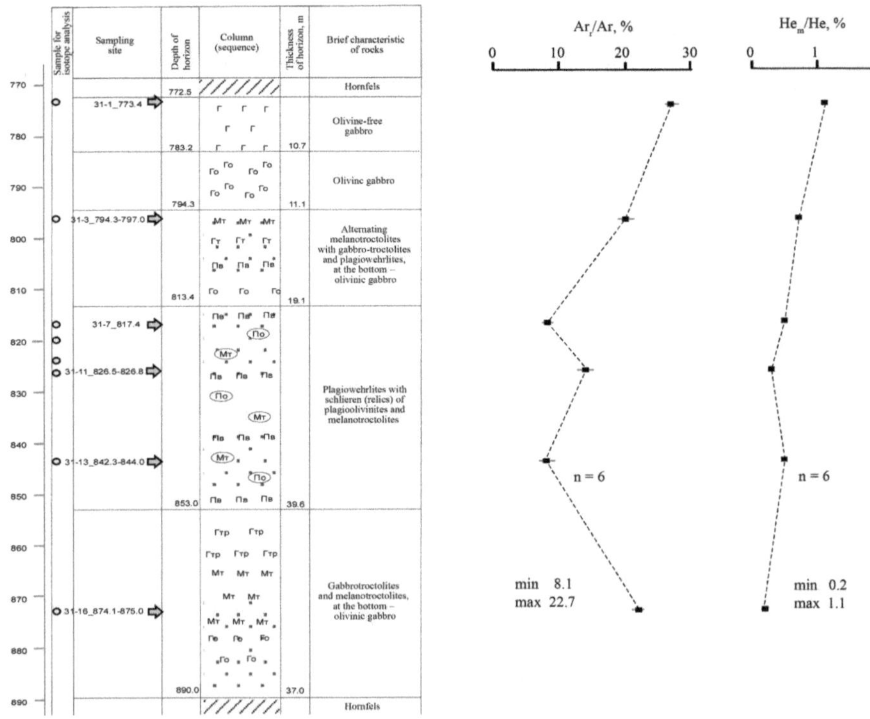

Fig. 21 Variations of He and Ar isotopic composition in rocks of Lower Talnakh intrusive, Bh. TG-31

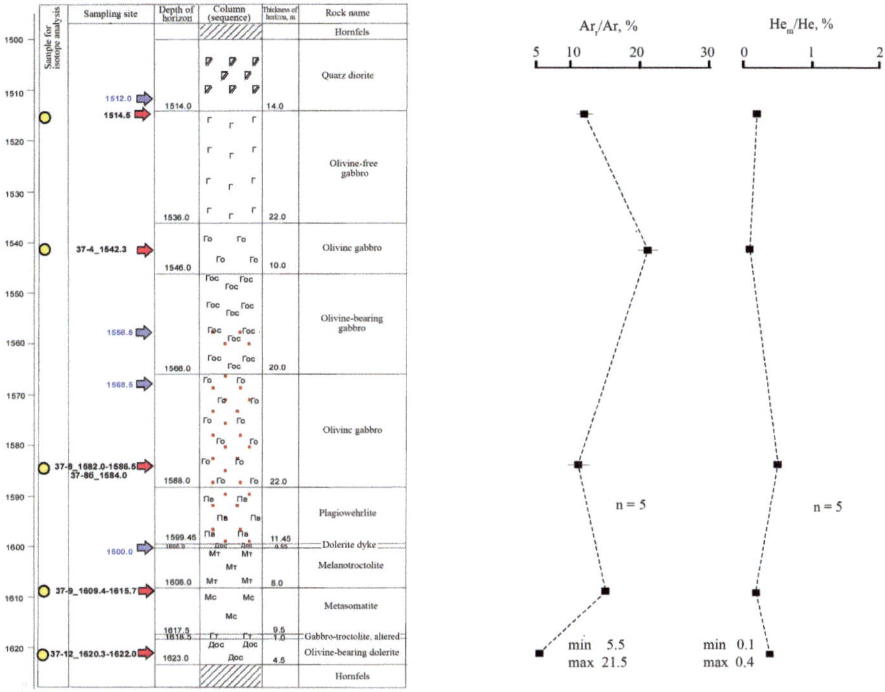

Fig. 22 Variations of He and Ar isotopic composition in rocks of Lower Norilsk intrusive, Bh. NP-37

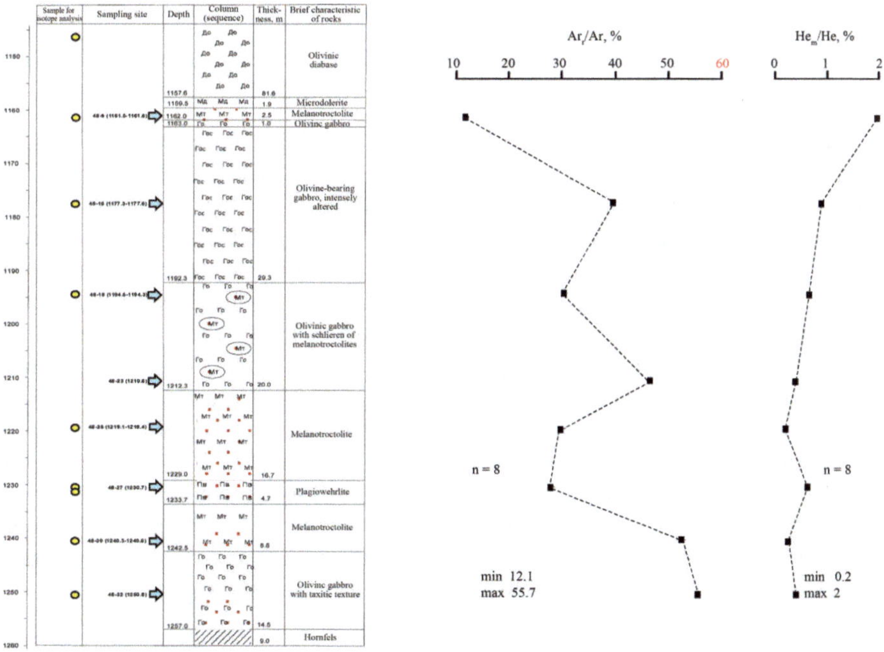

Fig. 23 Variations of He and Ar isotopic composition in rocks of Mikchangda intrusive, Bh. MD-48

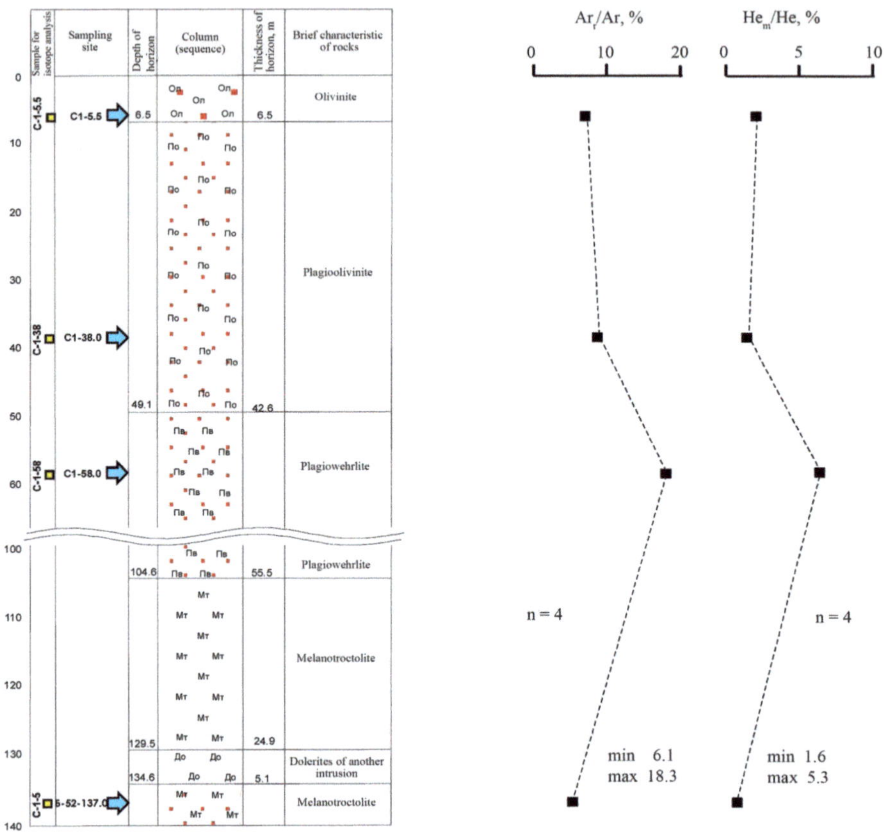

Fig. 24 Variations of He and Ar isotopic composition in rocks of Binyuda intrusive, Bh. S1

Extremely interesting data were obtained when studying argon from drill core of RT-2 located in the central Oktyabrsky deposit, but outside the projection of ore bodies (Table 13). The share of air argon here (in the 1371–1454 m interval) is low and ranges extensively (from 44 to 85%). At the top of the studied section, in hornfels, a value sets the minimum record—18.2%. Obviously, the participation of the near-surface fluids was also minimal, as well as of the mantle ones, since the share of mantle helium in this sample is negligible (**m** = 0.1%). Probably, there is an uncontaminated paleofluid, the source of which is at a great depth in the Earth's crust.

New data on the distribution of air argon are generally identical to those obtained earlier. A certain impoverishment of air argon is recorded against that adopted earlier for the commercially mineralized intrusions (**a** = 88%). Therefore, it is proposed to reduce the limit value of **a** to 80% used as an isotope criterion.

Isotope analysis of argon from gas-liquid inclusions gave a unique information: the participation of near-surface fluids in the formation of ore minerals is recorded; it is shown that it was most effective at the stage of ore genesis.

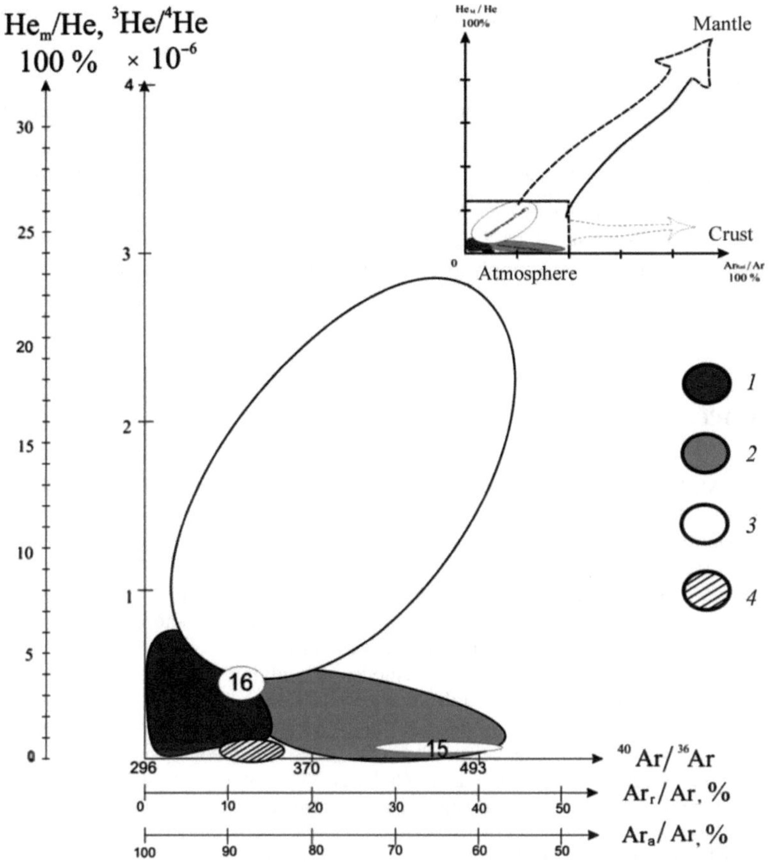

Fig. 25 He-Ar isotope criterion of ore presence degree in intrusives of Norilsk-Taimyr ore District. Occurrence zones of intrusive: 1—rich (commercial ore-bearing); 2—intermediate (ore-bearing); 3—poor (weakly ore-bearing); 4—satellites of commercial ore-bearing intrusives. Intrusives: 15—Mikchangda; 16—Binyuda

Table 18 Isotopic He–Ar criterion of ore presence

Geological economic group	Share of mantle helium **m**, %	Share of air argon **a**, %
Commercial ore-bearing (rich)	1–4	80–100
Satellites	0.2–1	81–90
Ore-bearing (average)	1–4	60–80
Weakly ore-bearing (poor)	5–22	66–95

Intrusions of all three geological economic types are different as regards the set of data on isotopes of noble gases. Rich and average intrusions differ from the poor ones in isotopic composition of helium and between themselves in argon isotope ratio. Ore-rich, intermediate and poor intrusions are segregated as definite non-overlapping areas in the diagram with $^{3}He/^{4}He$ and $^{40}Ar/^{36}Ar$ coordinates (Fig. 25). Helium-argon isotope criterion resulting from these empirical data, therefore, makes it possible to diagnose the degree of ore presence in mafic intrusions in the Norilsk district at the first stages of geological exploration by an efficient and cost-effective method of using diagrams with "criterial" areas and Table 18.

References

1. Ozima M, Podosek F (1987) Geochemistry of noble gases. Leningrad Nedra 344
2. Prasolov EM (1990) Isotope geochemistry and origin of natural gases. Leningrad Nedra 283
3. Tolstikhin IN (1986) Isotope geochemistry of helium, argon and rare gases. Leningrad Izdatel Nauka 200
4. Ikorsky SV, Kamensky IL (1998) Method of crushing rocks and minerals in glass vials in the course of isotopic studies of noble gases. In: XV symposium on isotope geochemistry, 24–27 November 1986. Abstracts of papers. Moscow, p 115
5. Tolstikhin IN, Prasolov EM (1971) Methods of studying isotopes of noble gases from microinclusions in rocks and minerals. Trans VNIISIMS XIV:86–98
6. Neruchev SS, Prasolov EM (1995) A fluid-geochemical model of platinoid deposits associated with trappean magmatism. Platin Russ 94–101 (Geoinformmark)
7. Petrov OV, Sergeev SA, Prasolov EM, Khalenev VO, Lokhov KI (2010) Geochronological and isotope-geochemical characteristics of mafic intrusions in the Norilsk district. Dokl RAS 434(3):388–390
8. Zavileisky DI, Prasolov EM (2004) Sources of ore matter and fluids in a low-sulphide platinum-metal horizon of Norilsk-1 intrusion. In: Proceedings of the XVII symposium on isotope geochemistry. Moscow, p 89
9. Petrov O et al (2011) Isotope correlations in rocks and ores of productive intrusions in the Norilsk district. Platin Russ 7:467–475
10. Khalenev VO (2009) Isotopic composition of helium and argon in paleofluids of Maslovskoye ore occurrence (Norilsk-Taimyr district). Reg Geol Metallogeny (39):85–99
11. Prasolov EM, Khalenev VO, Petrov OV (2007) Abilities of isotopic geochemistry of rare gas for differentiation of intrusions of the Norilsk area according to their ore potential. In: 7th international symposium on applied isotope geochemistry (AIG 7). Republic of South Africa, Stellenbosch. p 114
12. Prasolov EM, Khalenev VO, Gruzdov KA (2008) Noble gases isotopic features of mafic intrusions (Taimyr-Norilsk area) as the indicator of Cu-Ni-PGE ore accumulation scale. 33 IGC 14 August 2008, Oslo, Norway. Abstracts MPC 01220P
13. Aplonov VS (1995) Fluid regime and problems of platinum presence in differentiated mafic intrusions. Platin Russ 2(1):102–107
14. Aplonov VS (2001) Thermobarogeochemical model of the Talnakh platinoid copper-nickel deposit. SPb: VNIIOkeangeologia, 234 p
15. Grinenko LN (1966) Isotopic composition of sulphur in sulphides of the Talnakh copper-nickel deposit in connection with questions of its genesis. Geol Ore Deposits (4):15–31

16. Zolotukhin VV (1971) On the genesis of the so-called "liquation" copper-nickel ores in the light of new data (on the infiltration autometasomatic hypothesis). Geol Geophys (9):12–22
17. Mamyrin BA, Tolstikhin IN, Anufriev GS, Kamensky IA (1969) Anomalous isotopic composition of He in volcanic gases. DAN SSSR 184(15):1197
18. Naldrett AJ (2003) Magmatic sulphide deposits of copper-nickel and platinum-metal ores. SPb: SPBGU, 487 p
19. Arndt NT, Czamanske GK, Walker RJ et al (2003) Geochemistry and Origin of the Intrusion Hosts of the Norilsk-Talnakh Cu-Ni-PGE sulphide deposits. Econ Geol 98:495–515
20. Burnard PG, Hu R, Turner G, Bi XW (1999) Mantle, crustal and atmospheric noble gases in Ailaoshan gold deposits, Yunnan province, China. Geochim et Cosmochim Acta 63 (10):1595–1604
21. Dalrymple GB, Gramanske GK, Lanphere MA et al (1991) $^{40}Ar/^{39}Ar$ ages of samples from the Norilsk-Talnakh mineralized intrusions and the Siberian flood basalts, Siberia. EOS (Trans Amer Geophys Union) 72:570
22. Kamo SL, Czamanske GK, Amelin Y et al (2003) Rapid eruption of Siberian flood-volcanic rocks and evidence for coincidence with the Permian-Triassic boundary and mass extinction at 251 Ma. Earth Planet Sci Lett 214:75–91
23. Mamyrin BA, Tolstikhin IN (1984) Helium isotopes in nature, vol 3. Elsevier, Amsterdam, 273 p (Developm. in Geochem)
24. Tolstikhin IN, Marty B (1998) The evolution of terrestrial volatiles: a view from helium, neon, argon and nitrogen isotope modeling. Chem Geol 147:27–52

Sulphur Isotopes

Edward Prasolov, Vladimir Khalenev, Boris Belyatsky, Edward Prilepsky and Tatiana Nazarova

Abstract The chapter presents the results of sulphur isotope composition measurements ($\delta^{34}S$, in ‰), which showed that in rich intrusions sulphur is in most cases isotopically heavy ($\delta^{34}S$ = 9–13, average about 11‰), and in this parameter corresponds to the crustal source, apparently, in most cases, sedimentary anhydrite with $\delta^{34}S = 16.5 \pm 1.5$. The sulphur from medium and poor intrusions is indistinguishable by the isotopic composition and is characterized by a much lower $\delta^{34}S$ with a wide range of variations (0–9‰). Clear regularities in the distribution of sulphur isotope composition along the section of intrusions have not been revealed. The criterion should be considered as additional because of the possible fractionation of sulphur isotopes in the formation of ores. Thus, in the richest intrusions, sulphur, like helium, is of predominantly crustal origin; it appears to have been borrowed from anhydrites of the enclosing strata as a result of intensive migration (circulation) of water caused by injection. The origin of sulphur in poor and medium intrusions remains unclear. It is possible that some of the sulphur has a mantle genesis.

1 Methodology, Samples

Isotopic composition of sulphur in sulphides ($^{34}S/^{32}S$ isotopic ratio), probably, corresponds to that in fluid sulphur. It is important to find out, whether the genetic (fluid) models derived from the study of such different isotope systems—noble gases and sulphur—are adequate. However, unlike noble gases, when interpreting the data on sulphur, we cannot with the same confidence offer solutions due to a certain indefiniteness of genetic criteria. This indefiniteness is inherent in the systems of stable isotopes and is caused by isotopic fractionation accompanying chemical transformations. In reality, isotopic composition of the reaction products depends not only on the fractionation coefficient of isotopes between the initial

E. Prasolov (✉) · V. Khalenev · B. Belyatsky · E. Prilepsky · T. Nazarova
Russian Geological Research Institute (VSEGEI), St. Petersburg, Russia
e-mail: edward_prasolov@vsegei.ru

O. Petrov (ed.), *Isotope Geology of the Norilsk Deposits*, Springer Geology,
https://doi.org/10.1007/978-3-030-05216-4_2

matter (for example, sulphate or hydrogen sulphide) and the product (hydrogen sulphide or sulphides), but also on the degree of reaction completion. At full transition of the atoms of a certain element from the initial matter into the product, isotopic composition, naturally, remains unchanged. On the one hand, despite these difficulties, we can expect the corresponding correlations (in case of adequacy of fluid formation models). On the other hand, it is due to these difficulties that we are forced to consider sulphur isotopic composition as an additional fluid genetic criterion.

It is known that isotopically heavy sulphur ($\delta^{34}S$ ~15–20‰) corresponds to sulphates in sedimentary rocks; sulphides, including hydrothermal ones, are much lighter; $\delta^{34}S$ value in them reaches zero [6, 8–10]. It is believed that it is approximately the same and is typical of the mantle sulphur. This information is used in the interpretation of isotopic data.

Isotopic composition of sulphur here and below is presented as usual as $\delta^{34}S$ values, where the "standard" is sulphur of troilite in Canyon Diablo meteorite

$$\delta^{34}S(\text{‰}) = \left(\frac{\left({}^{34}S/{}^{32}S\right)_{\text{sample}}}{\left({}^{34}S/{}^{32}S\right)_{\text{standard}}} - 1 \right) \times 1000.$$

Earlier, Grinenko et al. [2–5] obtained very extensive data on isotopic composition of sulphur in different sulphides and anhydrite (overall, over 500 analyses) mainly in the Talnakh intrusion. Later, they were summarized and discussed in detail by V. S. Aplonov in [1] (Fig. 1).

Isotopically heavy sulphur was present in ores (Fig. 2) with $\delta^{34}S$ value 8–12‰, in the vast majority of cases (on the average, about 10‰) approaching the values characteristic of anhydrite (15–18, averaging at 16.5‰). These data renounced the mantle origin of sulphur in sulphide ores and pointed to a significant contribution of sulphur of sulphates in the sedimentary sequence. The conclusions about the genesis of fluids based on the data of sulphur isotopes agree well with the above results of isotopic studies of noble gases.

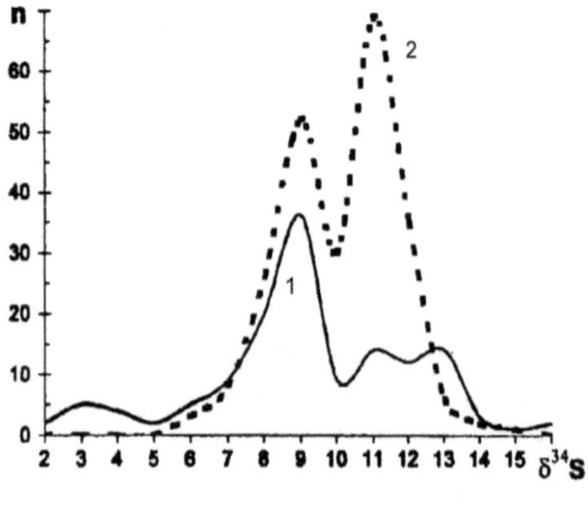

Fig. 1 Isotopic composition of sulphur in sulphides of solid (1) and disseminated (2) ores of Talnakh deposit (according to Aplonov [1])

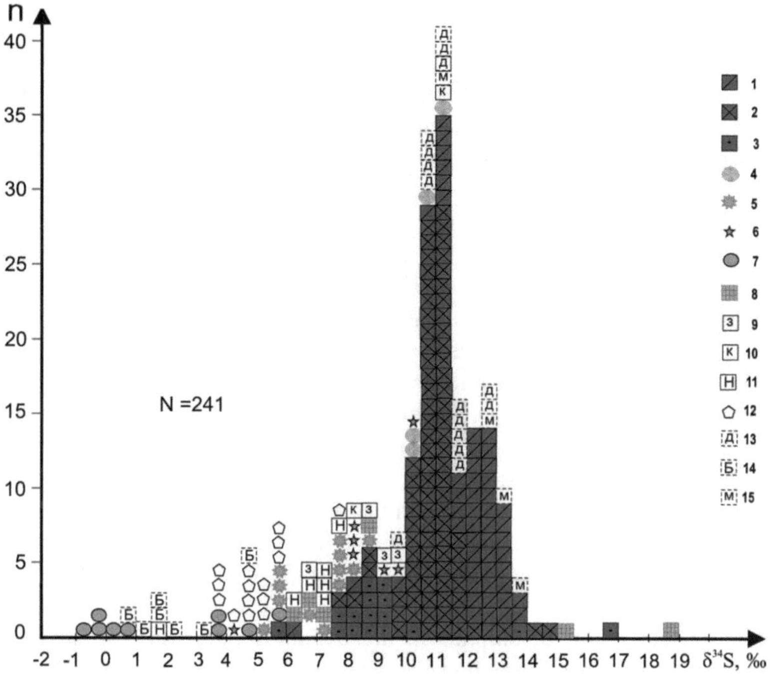

Fig. 2 Distribution of $\delta^{34}S$ value in intrusives with different degrees of ore presence (stage I). Intrusives: 1–3, rich (1—Kharaelakh; 2—Talnakh; 3—Norilsk-1); 4–8—intermediate (4—Chernogorsk; 5—Vologochan; 6—South Pyasina; 7—Zub-Marksheidersky; 8—Imangda); 9, 10—poor (9—Zelenaya Griva; 10—Kruglogorsky); 11, 12—satellites (11—Lower Talnakh; 12—Lower Norilsk); 13–15—prospective; (13—Dyumptalej; 14—Binyuda; 15—Mikchangda)

Conducting our own study of sulphur isotopes—one of the main components of ore minerals—at the **I stage** (2006–2008) we had the following objectives:

- to find the relationships (correlation) between the characteristics of the largest possible number of isotope systematics;
- to obtain data on isotopic composition of sulphur in a large number of intrusions with varying ore content.

At the II stage (2012–2014), we tried to obtain data on isotopic homogeneity of sulphur in an isolated intrusion.

Isotopic composition of sulphur in sulphides (in total) was analysed using mass spectrometer DELTAplusXL with an attachment EA-ConFlo III following IRM-MS procedure with reproducibility (1σ) about 0.2‰. In most cases, sulphur was investigated in the same objects and even samples as noble gases. At the I stage, 240 samples from 14 intrusions were studied; at the II stage, 80 samples from 4 intrusions.

2 Results and Discussion

Our data obtained at the I stage (Table 1, Fig. 2) confirmed a high value of $^{34}S/^{32}S$ ratio in rich intrusions as well as the absence of isotopic differences in disseminated and massive ores. Measurements showed that in rich intrusions, indeed, in most cases, sulphur was isotopically heavy ($\delta^3 4S$ = 9–13, on the average, about 11‰), and by this parameter, it corresponds to a crustal source, apparently, to sedimentary anhydrites with $\delta^{34}S = 16.5 \pm 1.5$ in most cases [1]. This interval comprises not only the three richest intrusions (Norilsk-1, Talnakh, Kharaelakh), but also the intermediate Chernogorsky and Imangda as well as Dyumptalei and Mikchangda intrusions with unknown prospects.

Sulphur of the intermediate and poor intrusions is indistinguishable by its isotopic composition, has a much lower $\delta^{34}S$ value with a wide range of variations (0–9‰). Isotopic composition of sulphur can be used as an additional criterion for assessing the degree of intrusion mineralization.

No clear pattern has been revealed in the distribution of isotopic composition of sulphur in the section of intrusions. This is evident from the results obtained at the I stage for individual objects (Table 1 and Figs. 3, 4, 5, 6, 7, 8, 9, 10, 11, 12, 13 and 14).

Isotopic composition of sulphur in the **Talnakh** intrusion was studied in 10 samples from OUG-2 drill core. It is characterized by a high average value of $\delta^{34}S$ – 10.6‰, which is typical of the rich platinum-copper-nickel deposits of the Norilsk group. The heaviest sulphur (to 15‰) was found in the upper part of the section in gabbrodiorites. Lower value of $\delta^{34}S$, 8.5‰, is recorded in the middle part, in olivine-free and olivine-bearing gabbro, and about 12‰ in the lower part, in mineralized plagiowehrlites and taxites at the contact with massive sulphide ore.

Sulphur isotopes in the **Kharaelakh** intrusion were studied in 19 samples from KZ-963 drill core. Similar to other rich intrusions of the Norilsk district, the average value is high (12.5‰). The lightest sulphur was found at the top of the section (to 11.4‰) in hornfels and metasomatic rocks with anhydrite and calcite. Sulphur becomes heavier in gabbro-diorites to 12.5‰, and only occasionally becomes lighter than 12‰ in plagiowehrlites in the middle of the section and in massive sulphide ore at the base of the section.

Sulphur of the **Chernogorsky** intrusion was studied only in two samples from MP-2bis drill core in the horizon of gabbro-troctolites with schlieren of melanotroctolites and leucogabbro. These samples are characterized by the average $\delta^{34}S$ value of 10.7‰.

Isotopic composition of sulphur in the **Vologochan** intrusion was studied in 5 samples from OV-29 drill core and has an average value of 7.1‰. The lightest sulphur is recorded in olivine gabbro (to 5.7‰). In gabbro-troctolites with schlieren of melanotroctolites and troctolites with a taxitic texture and metasomatic rocks sulphur becomes heavier to 8.5 and 7.8‰.

$\delta^{34}S$ value in the **South Pyasino** intrusion was determined in 7 samples from OV-25 drill core and averages at 8.3‰. In olivine-bearing gabbro at the top of the

Table 1 Sulphur isotopes in ore minerals of mafitic intrusives in Norilsk and Taimyr provinces (I stage)

Sample	Depth	Type of mineralization	Minerals	$\delta^{34}S$, ‰
Norilsk-1, Bh. MN-2				
N1-1	330.6	Disseminated	Cu–Ni sulphides	9.2
N1-2	338.5			16.7
N1-3	339.4	Low-sulphide		8.3
N1-6	359	Disseminated		13.8
N1-7	365			9.4
N1-8	367.5			9.1
N1-9	371			7.9
N1-10	380			7.5
Norilsk-1, Bh. MS-18				
18 1		Low-sulphide	Cu–Ni sulphides	5.5
18 2				10.4
Norilsk-1, Medvezhij Ruchej open pit				
Kn-97-1		Disseminated	Cu–Ni sulphides	8.6
Kn-97-2				8.3
Kn-97-3				8.9
Kn-97-4				8.9
Kn-97-5				8.8
Kn-97-6				9.0
Talnakh intrusive, Bh. OUG-2				
T-1		Disseminated	Cu–Ni sulphides	9.1
T-2				15.0
T-3				8.9
T-5,6				8.7
T-13				9.9
T-14				11.4
T-15				11.0
T-16				11.1
T-17				11.0
T-18				10.7
T-19 (T-14d.)				11.1
T-20 (T-18d.)				10.4
72r	1233.3	Massive	Pyrrhotite	7.8
78r	1238.6			10.5
82r	1242.9			10.7
87r	1248.2			11.1
89r	1250.4			11.0
102r (d.89r)	1250.4			11.1
96r	1257.7			10.9
87sr	1248.2		Chalcopyrite	11.2
89sr	1250.4			11.2
103sr (d.89sr)	1250.4			11.0

(continued)

Table 1 (continued)

Sample	Depth	Type of mineralization	Minerals	$\delta^{34}S$, ‰
26_V	1208.4	Disseminated	Cu–Ni sulphides	10.4
27_V	1208.9			10.8
21_ V (d.27B)	1208.9			10.8
28_V	1209.4			10.6
29_V	1210			10.3
30_V	1210.5			10.5
31_V	1211			10.8
32_V	1211.5			10.8
33_V	1212			11.1
34_V (d.33V)	1212	Disseminated	Cu–Ni sulphides	11.7
35_V	1213			11.5
37_V	1214			10.8
38_V	1215			10.6
39_V	1215.9			10.6
43_V	1218.1			11.1
44_V	1218.7			11.3
45_V	1219.2			11.3
47_V	1220.2			11.0
48_V	1220.7			11.2
49_V	1221.4			10.3
46_V (d.49V)	1221.4			11.0
50_V	1221.9			11.3
51_V	1222.1			11.2
52_V	1222.6			11.0
53_V	1223.2			11.1
54_V	1223.7			11.4
55_V	1224.3			12.0
56_V	1224.8			11.8
57_V	1225.3			11.6
60_V	1227.3			11.6
61_V (d.60V)	1227.3			11.2
63_V	1229.5			11.1
65_V	1230.5			10.5
66_V	1231.1			10.7
68_V	1231.8			10.8
69_V	1232.1			10.3
67_V (d.69V)	1232.1			10.5
71_S	1232.8	Massive		10.3
70_S (d.71S)	1232.8			10.7
72_S	1233.3			8.2
73_S	1233.8			10.5
74_S	1234.4			10.6
75_S	1235.3			9.7

(continued)

Table 1 (continued)

Sample	Depth	Type of mineralization	Minerals	$\delta^{34}S$, ‰
76_S	1236.4			11.1
77_S	1237.5			11.6
107_S (d.77S)	1237.5			11.2
78_S	1238.6			11.4
79_S	1239.6			10.7
80_S	1240.7			10.8
81_S	1241.8			10.8
82_S	1242.9			10.5
83_S	1244			11.3
105_S (d.83S)	1244			11.2
84_S	1245.1			10.9
85_S	1246.2			11.1
86_S	1247.1			11.2
87_S	1248.2			11.3
88_S	1249.3			11.3
89_S	1250.4			11.6
104_S (d. 89S)	1250.4			11.3
90_S	1251.5			11.5
91_S	1252.6			11.4
92_S	1253.7			11.1
93_S	1254.5			11.1
94_S	1255.6			12.1
95_S	1256.7	Massive	Cu–Ni sulphides	10.9
98_S (d. 95S)	1256.7			11.5
96_S	1257.7			10.8
97_S	1258.3			10.9
99_S	1259.5			10.8
100_S	1261			10.8
101_S	1262.3			11.1
106_S (d. 101S)	1262.3			11.1
Kharaelakh intrusive, Bh. KZ-844				
844-2V		Disseminated	Cu–Ni sulphides	11.4
844-2V	949.5			12.5
844-3,4V				13.1
844-3,4V	955.0–956.0			13.5
844-6V				12.4
844-7V				13.3
844-10,11V				12.2
844-11V	1021			12.5
844-15V				6.3
844-18S	1046	Massive		13.0
844-19S	1055			13.0
844-19S				11.7
844-20S	1063			12.7

(continued)

Table 1 (continued)

Sample	Depth	Type of mineralization	Minerals	$\delta^{34}S$, ‰
Kharaelakh intrusive, Bh. KZ-963				
963-5		Veinlet	Cu–Ni sulphides	11.4
963-12		Massive		13.0
963-17				12.8
963-17 (d.)				12.7
963-18				12.8
963-25				12.4
963-30		Disseminated		11.5
963-30				13.0
963-30 (d.)				11.9
963-31				11.9
963-37				12.6
963-37				12.5
963-38				13.2
963-54				13.1
963-60				12.8
963-71				13.1
963-75		Massive		13.2
963-75 (d.)				13.6
963-78				11.8
963-86				12.6
963-88				12.3
963-88 (d.)				12.6
963-89				12.4
963-95				12.8
Kharaelakh intrusive, Oktyabrskoe deposit				
1/S-08-16		Malachite ore	Cu–Ni sulphides	11.3
2/S-08-16		Cubanite-chalcopyrite ore		12.1
2/SER-02		Galena-talnachite with PGM	Cu–Ni sulphides	12.0
3/S-08-17		Talnachite ore	Talnachite	12.5
3/S-08-17		Talnachite ore		12.5
500		Galena-chalcopyrite ore	Cu–Ni sulphides	11.3
3765		Bornite ore		12.9
5481		Chalcopyrite ore		10.5
5487		Bornite ore		12.6
5499		Chalcopyrite ore		11.2
6001		Galena-bornite-chalcopyrite ore		12.5
6027		Galena-bornite ore		13.6
4370-10		Galena-chalcopyrite ore		12.2
4370-11				12.7
4370-8				12.1

(continued)

Table 1 (continued)

Sample	Depth	Type of mineralization	Minerals	$\delta^{34}S$, ‰
3a/S-08-16		Chalcocite-calcite vein	Chalcocite	8.4
1/S-08-17	242–3530	Massive	Cubanite	9.6
5S-08 c/g			Anhydrite D_1zb	17.7
5S-08 f/g				17.9
Talnakh deposit				
6b/S-08-16		Pyrite-pyrrhotite ore	Pyrrhotite	14.0
4a/S-08-16		Sulphide-anhydrite ore	Cu–Ni sulphides	10.1
6a/S-08-16		Pyrite-pyrrhotite ore	Pyrite	13.4
Chernogorsk intrusive, Bh. MP-2bis				
CH-11		Disseminated	Cu–Ni sulphides	10.4
CH-13				11.0
CH 11_d.				10.9
CH-13_d.				11.2
Zub-Marksheidersky intrusive, Bh. MP-27				
27-13_d.		Disseminated	Cu–Ni sulphides	–0.1
27_1				5.7
27_3				0.8
27_4				3.7
27_5				4.7
27_10				3.9
27_13				–0.4
27_14				0.2
MP-27/96.9				0.0
MP-27/97.4				–0.7
MP-27/97.5				–0.1
Vologochan intrusive, Bh. OV-25				
OV-29/814.2		Disseminated	Cu–Ni sulphides	7.7
OV-29/814.7				6.6
OV-29/852.3				–4.0
OV-29/853.9				5.1
OV-29/854.3				8.5
OV-29/862.4				7.9
OV-29/864.5				5.9
OV-29/867.5				7.2
29-24				8.5
29_9		Disseminated	Cu–Ni sulphides	5.7
29-16				7.9
29-17				5.6
29-19				7.8

(continued)

Table 1 (continued)

Sample	Depth	Type of mineralization	Minerals	$\delta^{34}S$, ‰
South Pyasina intrusive, Bh. OV-25				
25_4		Disseminated	Cu–Ni sulphides	8.0
25-20				4.3
25-31				8.1
25-35				8.5
25-36				10.5
25-41				9.6
25-44				9.5
Imangda intrusive, Bh. KP-4				
4_2		Disseminated	Cu–Ni sulphides	6.4
4_5				7.0
4_6				6.7
4_8				8.7
4_9				18.8
4_10				15.4
Mikchangda intrusive, Bh. MD-48				
48-9		Disseminated	Cu-Ni sulphides	12.8
48-25				11.0
48-27, 28				13.3
48-32, 33				14.0
Binyuda intrusive, Bh. S-1, 2, 3				
S1-1		Disseminated	Cu–Ni sulphides	2.0
S1-2				1.6
S1-3				1.7
S-1-5 (N 52)				0.7
S2_1				1.3
S2_2				4.7
S3-2				3.0
Dyumptalej intrusive, Bh. TP-43				
43_1	750	Disseminated	Cu–Ni sulphides	11.7
43_2	848.5			9.9
43_3	858.5			11.2
43_4	874			11.0
43_5	883.2			10.8
43_6	892.8			10.9
43_8	901.4			11.2
43_9	913			11.7
43_10	915			11.6
43_11	915.5			10.7
43_12	917.8			11.7
43_13	918			11.0
43_14	920			11.6
43_15	937			12.9
43_16	939.5			12.7

(continued)

Table 1 (continued)

Sample	Depth	Type of mineralization	Minerals	$\delta^{34}S$, ‰
Lower Talnakh intrusive, Bh. TG-31				
31_1		Disseminated	Cu–Ni sulphides	1.8
31_3				6.5
31_9				7.0
31_10				7.3
31_11				7.3
31_11 (d.)				7.0
31_13				6.6
31_16				8.0
31-16 (d.)				7.6
Lower Norilsk intrusive, Bh. NP-37				
37_1	1609.4	Disseminated	Cu–Ni sulphides	3.8
37_2	1612.8			3.9
37_3	1613.4			4.9
37_4	1614			5.1
37_5	1615.6			3.9
37_6	1615.7			5.6
37_7	1617.9			4.0
37_8	1619.1			4.6
37_9	1620.3			4.9
37-9				7.7
37_10	1621.5			5.2
37_11	1622			5.7
37_12	1622.2			5.2
37-12				4.8
37_13	1622.9			5.1
Zelenaya Griva intrusive, Bh. F-233				
F-233-2		Disseminated	Cu–Ni sulphides	9.5
F-233-7				8.5
F-233-10				6.8
F-233-10_d.				7.5
F-233-11				9.7
Kruglogorsky intrusive, Bh. MP-2 bis				
K-4		Disseminated	Cu–Ni sulphides	11.4
K-6				8.0
Agatsky intrusive				
25-25, 26		Disseminated	Cu–Ni sulphides	15.5
Igarka intrusive				
4/S-08-17		Disseminated	Native copper	1.1

Notes Samples Kn-97-1, Kn-97-2 and Kn-97-3 have duplicates (Kn-97-1 = Kn-97-4; Kn-97-2 = = Kn-97-5; Kn-97-3 = Kn-97-6), N1-4—olivine-free gabbro and N1-5—olivine-bearing gabbro (NDA)

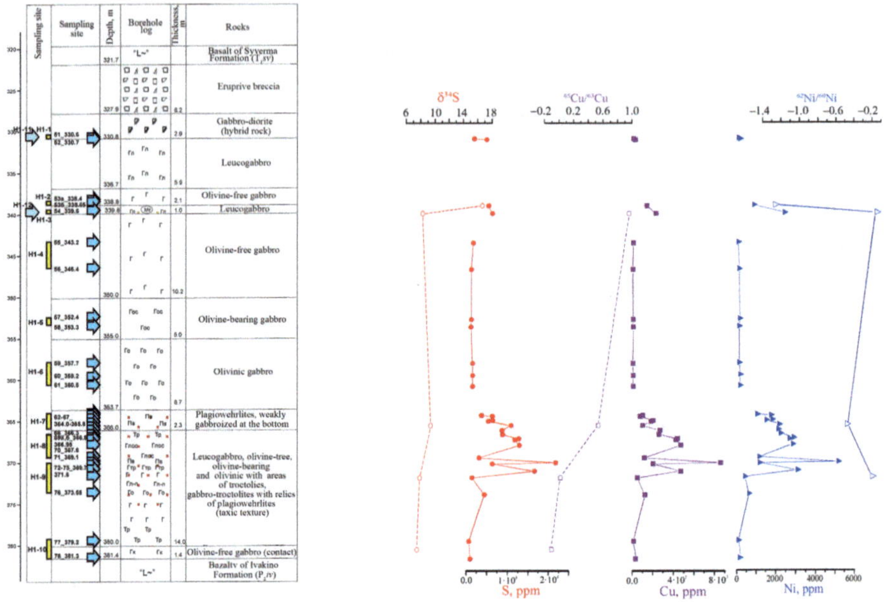

Fig. 3 Variations of isotopic composition of sulphur, copper, nickel in Cu-Ni sulphides and concentrations of sulphur, copper, nickel in rocks of Norilsk-1 intrusive, Bh. MN-2

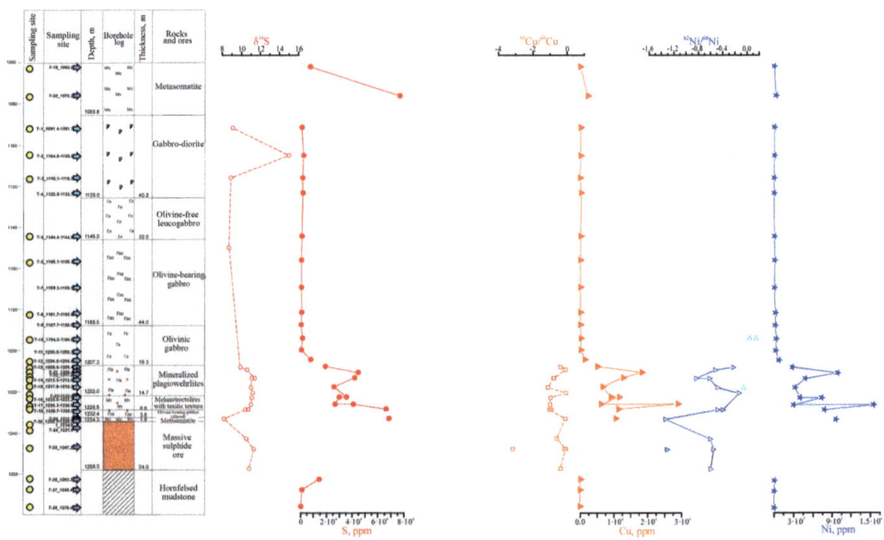

Fig. 4 Variations of isotopic composition of sulphur, copper, nickel in Cu-Ni sulphides and concentrations of sulphur, copper, nickel in rocks of Talnakh intrusive, Bh. OUG-2

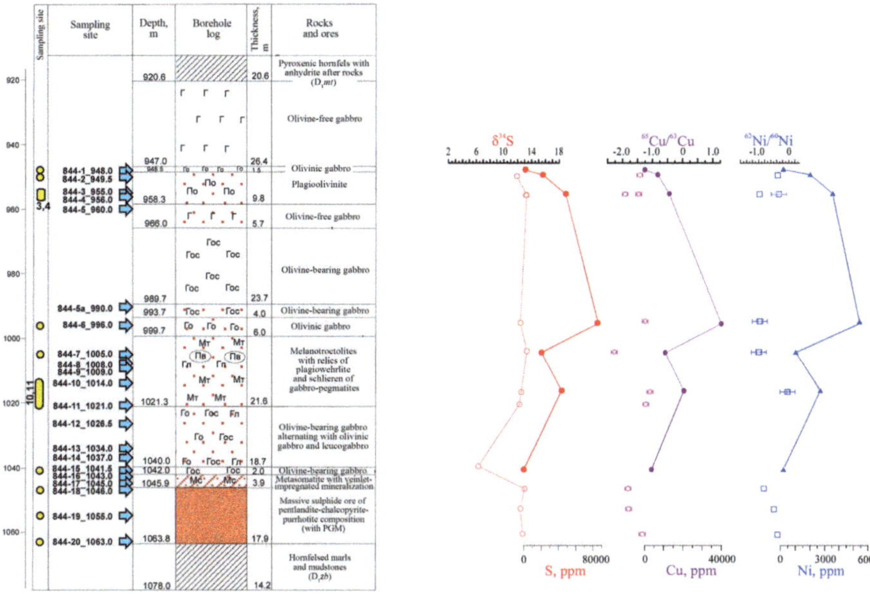

Fig. 5 Variations of isotopic composition of sulphur, copper, nickel in Cu-Ni sulphides and concentrations of sulphur, copper, nickel in rocks of Kharaelakh intrusive, Bh. KZ-844

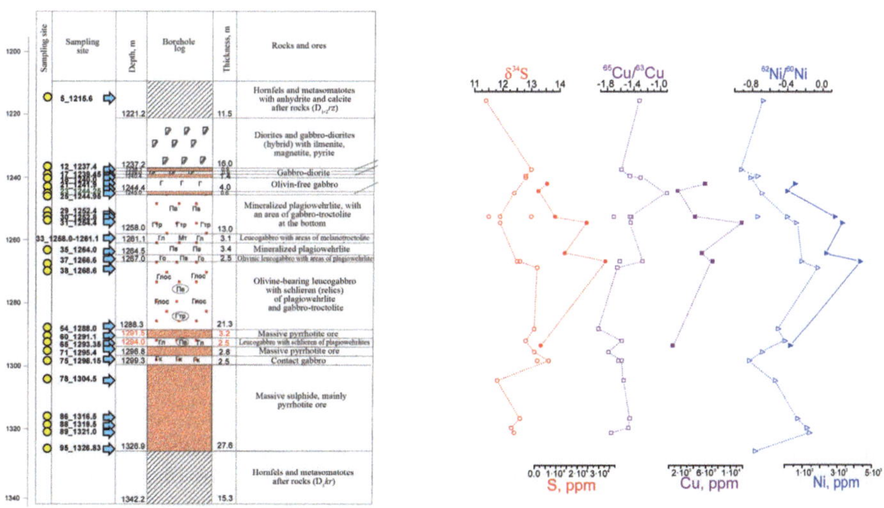

Fig. 6 Variations of isotopic composition of sulphur, copper, nickel in Cu-Ni sulphides and concentrations of sulphur, copper, nickel in rocks of Kharaelakh intrusive, Bh. KZ-963

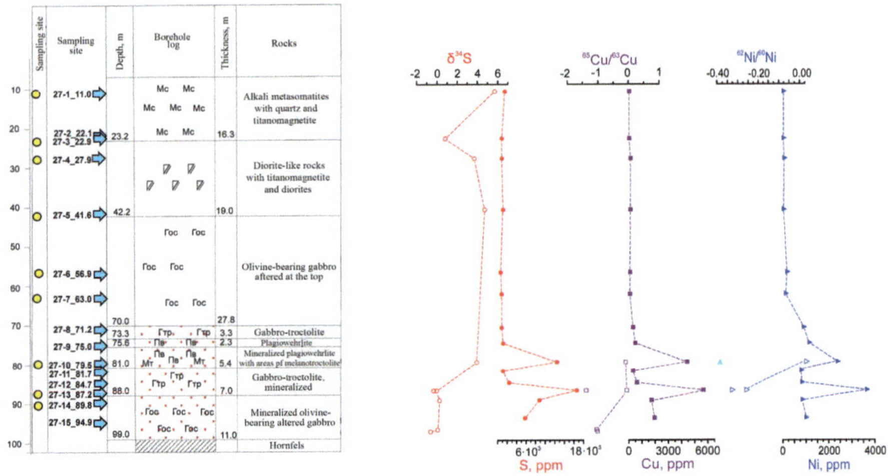

Fig. 7 Variations of isotopic composition of sulphur, copper, nickel in Cu-Ni sulphides and concentrations of sulphur, copper, nickel in rocks of Zub-Marksheidersky intrusive, Bh. MP-27

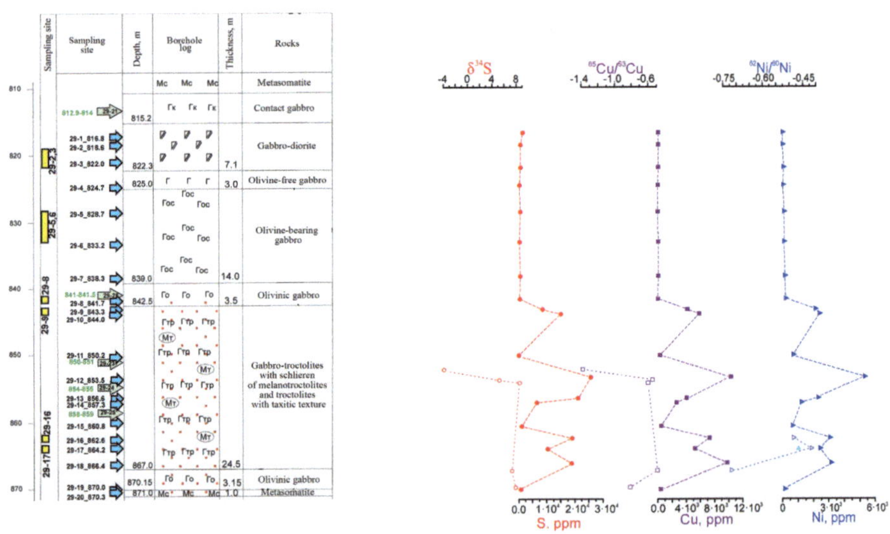

Fig. 8 Variations of isotopic composition of sulphur, copper, nickel in Cu-Ni sulphides and concentrations of sulphur, copper, nickel in rocks of Vologochan intrusive, Bh. OV-29

section, $\delta^{34}S$ is 8.0‰; in olivine leucogabbro, sulphur is much lighter (4.3‰). In melanotroctolites, $\delta^{34}S = 8.5$‰; in gabbro-troctolites sulphur is the heaviest (10.5‰). At the base of the section. in gabbro-troctolites and olivine gabbro, isotopic sulphur is almost similar ($\delta^{34}S = 9.5$‰).

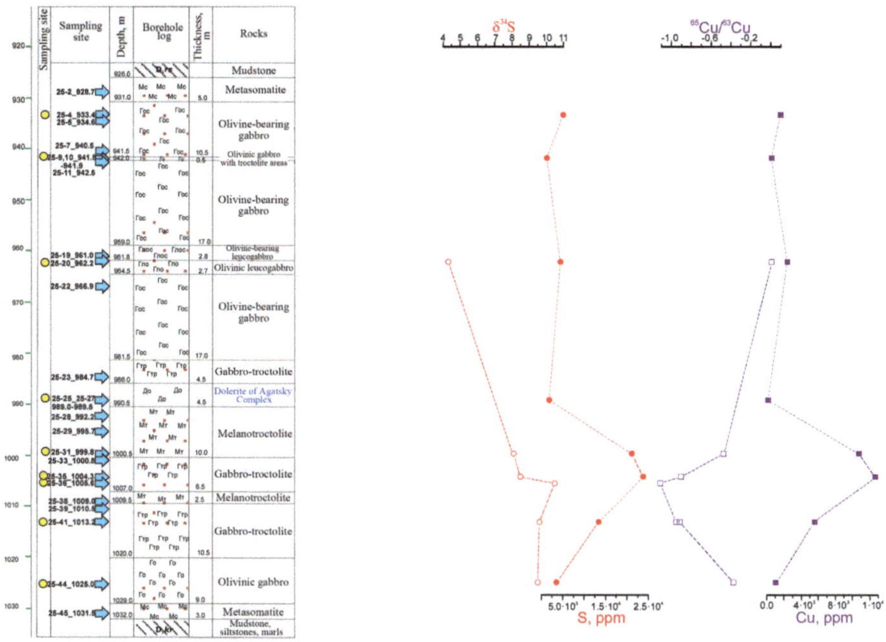

Fig. 9 Variations of isotopic composition of sulphur, copper in Cu-Ni sulphides and concentrations of sulphur, copper in rocks of South Pyasina intrusive, Bh. OV-25

Isotopic composition of sulphur in the **Zub-Marksheidersky** intrusion was studied in 7 samples from MP-27 drill core and is characterized by a low average value of $\delta^{34}S$, 6‰. Sulphur, which is heaviest at the top of the section (to 5.7‰), in alkaline metasomatic rocks with quartz and in diorites with titanomagnetite becomes much lighter (0.8–3.7‰). In the mineralized plagiowherlite with areas of melanotroctolites, $\delta^{34}S$ is 4.7‰. $\delta^{34}S$ value decreases markedly in mineralized gabbro-troctolite and olivine-bearing gabbro (to −0.4 and 0.2‰).

In **Zelenaya Griva** intrusion (four samples from F-233 drill core), the average value of $\delta^{34}S = 8.6$‰. Sulphur becomes lighter from olivine-bearing gabbro (9.5‰) to gabbro-troctolites (8.0‰), and then to the mineralized melanotroctolite (6.5‰), and becomes heavier in olivine gabbro (10‰).

Isotopic composition of sulphur in the **Lower Talnakh** intrusion was determined in 7 samples from TG-31 drill core and has an average $\delta^{34}S$ value of 6.3‰. The lightest sulphur is recorded at the top of the section in olivine-free gabbro at the contact with hornfelsed rocks (1.8‰). In the rest of the rocks down the section, sulphur is relatively homogeneous. Thus, in melanotroctolites alternating with gabbro-troctolites, $\delta^{34}S = 6.5$; in plagiowehrlites, 7.3; and down the section in gabbro-troctolites and melanotroctolites, 8.0‰.

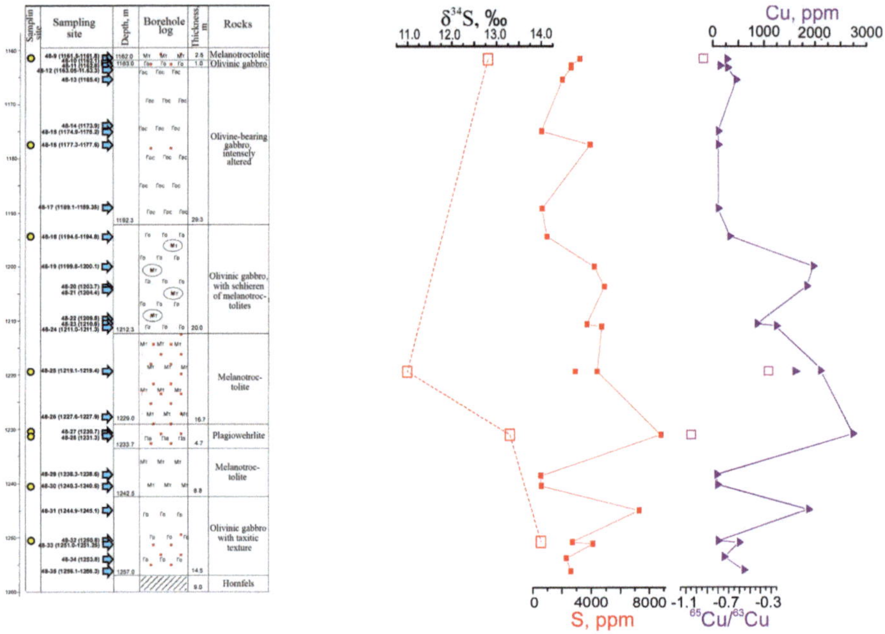

Fig. 10 Variations of isotopic composition of sulphur, copper in Cu-Ni sulphides and concentrations of sulphur, copper in rocks of Mikchangda intrusive, Bh. MD-48

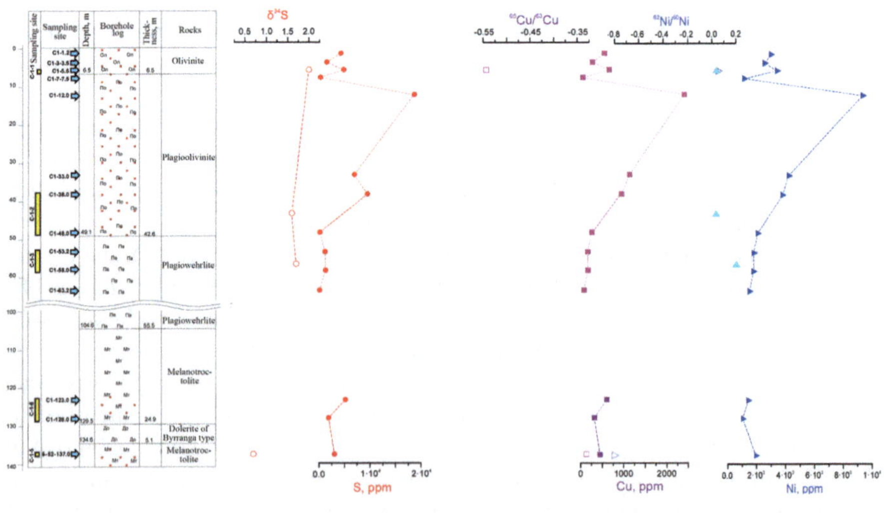

Fig. 11 Variations of isotopic composition of sulphur, copper, nickel in Cu-Ni sulphides and concentrations of sulphur, copper, nickel in rocks of Binyuda intrusive, Bh. S-1

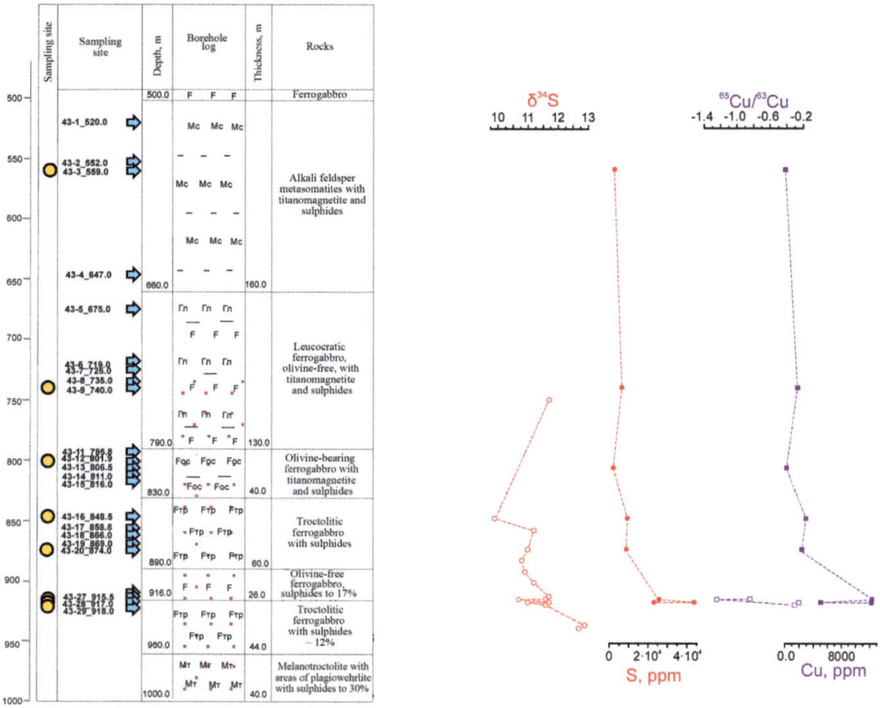

Fig. 12 Variations of isotopic composition of sulphur, copper in Cu-Ni sulphides and concentrations of sulphur, copper in rocks of Dyumptalej intrusive, Bh. TP-43

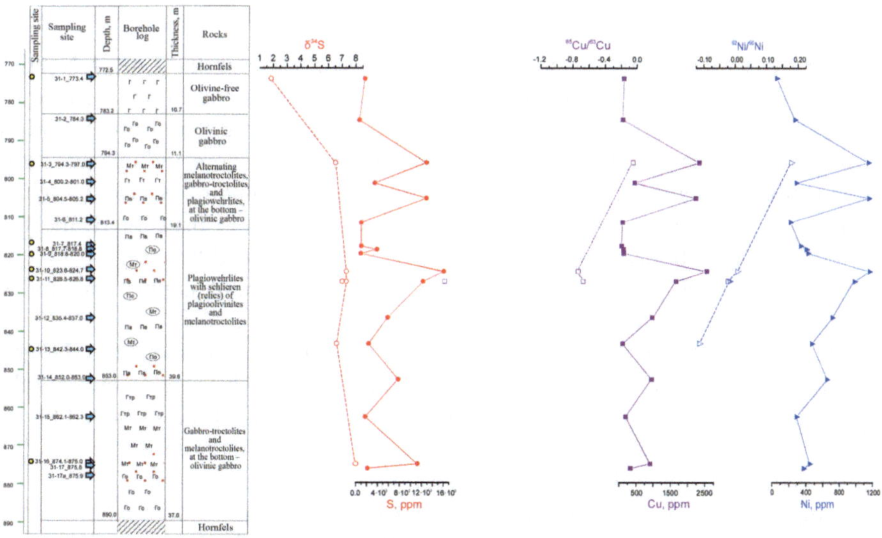

Fig. 13 Variations of isotopic composition of sulphur, copper, nickel in Cu-Ni sulphides and concentrations of sulphur, copper, nickel in rocks of Lower Talnakh intrusive, Bh. TG-31

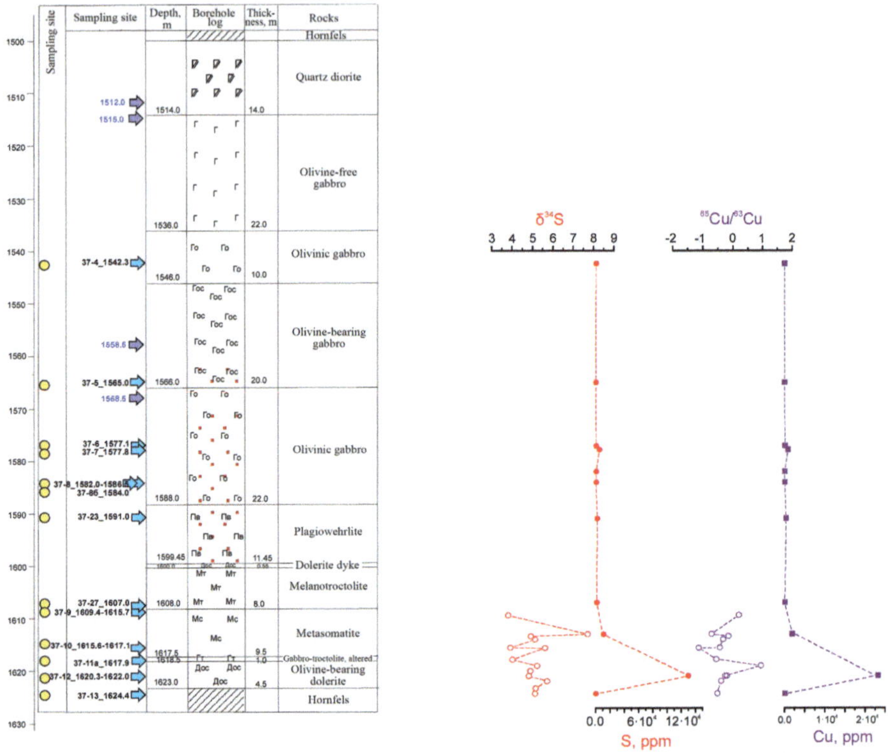

Fig. 14 Variations of isotopic composition of sulphur, copper, nickel in Cu–Ni sulphides and concentrations of sulphur, copper, nickel in rocks of Lower Norilsk intrusive, Bh. NP-37

Average $\delta^{34}S$ value in 13 samples from NP-37 drill core in the **Lower Norilsk** intrusion equals 4.7‰. The heaviest sulphur (5.7‰) was found in olivine-bearing dolerite at a depth of 1622 m. The Lower Norilsk intrusion differs from Norilsk-1 intrusion by a less mantle composition of helium and a more radiogenic argon composition as well as a light isotopic composition of sulphur. By isotope characteristics of helium and argon, the Lower Norilsk intrusion is virtually identical to the Lower Talnakh intrusion, and as for $\delta^{34}S$ value, the difference is, on the average, 2‰, which, most likely, points to a common source of sulphur and similar conditions of their formation.

Sulphur studied in 4 samples from MD-48 drill core in the **Mikchangda** intrusion has a high average $\delta^{34}S$ value of 12.7‰. Similar values are recorded at the top of the section, 12.8‰; in melanotroctolites in the middle of the section, 11.0; in plagiowehrlites, 13.3; and the heaviest sulphur, 14.0‰, occurs in olivine gabbro with a taxitic texture at the base of the section.

Isotopic composition of sulphur in the **Binyuda** intrusion was studied in 4 samples from C-1 drill core and has a low average value (1.5‰), $\delta^{34}S$ is 2‰ at the top of the section in olivinites, 1.6 in plagioolivinites, 1.7 in the middle of the section in plagiowehrlites and 0.7‰ at the base of the section. It was also necessary to determine the assignment of the understudied Binyuda intrusion to a certain mineralized group. On the helium-argon diagram, it is between the rich and poor criterial areas. Taking into account that $\delta^{34}S$ value is so low, this intrusion should be assigned to weakly mineralized (poor).

Decrease and extensive variations of $\delta^{34}S$ values in poor and average intrusions can be caused by three factors. Firstly, multiple sources of sulphur with different isotopic composition; secondly, mixing of sulphur from two sources in varying proportions; thirdly, isotopic fractionation of sulphur from one source with varying degrees of Rayleigh depletion of source material. The first cause is unlikely: it is difficult to imagine the presence of 8–9 sources in a confined space. The second cause cannot be excluded. The entire range of $\delta^{34}S$ variation in sulphides can actually be caused by mixing of isotope-heavy "sulphate" sulphur of sediments and deep, possibly, mantle sulphur with $\delta^{34}S$ ~0‰. While sharing this approach, one should acknowledge the maximum contribution of crustal sulphur in rich intrusions. The third cause can also be considered acceptable. The degree of isotope fractionation during partial transition of atoms from one chemical form to another depends on temperature. In addition, isotopic composition of the product is determined by the reaction completion degree. Natural variability of these two factors is quite sufficient for such major changes in isotopic composition of sulphide sulphur.

Let us consider the correlations in the distribution of sulphur isotopes and noble gases making up the diagnosed mixture of mantle, crustal and atmospheric components.

In the diagram showing the association of $\delta^{34}S$ and **m** (the share of mantle helium), there is a clear correlation (Fig. 15a) in three richest intrusions (similar to **m** and **a** in Fig. 9). In these three objects, the distribution of sulphur isotopes as well as of helium and argon (Fig. 15b) may be caused by mixing of two varieties of shallow fluids, in one of which $\delta^{34}S$ ~12‰ and **a** ~90%; and in the other one, 10 and 100%. The regression line is also adjoined by less rich intrusions—Chernogorsky, Zelenaya Griva, and Mikchanda.

Distribution of sulphur and helium isotopes in six other intrusions renounces the idea of a decisive influence of mixing. Here, considerable $\delta^{34}S$ variations from 1 to 9‰ are, probably, caused by the third group of factors associated with isotope fractionation.

Thus, in the richest intrusions, sulphur as well as helium is mainly of crustal origin; sulphur is, apparently, derived from the host rock anhydrites due to intense water migration (circulation). The origin of sulphur in poor and intermediate intrusions remains unclear.

Attempts were taken to determine ore formation temperature on the basis of isotopic differences between coexisting sulphides [7]. When using pyrrhotite-chalcopyrite isotope geothermometer, the values of 68–152 °C were obtained. These values are, apparently, underestimated due to the lack of isotopic equilibrium (for sulphur) between minerals which crystallized at different stages of existence of the cooling intrusion.

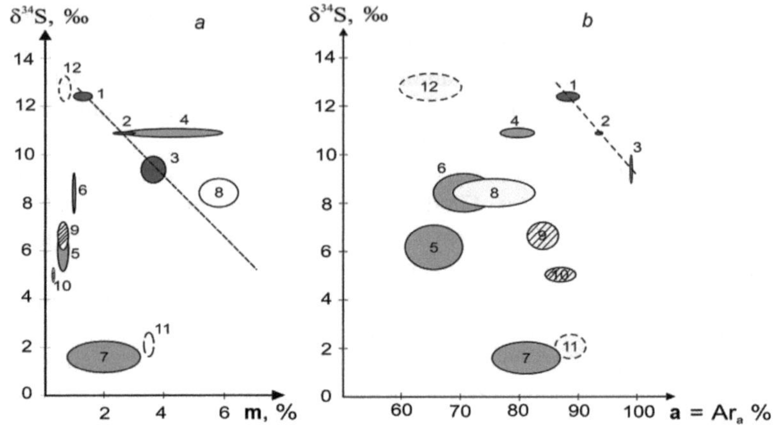

Fig. 15 Relationship between isotopic composition of sulphur and helium (**a**), sulphur and argon (**b**) in intrusives of Norilsk ore District. Correlation lines are drawn for intrusives of the rich group. Rich intrusives (1—Kharaelakh; 2—Talnakh; 3—Norilsk-1); average (4—Chernogorsk; 5—Vologochan; 6—South Pyasina; 7—Zub-Marksheidersky); poor (8—Zelenaya Griva), satellites (9—Lower Talnakh; 10—Lower Norilsk); prospective (11—Binyuda; 12—Mikchangda)

At the **II stage** (2012–2014), isotopic heterogeneity of an individual intrusion was investigated. On top of everything else, this was needed for justifying the required number of samples in the evaluation of ore presence in individual intrusions. About 60 samples from 15 wells on the Oktyabrsky (O) deposit (Kharaelakh intrusion) were analysed, 10 samples from two wells in the Talnakh intrusion, 6 samples from the Norilsk-1 massif and 4 samples from the Maslovskoye intrusion (total 80 samples). To identify the isotope differentiation between different types of sulphides, in each specimen two samples were distinguished with magnetic (electromagnetic) and nonmagnetic fractions.

The results have confirmed the previous measurements. Thus, in sulphides of the Kharaelakh intrusion, earlier in 19 samples from one well, the average value of $\delta^{34}S$ = 12.5‰ was obtained. At the II stage, averaging of the results of analyses of 59 samples from 14 wells gave the value $\delta^{34}S$ = 12.45‰ (Table 2). The analysis of spatial distribution (Table 3, Fig. 16) shows that from the north-west (from FJL) to the south-east (SFD), $\delta^{34}S$ value in sulphides slightly decreases (from 12.8 to 11.5‰) approaching the value typical for the Talnakh intrusion (11.2‰ based on 6 samples).

At the same time, there is a very large spread of $\delta^{34}S$ values in samples from some wells on the western flank of the Oktyabrsky deposit—from 7 to 15.8‰. It is interesting to note that when averaging the data on wells, positive anomalies compensate the negative ones, and, as a result, the average values do not differ much from those characteristic of this district (about 12.5‰). Such large variations are, apparently, caused by isotopic fractionation of sulphur from one source with varying degrees of Rayleigh depletion of the source material resulting in increasing

Table 2 Sulphur isotopes in sulphides of mafitic intrusives in Norilsk and Taimyr provinces (II stage)

Sample number	$\delta^{34}S$, ‰	Sample number	$\delta^{34}S$, ‰
Kharaelakh intrusive		ZF-12/398 em/fr	14.5
KZ361bis-1062,8 s/fr	11.5	KZ981-1122.4-1123.0	13.3
KZ361bis-1062,8 em/fr	11.5	KZ981-1122.4-1123.0	12.3
KZ361bis-1063,4 s/fr	11.6	KZ1112-1098.4	13.0
KZ361bis-1063,4 em/fr	11.3	ZF19-407.9-408.5	15.4
KZ361bis-1074,4 s/fr	11.7	ZF19-407.9-408.5	15.4
KZ361bis-1074,4 em/fr	11.9	ZF13-441.9-442.5	15.8
KZ361bis-1075,6 s/fr	11.6	ZF13-441.9-442.5	15.5
KZ361bis-1075,6 em/fr	11.4	ZF13-448.0	15.5
KZ931-643,9 s/fr	12.7	ZF13-448.0	15.3
KZ931-643,9 em/fr	12.7	ZF13-452.3-452.9	11.1
KZ931-645 s/fr	12.2	ZF13-452.3-452.9	11.2
KZ931-645 em/fr	12.2	ZF13-478.4-479.0	12.8
K952-971,1 s/fr	12.2	ZF13-478.4-479.0	12.6
K952-971,1 em/fr	12.3	ZF18-467.5	7.0
K952-1010,4 s/fr	12.3	ZF18-467.5	9.9
K952-1010,4 em/fr	12.3	PT2-1432.8-1433.4	11.6
K1089-1155,6 s/fr	12.0	PT2-1432.8-1433.4	11.1
K1089-1155,6 em/fr	12.0	*Talnakh intrusive*	
K1089-1156,7 s/fr	12.6	OUG-2 s/fr	11.1
K1089-1156,7 em/fr	12.0	OUG-2 em/fr	10.5
K1112-1092,4 s/fr	12.6	KZ 774-1023.7	8.5
K1112-1092,4 em/fr	12.6	KZ 774-1029.0-1029.6	13.0
K1112-1094,8 em/fr	12.8	KZ 774-1032.4-1033.0	11.7
K1112-1094,8 s/fr	12.8	KZ 774-1032.4-1033.0	11.4
K1112-1098,4 s/fr	12.9	KZ774-1049.2-1049.7	14.2
K1112-1098,4 em/fr	12.8	KZ774-1049.2-1049.7	14.4
K1112-1100,4 s/fr	12.7	KZ774-1057.8-1058.2	9.2
K1112-1100,4 em/fr	12.7	KZ774-1057.8-1058.2	9.3
K1112-11022,4 s/fr	13.0	*Norilsk-1*	
K1112-1102,4 em/fr	12.9	N-1/0 s/fr	7.3
K1112-1103,8 s/fr	12.8	N-1/0 em/fr	7.7
K1112-1103,8 em/fr	12.8	N-1/2 s/fr	7.4
KZ1319-598,4 s/fr	11.7	N-1/2 em/fr	7.7
KZ1319-598,4 em/fr	11.8	N-1/6 s/fr	9.0
KZ952-1013,5 s/fr	12.9	N-1/6 em/fr	8.1
KZ952-1013,5 em/fr	12.5	*Maslovsky intrusive*	
K1084-1146,9 d/sulphides	11.5	OM10-1072.9-1073.5	9.0
K1084-1146,9 em/fr	11.5	OM10-1072.9-1073.5	8.8
K1089-1154,4 s/fr	12.3	OM10-1082.4-1083.0	9.3
K1089-1154,4 em/fr	11.9	OM10-1082.4-1083.0	9.6
ZF-12/398 s/fr	14.6		

Table 3 $\delta^{34}S$ value in sulphides of Oktyabrskoe deposit (mean values)

Parameter	Western flank	Central part	Southern flank	In general
Mean for samples, ‰	12.81	12.45	11.45	12.45
Number of samples	26	23	10	59
Mean for boreholes, ‰	12.68	12.36	11.53	12.41
Number of boreholes	8	4	2	14

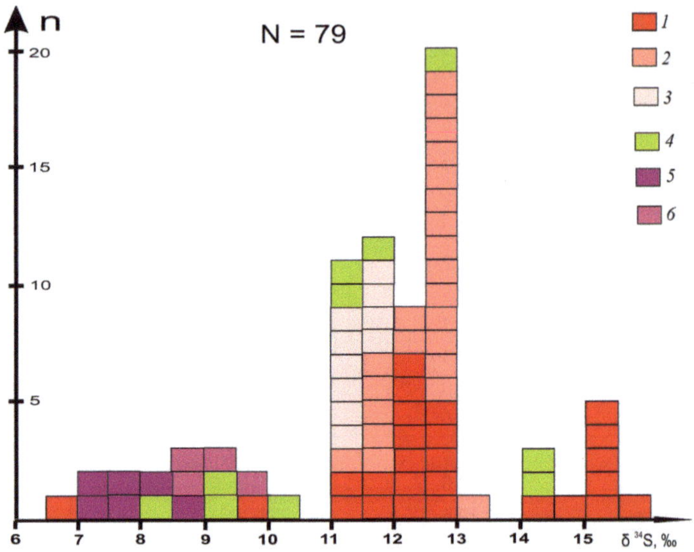

Fig. 16 Distribution of $\delta^{34}S$ value of sulphides in some intrusives of the Norilsk district. 1–3—Oktyabrskoe (1—western flank; 2—central part; 3—southern flank); 4—Talnakh; 5—Norilsk-1; 6—Maslovsky deposit

weight of sulphur in some fractions and its decreasing weight in the other ones. It is important that by measuring isotopic composition of sulphur in several parts of the section and averaging it, we can avoid the impact of heterogeneity and obtain isotopic characteristics of the source material. The latter are necessary for the diagnosis of mineralization degree of intrusions on the basis of isotopic composition of sulphur.

In the study of isotopic composition of sulphides in each of the 40 samples they were preliminarily divided into magnetic, electromagnetic and non-magnetic mineral fractions. In illustrations and statistical buildups, each fraction was regarded as a separate sample. It appeared that, in most cases, the samples (76% of 80 samples) are indistinguishable by isotopic composition of sulphur (differences do not exceed 0.3‰). Significant differences (0.9, 1.0, 2.9‰) were recorded only in three samples. The causes for abnormality are investigated. Table 4 shows the average values of isotopic composition of sulphur in the group of intrusions in the Norilsk-Taimyr

Table 4 Sulphur isotopes in intrusives of Norilsk and Taimyr provinces (averaged values)

Intrusive	Borehole	Type of mineralization	$\delta^{34}S \pm \sigma$, ‰	MSWD	Number of samples
Norilsk-1	MN-2	Disseminated	9.7 ± 0.7	2.6	13
	MN-2	Low-sulphide	8.1 ± 1.4	2.5	3
Talnakh	OUG-2	Disseminated	10.9 ± 0.1	0.9	49
	OUG-2	Massive	10.9 ± 0.1	0.7	46
Kharaelakh	KZ-844	Disseminated	11.9 ± 0.7	2.2	9
	KZ-844	Massive	12.6 ± 0.3	0.6	4
	KZ-963	Veinlet	11.4	0.0	1
	KZ-963	Disseminated	12.6 ± 0.2	0.6	10
	KZ-963	Massive	12.7 ± 0.1	0.2	5
	MP-2bis		10.9 ± 0.2	0.3	4
Zub-Marksheidersky	MP-27	Disseminated	1.6 ± 0.7	2.4	11
Vologochan	OV-29		6.2 ± 0.9	3.3	13
South Pyasina	OV-25		8.4 ± 0.8	2.0	7
Imangda	KP-4		10.5 ± 2.2	5.3	6
Mikchangda	MD-48		12.8 ± 0.6	1.3	4
Binyuda	S-1		2.1 ± 0.5	1.3	7
Dyumptalej	TP-43		11.4 ± 0.2	0.8	15
Lower Talnakh	TG-31		6.6 ± 0.6	1.8	9
Lower Norilsk	NP-37		5.0 ± 0.3	1.0	15
Zelenaya Griva	F-233		8.4 ± 0.6	1.3	5
Kruglogorsky	MP-2bis		9.7 ± 1.7	2.4	2
Agatsky	OV-25		15.5	0.0	1

Notes MSWD—root-mean-square deviation (‰) in a single sample, σ—root-mean square deviation of mean value

Province obtained at both stages of the study. They give an idea of the variations of isotopic composition of sulphur in different objects.

Thus, in the richest intrusions, sulphur, similar to helium, is predominantly of crustal origin; it was, apparently, derived from anhydrites of the enclosing sequences as a result of intense water migration (circulation) caused by intrusion of a lava body. The origin of sulphur in poor and intermediate intrusions remains unclear. Possibly, a certain part of sulphur is of mantle origin.

Decrease and extensive variations of $\delta^{34}S$ values in poor and intermediate intrusions can be accounted for by a number of reasons: several sources of sulphur with different isotopic composition; mixing of sulphur from two sources in varying proportions; isotopic fractionation of sulphur from one source with varying degrees of Rayleigh depletion of the source material. The first cause is unlikely: it is difficult to imagine the presence of 8–9 sources in a confined space. The second cause cannot be excluded: the entire range of $\delta^{34}S$ variation in sulphides can actually be accounted for by mixing of isotopically heavy "sulphate" sulphur of sediments and deep, possibly, mantle sulphur with $\delta^{34}S \sim 0‰$. From this viewpoint, one should acknowledge the maximum contribution of crustal sulphur in rich intrusions. The third cause can also be considered acceptable. The degree of isotope fractionation during partial transition of atoms from one chemical form to another depends on temperature. In addition, isotopic composition of the product is determined by the reaction completion degree. Natural variability of these two factors is quite sufficient for such major changes in isotopic composition of sulphide sulphur.

S-isotope criterion. The intrusions rich in ore (and some others) are characterized by isotopically heavy sulphur with $\delta^{34}S$ values ranging from 8 to 13‰. In the intermediate and poor intrusions, it is lighter (0–8‰), and in isotopic composition, it is indistinguishable in these two types of objects. This criterion should be considered as an additional one due to a possible impact of sulphur isotope fractionation during ore formation. In addition, for isotopic analysis, ore minerals—sulphides are required.

References

1. Aplonov VS (2001) Termobarogeochemical model of the Talnakh platinoid-copper-nickel deposit. SPb.: VNIIOkeangeologia, 234 p
2. Vinogradov AP, Grinenko LN (1964) On the influence of enclosing rocks on isotopic composition of sulphur. Geochemistry (6):491–499
3. Gorbachev NS, Grinenko LN (1973) Isotopic composition of sulphide sulphur and sulphates in the Oktyabrsky deposit of sulphide ores (Norilsk district) with regard to the questions of its genesis. Geochemistry (8):1127–1136
4. Grinenko LN (1966) Isotopic composition of sulphur in sulphides of the Talnakh copper-nickel deposit in connection with questions of its genesis. Geology of ore deposits (4):15–30
5. Grinenko LN (1988) Methodology of isotope studies of the deposits of copper-nickel formation. Isotope geochemistry of ore-formation processes. Nauka, pp 135–149
6. Grinenko VA, Grinenko LN (1974) Geochemistry of sulphur isotopes. Nauka, 274 p
7. Kovalenker VA, Gladyshev GD, Nosik LP (1974) Isotopic composition of sulphur in sulphides of the Talnakh ore cluster deposits in connection with their salt content. Izv USSR Acad Sci Ser Geol (2):80–91
8. Faure G (1989) Fundamentals of isotope geology. Mir, 590 p
9. Hofs J (1983) Geochemistry of stable isotopes. Mir, 200 p
10. Faure G, Mensing TM (2005) Isotopes: principles and applications. Wiley, New-Jersey, 897 p

Copper and Nickel Isotopes

Sergey Sergeev, Igor Kapitonov, Robert Krymsky, Dmitriy Sergeev, Elena Adamskaya and Nikolay Goltsin

Abstract The chapter describes a technique of analysing copper and nickel isotopes in ores and magmatic rocks of deposits. Average δ^{65}Cu value strongly diverges in rich intrusions: Kharaelakh (−1.55‰), Talnakh (−0.7‰), and Norilsk-1 (+0.25‰). Perhaps this is due to the mixing of the crustal and mantle sources of matter in the formation of these massifs. The Kharaelakh intrusion by δ^{65}Cu has a large share of crustal matter. In the remaining massifs, the average value of δ^{65}Cu is almost the same (−0.8 to −0.3‰) and does not correlate with the isotope composition of sulphur. The average values of nickel ^{61}Ni/^{60}Ni isotope composition vary within narrow limits from −0.6 to 0‰ in all intrusions of the region and point to a single source of this metal. Data on copper and nickel isotopes in the products of mining and metallurgical companies in the world are given. Variations in the isotopic composition of copper are more associated with raw materials sources of companies, and the isotope composition of nickel, with different technological processes. A possible change in the isotopic composition of nickel during carbonylation (a mond process) is considered.

1 Methodology, Samples

For the first time in national geology, isotopic studies of the second main component of minerals—metals—were conducted, namely copper and nickel from sulphides. So far, there is too little evidence of the geochemistry of isotopes of these elements to use any genetic criteria. Finding of correlations with isotopic variations in other systematics should ensure progress in this sphere.

At the I stage (2006–2008), bulk samples of sulphides were studied, isotopic compositions of copper were determined (^{65}Cu/^{63}Cu ratio) in 180 samples from 16 intrusions as well as of nickel (^{62}Ni/^{60}Ni and ^{61}Ni/^{60}Ni ratios) in 70 samples from 9

S. Sergeev (✉) · I. Kapitonov · R. Krymsky · D. Sergeev · E. Adamskaya · N. Goltsin
Russian Geological Research Institute (VSEGEI), St. Petersberg, Russia
e-mail: sergey_sergeev@vsegei.ru

O. Petrov (ed.), *Isotope Geology of the Norilsk Deposits*, Springer Geology,
https://doi.org/10.1007/978-3-030-05216-4_3

intrusions. Besides, the composition of nickel from olivine was investigated in 8 samples from 4 intrusions (Tables 1, 2). At the II stage (2012–2014), isotopic composition of copper was studied in 53 samples of rocks and ores from four intrusions (Table 3) including 37 samples from the Oktyabrsky deposit (Kharaelakh intrusion)—the results are partly presented in the work by O. V. Petrov et al. "Isotope correlations in rocks and ores of productive intrusions in the Norilsk District" (Platinum of the USSR. 2011. Vol. 7, pp. 467–475).

Preparation of specimens for mass spectral analysis included a number of procedures. Decomposition of samples (weighted portions of 100–130 mg) was performed in a mixture of acids (HCl, HF, HNO_3, $HClO_4$) using extremely pure acids and Teflon dishes. Further, copper and nickel were distinguished by chromatography. Measurements were taken in 3% HNO_3 solutions with Ni content 1 ppm and 0.5 ppm Cu with an error ~0.15‰. Isotopic composition of these elements is expressed in the value of isotopic ratio shifts in per mill (δ value is calculated similar to δS) with respect to the international standards for copper NIST-976 and nickel NIST-986.

For measuring the isotopic composition of copper and nickel a multicollector inductively coupled plasma mass spectrometer was used (ICP-MS) ThermoFinnigan Neptune.

Isotopic composition of nickel was measured using a scheme with normalization after the isotopic copper standard NIST-976; and isotopic composition of copper, with normalization after the isotopic standard of nickel NIST-986.

The obtained samples were placed into the autosampler Cetac-100 connected to the Neptune device. Then, a sequence of measurements was run according to the scheme: Blank sample 1—Standard 1—Blank sample 2—Sample 1—Blank sample 3—Standard 2—Blank sample 4—Sample 2—Blank sample 5—Standard 3—Blank sample 6—Measurement process control and primary processing of the results were carried out using the software of the mass spectrometer.

2 Results and Discussion

Results of the studies at the II stage are presented in Tables 1 and 2; at the II stage, in Table 3. In most cases, δ^{65}Cu value ranges from −1.4 to 0‰ (Fig. 1). The average δ^{65}Cu value in sulphides of the Kharaelakh intrusion at the I stage of the studies virtually coincides with that of the II stage (about −1.5 to −1.6‰). No spatial differences have been revealed within the Oktyabrsky deposit (Kharaelakh intrusion) (Tables 1 and 2). Variations within the western front of the deposit are the largest. It should be noted that there is an isotopic anomaly of copper from RT-2 drill core (central part of the deposit), in which δ^{65}Cu = −0.83‰. As noted above, a sample from this well contained the most radiogenic argon (Fig. 2).

Table 1 Isotopic composition of copper (I stage)

Intrusive, borehole number	Sample number	Types of mineralization	δ^{65}Cu, ‰
Norilsk-1, MN-2	N1-2	Disseminated	NDA
	N1-3	Low-sulphide	0.96
	N1-7	Disseminated	0.54
	N1-9		0.02
	N1-10		−0.10
Norilsk-1, MS-18	18-1	Low-sulphide	0.19
	18-2		0.73
Norilsk-1, Medvezhij Ruchej open pit	Kn-97-1	Disseminated	0.61
	Kn-97-2		0.09
	Kn-97-3		0.33
Talnakh, OUG-2	28		−0.07
	48		−0.04
	72	Massive	−0.08
	82		−0.59
	87	Chalcopyrite	−0.05
	87	Pyrrhotite	−0.13
	87	Massive	−3.16
	96		−0.35
	T-13	Disseminated	−0.38
	T-14		−0.74
	T-19 (d T-14)		−0.77
	T-15		−1.10
	T-16		−0.95
	T-17		−0.97
	T-18		−0.90
	T-20 (d T-18)		−1.01
Kharaelakh, KZ-844	8442		−1.43
	844-3,4		−1.47
	844-11		−1.92
	844-10,11		−1.09
	844-18	Massive	−1.82
	844-19		−1.80
	844-20		−1.35
	844-6	Disseminated	−1.25
	844-7		−2.28
	844-11		−1.23
Kharaelakh, KZ-963 (2006)	963-5	Veinlet	−1.32
	963-12	Massive	−1.59
	963-17		−1.46
	963-18		−1.30
	963-25		−0.90
	963-30	Disseminated	−1.44
	963-30		−1.46
	963-30 (d)		−1.70

(continued)

Table 1 (continued)

Intrusive, borehole number	Sample number	Types of mineralization	$\delta^{65}Cu$, ‰
Kharaelakh, KZ-963 (2006)	963-31	Disseminated	−1.45
	963-37		−1.61
	963-37		−1.27
	963-38		−1.65
	963-54		−1.93
	963-60		−1.58
	963-71		−1.78
Kharaelakh, KZ-963 (2006)	963-75	Massive	−1.58
	963-75 (d)		−1.63
	963-78		−1.55
	963-86		−1.46
	963-88		−1.48
	963-89		−1.74
Kharaelakh	2S-08/17 (d)		−1.86
Chernogorsk, MP-2 bis	CH-11	Disseminated	−0.01
	CH-11 (d)		−0.04
	CH-13		−0.27
	CH-13 (d)		0.00
Zub-Marksheidersky, MP-27	27 10		−0.11
	27-13		−0.07
	27-13 (d)		−1.40
	MP-27/96,9		−1.06
	MP-27/97,4		−1.03
Vologochan, OV-29	29-16		−1.03
	29-17		−0.40
	29-19		−0.82
	29-17 (d)		−0.19
	29-24		−1.07
	OV-29/852.3		−1.39
	OV-29/853.9		−0.55
	OV-29/854.3		−0.61
	OV-29/867.5		−0.50
South Pyasina, OV-25	25-20		0.01
	25-31		−0.47
	25-35		−0.90
	25-36		−1.11
	25-41		−0.95
	25-41 (d)		−0.91
	25-44		−0.37

(continued)

Table 1 (continued)

Intrusive, borehole number	Sample number	Types of mineralization	$\delta^{65}Cu$, ‰
Vologochan, OV-29	S1-1		0.06
Imangda, KP-4	4-5		0.27
Mikchangda, MD-48	48-9		−0.94
	48-25		−0.33
	48-27, 28		−1.05
Binyuda, Bh. S-1	S1-1		−0.54
	S1-5		−0.33
Binyuda, Bh. S-2	S2-2		−0.62
	S2-1		−0.22
Dyumptalej, TP-43	43-26		−0.83
	43-27		−1.24
	43-29		−0.25
	43-30		−0.30
Lower Talnakh, TG-31	31-3		−0.05
	31-10		−0.73
	31-11		−1.09
	31-11 (d)		−0.67
	31-13		−0.11
Lower Norilsk, NP-37	37-9		−0.71
	37-9a		0.19
Lower Norilsk, NP-37	37-9v		−0.16
	37-9 g		−0.33
	37-10a		−0.44
	37-10a-2		−1.14
	37-11a		−0.56
	37-11b		0.95
	37-12		−0.27
			−0.21
	37-12v		−0.40
	37-13		−0.53
Zelenaya Griva, F-233	F-233-2		−0.10
	F-233-10		−1.21
	F-233-10 (d)		−0.26
Kruglogorsky, MP-2bis	K-6		−0.26
Igarka	IP-353/8	Chalcocite	−0.09
	110-3499	Native copper	0.08
Arylakh	2/S-08-17		−1.85
Kharaelakh	3/S-08-17	Talnakhite	−1.27
	1/S-08-17	Cubanite	−0.01
Stillwater	9-4	Disseminated	−0.13

Table 2 Isotopic composition of nickel (I stage)

Intrusive, borehole number (year of meas.)	Sample number	Types of mineralization	$\delta^{62}Ni/^{60}Ni$, ‰	$\delta^{61}Ni/^{60}Ni$, ‰
Norilsk-1, MN-2 (2005)	N1-2	Disseminated	−1.260	−0.66
	N1-3	Low-sulphide	−0.15	−0.10
	N1-7	Disseminated	−0.47	−0.22
	N1-9		−0.19	−0.11
Norilsk-1, MS-18 (2007)	18 1	Low-sulphide	0.63	1.180
Talnakh, OUG-2 (2005)	28	Disseminated	−0.52	−0.24
	48		−0.13	−0.04
	72	Massive	−1.330	−0.65
	82		−0.61	−0.32
	87	Chalcopyrite	−1.300	−0.67
	87	Pyrrhotite	−0.55	−0.27
	87	Massive	−1.310	−0.67
	96		−0.59	−0.28
Talnakh, OUG-2 (2006)	T-13	Disseminated	−0.23	−0.08
	T-14		−0.83	−0.37
	T-19 (=T-14)		−0.62	−0.28
	T-15		−0.49	−0.27
	T-18		−0.40	−0.18
	T-20 (=T-18)		−0.48	−0.30
Kharaelakh, KZ-844 (2005)	8442 V		−0.35	−0.15
	844-3,4		−0.96	−0.52
	844-11		−0.00	−0.02
	844-18	Massive	−0.81	−0.44
	844-19		−0.47	−0.20
	844-20		−0.35	−0.17
Kharaelakh, KZ-844 (2006)	844-3,4	Disseminated	−0.32	−0.16
	844-6		−0.95	−0.40
	844-7		−0.98	−0.45
	844-10,11		0.02	0.02
Kharaelakh, KZ-963 (2006)	963-5	Veinlet	−0.68	−0.23
	963-12	Massive	−0.93	−0.45
	963-17		−0.74	−0.34
	963-18		−0.82	−0.37
	963-25		−0.69	−0.32
	963-30	Disseminated	−0.74	−0.38
	963-30 (d)		−0.40	−0.23
	963-31		−0.31	−0.10
	963-37		−0.24	−0.12

(continued)

Table 2 (continued)

Intrusive, borehole number (year of meas.)	Sample number	Types of mineralization	$\delta^{62}Ni/^{60}Ni$, ‰	$\delta^{61}Ni/^{60}Ni$, ‰
Kharaelakh, KZ-963 (2006)	963-38	Disseminated	−0.06	−0.02
	963-54		−0.51	−0.19
	963-60		−0.43	−0.20
	963-71	Massive	−0.69	−0.31
	963-75 (d)		−0.83	−0.45
	963-78		−0.54	−0.25
	963-86		−0.29	−0.15
	963-88		−0.18	−0.10
	963-89		−0.15	−0.09
	963-95		−0.77	−0.40
Chernogorsk, MP-2bis (2007)	CH-11	Disseminated	0.01	−0.07
	CH-11 (d)		0.04	0.03
	CH-13		−0.20	−0.37
	CH-13 (d)		−0.07	−0.14
Zub-Marksheidersky, MP-27 (2007)	27 10		0.10	0.03
	27-13		−0.13	−0.33
	27-13 (d)		−0.09	−0.26
Vologochan, OV-29 (2006)	29-16		−0.49	−0.23
	29-17		−0.42	−0.19
	29-19		−0.72	−0.34
Vologochan, OV-29 (2007)	29-17 (d)		−0.14	−0.35
Binyuda, C1 (2007)	C1-1		0.06	0.05
	C1-5		−0.36	−0.80
Lower Talnakh, TG-31 (2006)	31 3		0.18	0.15
	31 10		0.01	−0.02
	31-11		−0.02	0.07
	31-11 (d)		−0.02	0.10
	31-13		−0.11	−0.06
Zelenaya Griva, F-233 (2007)	F-233-2		0.03	0.15
	F-233-10		−0.11	−0.21
	F-233-10 (d)		−0.09	−0.22
Kruglogorsky, MP-2bis (2007)	К-6		−0.09	−0.24

Table 3 Isotopic composition of copper (II stage)

Sample number	Horizon depth (m)/ Fraction	Rock name	$\delta^{34}S$, ‰	$\delta^{65}Cu$, ‰	Standard error, ‰
Oktyabrskoe deposit					
Western flank					
ZF-13	452.3	Gabbro, coarse-grained, olivinic, melanocratic. Ore: impregnations 8–10%, size 0.3–2 mm	11.24	−1.72	0.54
ZF-13	452.3–452.9/em	Gabbro, coarse-grained, olivinic, melanocratic. Ore: impregnations 8–10%, size 0.3–2 mm	11.24	−1.72	0.54
KZ-931	643.9/em	Disseminated ore	12.70	−1.01	0.03
KZ-931	643.9/em	Rich ore	12.20	−2.26	0.06
KZ-931	643.9/s	Disseminated ore	12.70	−0.68	0.21
KZ-931	643.9	The same	12.70	−0.99	0.05
KZ-931	645	Rich ore	12.20	−2.34	0.15
KZ-931	645	The same	12.20	−2.29	0.12
KZ-931	645/s	–	12.20	−2.16	0.08
KZ-952	1010.4	Disseminated ore	12.30	−1.07	0.06
KZ-952	1010.4	The same	12.30	−1.04	0.10
KZ-952	1013.5		12.50	−1.09	0.07
KZ-952	1013.5	Disseminated ore	12.50	−0.95	0.07
KZ-1319	598.4/em	Rich ore	11.80	−1.22	0.07
ZF-19	407.9–408.5/n/s	Disseminated ore, 50, %, in biotite-chlorite-feldspar metasomatite	15.35	−1.46	0.10
ZF-19	407.9–408.5/em/s	The same	15.40	−1.08	0.09
ZF-13	448.0/em/s	Gabbro, coarse-grained, olivinic, melanocratic. Secondary: serpentine. Ore: segregations over 30 mm	15.30	−1.88	0.08
ZF-13	478.4–479.0/em/s	Disseminated ore 40–50% in carbonate-epidote-pyroxene metasomatite	12.60	0.01	0.09
ZF-18	467.5/em/s	Gabbro, coarse-grained,, inequigranular, olivine-bearing. Ore: segregations to 6 mm	9.90	−0.29	0.10
Central part					
KZ-1089	1154.4/em (50.75)	Rich ore	11.90	−1.26	0.06
KZ-1089	1155.6/em	The same	12.00	−1.38	0.10
KZ-1089	1155.6/em		12.0	−1.38	0.08
KZ-1089	1155.6/em		12.00	−1.45	0.10

(continued)

Table 3 (continued)

Sample number	Horizon depth (m)/ Fraction	Rock name	$\delta^{34}S$, ‰	$\delta^{65}Cu$, ‰	Standard error, ‰
KZ-1112	1092.4/em (48.56)	Disseminated ore	12.60	−1.83	0.08
KZ-1112	1094.8/em (50.88)	Rich ore	12.80	−1.78	0.04
KZ-1112	1098.4/em	Rich ore	12.80	−1.89	0.09
KZ-1112	1098.4/ em/s	Ore zone, fine-crystalline, rich sulphide ore with medium-crystalline chalcopyrite areas	13.00	−1.56	0.08
KZ-1112	1100.4/em	Rich ore	12.70	−1.70	0.05
KZ-1112	1102.4/em	The same	12.90	−1.74	0.08
KZ-1112	1103.8/em		12.80	−1.93	0.06
Southern flank					
RT-2	1432.8–1433.4/n/s	Plagiowehrlite, coarse-grained, poikilitic (f/g olivine in pyroxene). Secondary: iddingsite, chlorite, serpentine. Ore: to 15% of impregnations in interstices and segregations to 3 mm	11.60	−0.83	0.11
KZ-361bis	1063.4	Rich ore	11.30	−1.74	0.07
KZ-361bis	1074.4	The same	11.90	−1.98	0.06
KZ-361bis	1075.6		11.40	−1.42	0.05
KZ-1084	1146.9	Disseminated ore	11.50	−1.57	0.07
KZ-1084	1156.7/em	The same	11.50	−1.42	0.08
ZF-12	398/em	No description	14.50	−1.96	0.12
Norilsk-1					
N1-10		Sulphide vein in picrites	7.30	0.26	0.10
N1-0	м	The same	7.30	0.21	0.20
N1-0	em		7.70	0.24	0.20
N1-2	м	Cubanite	7.40	−1.44	0.10
N1-2	em		7.70	−1.44	0.10
N1-6	м	Chalcopyrite disseminated ore	9.00	−0.66	0.12
N1-6	em	The same	8.10	−0.56	0.09
Maslovsky deposit					
OM-10	1072.9–1073.5/n/s	Dolerite, coarse-medium-grained, olivine, poikilitic (olivine in pyroxene). Secondary: chlorite, serpentine, iddingsite, talc, amphibole. Ore: to 5% of impregnations to 6 mm	8.90	0.10	0.10

(continued)

Table 3 (continued)

Sample number	Horizon depth (m)/ Fraction	Rock name	$\delta^{34}S$, ‰	$\delta^{65}Cu$, ‰	Standard error, ‰
OM-10	1082.4–1083.0/n/s	Dolerite, coarse-medium-grained, olivine-bearing, poikiloophitic. Secondary: chlorite, serpentine, iddingsite, talc, amphibole. Ore: to 3% impregnations to 6 mm	9.30	−0.92	0.12
Talnakh intrusive					
OUG-2	em	Cubanite, pyrrhotite, chalcopyrite	10.50	−1.10	0.10
OUG-2	M	The same	11.10	−0.96	0.13
KZ-774	1029.0–1029.6/n/s	Dolerite, poikiloophitic, olivinic, medium-coarse-grained. Secondary: iddingsite, amphibole, chlorite. Ore: impregnations 5–10% to 1 mm	12.95	−0.31	0.10
KZ-774	1032.4–1033.0/ em/s	Wehrlite, fine-medium-grained, serpentinized. Secondary: anhydrite, carbonate. Ore: 5% of segregations to 2–6 mm	11.36	−0.72	0.11
KZ-774	1049.2–1049.7/n/s	Gabbro, medium-coarse-grained, olivinic. Secondary: serpentine, iddingsite, chlorite, anhydrite, carbonate. Ore: to 5–10% of impregnations and accumulations to 10 mm	14.19	−0.97	0.17
KZ-774	1057.8–1058.2/n/s	Dolerite, poikiloophitic, olivinic, medium-coarse-grained. Secondary: serpentine, iddingsite, chlorite. Ore: 15% of accumulations to 10 mm	9.21	0.28	0.12
KZ-981	1122.4–1123.0/n/s	Troctolite, pegmatoid, pyroxenic, porphyric, serpentinized. Ore: rare accumulations to 3 mm	13.31	−0.95	0.11

Comparison of isotopic composition of copper and sulphur (Fig. 3):

- in the vast majority of cases (10 out of 13 objects), the average value of copper isotope ratio is practically the same ($\delta^{65}Cu$ from −0.8 to −0.3‰) and is independent of sulphur isotopic composition;
- $\delta^{65}Cu$ value differs significantly in two rich intrusions—Kharaelakh (–1.55‰) and Norilsk-1 (+0.25‰);
- similar to other systematics considered above, there is a clear correlation for the three richest intrusions (as well as for the Imangda and Dyumptalei). A negative correlation shows up in copper weight reduction and sulphur weight increase; therefore, mixing of copper from different sources as a cause of variations seems more likely. However, one cannot completely exclude isotopic fractionation of copper, especially, if it occurred before the formation of its compounds with

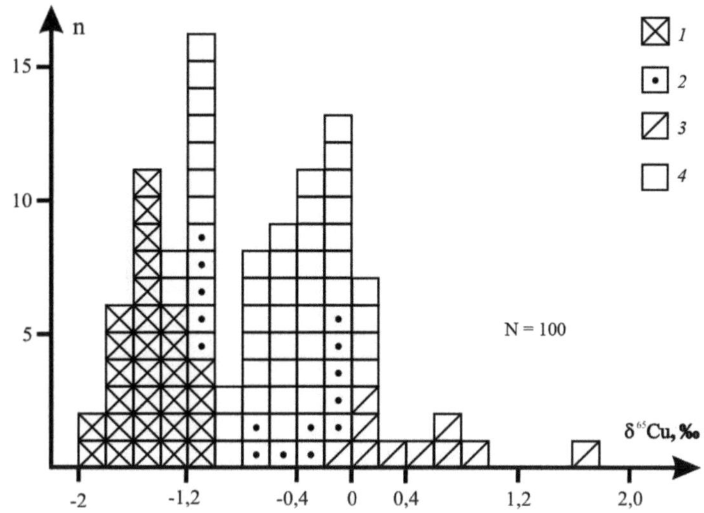

Fig. 1 Distribution of δ^{65}Cu value in intrusives with different degree of ore presence (I stage). *Intrusives* 1—Kharaelakh; 2—Talnakh; 3—Norilsk-1; 4—The rest

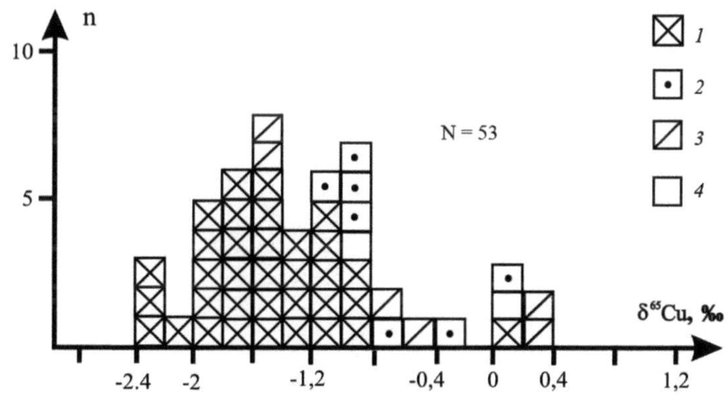

Fig. 2 Distribution of δ^{65}Cu value in intrusives with different degree of ore presence (II stage of work). See Legend in Fig. 1

sulphur. However, in the light of the idea of mixing fluids from different sources, this phenomenon appears to be more likely. Then, copper from the Kharaelakh intrusion should be more "crustal" (Fig. 4).

Isotopic composition of nickel (average values) varies within a narrow range (from −0.6 to 0‰ for δ ^{61}Ni/^{60}Ni ratio) pointing to a single source of this metal (Table 2). No association is found with isotopic composition of copper and sulphur. Nickel (unlike copper) has several isotopes, which allows diagnosing mass fractionation. A slight fractionation of isotopes is revealed giving ^{62}Ni/^{60}Ni ratio

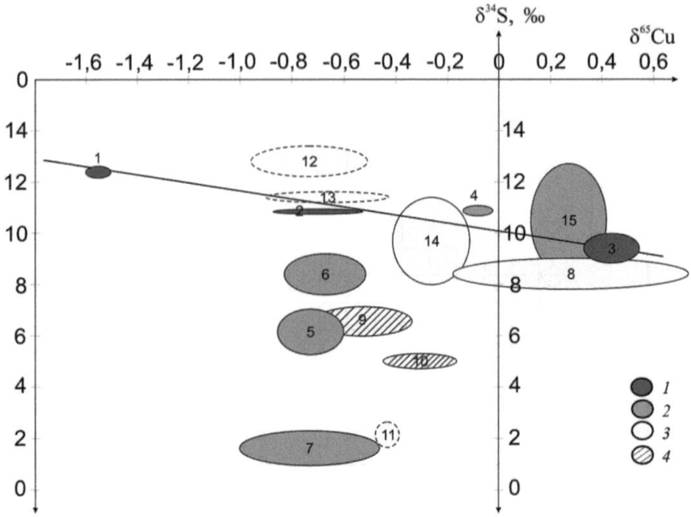

Fig. 3 Relationship of isotopic composition of Cu and S in ores of intrusives in the Norilsk District. Intrusives: 1—rich (1—Kharaelakh; 2—Talnakh; 3—Norilsk-1); 2—average (4—Chernogorsk; 5—Vologochan; 6—South Pyasina; 7—Zub-Marksheidersky; 15—Imangda); 3—poor (8—Zelenaya Griva; 14—Kruglogorsky); 4—satellites (9—Lower Talnakh; 10—Lower Norilsk) and prospective (11—Binyuda; 12—Mikchangda; 13—Dyumptalej)

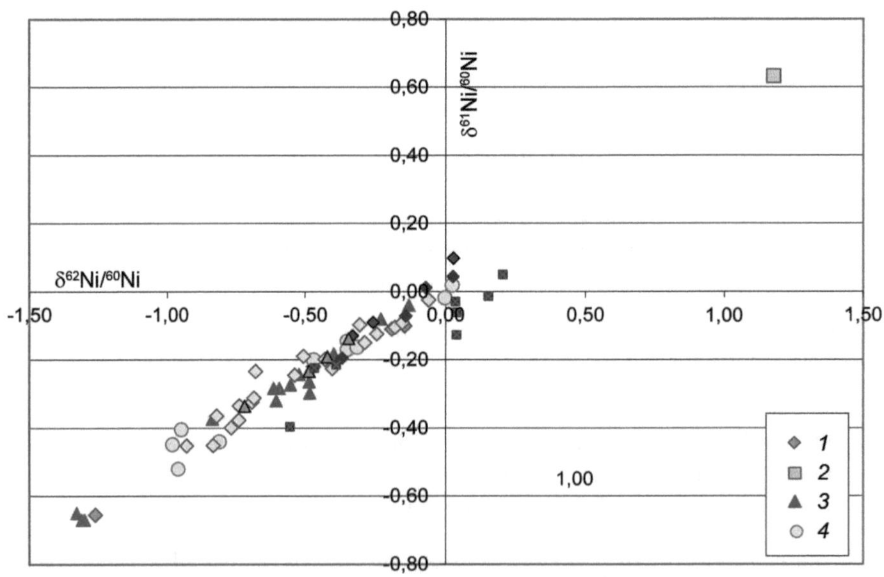

Fig. 4 Isotopic composition of Ni in sulphides and olivine. Intrusives: 1, 2—Norilsk-1 (1—MN-2, 2005; 2—MS-18, 2007); 3—Talnakh, OUG-2; 4—Kharaelakh, KZ-844 (2005)

variations to 1.3% in certain samples (Table 2). Generally, separation of isotopes could occur in different situations, in particular, when extracting nickel from silicate material or during formation of nickel compounds (and other metals) with sulphur. Study of nickel isotope distribution in silicates (olivine) shows that the first process is more likely, as nickel isotopes from olivine are also fractionated, and δ^{62}Ni value in it is slightly higher than in ore minerals. However, this is the first result; further extensive and more accurate research is required for reliable conclusions. Thus, it can be presumed that there is a single source of nickel in intrusions, and this source may be mafic and ultramafic rocks of the Norilsk district, which are, apparently, not the main source of copper.

The obtained data on isotopic composition of copper and nickel will be necessary in future in the search for copper and nickel sources directly from isotopic label.

3 Cu and Ni Isotopes in Technogenic Materials of the Norilsk MMC

Possibilities of determining the regional affiliation of the products of Ni–Cu–PGE production. At CIR, the study of copper and nickel isotopic composition was conducted for solving certain geological, geochemical, and environmental tasks. It was aimed at finding differences in isotopic composition of Cu and Ni in the products of Ni–Cu–PGE producing facilities using different materials and different technological cycles for regional identification. The collection comprises samples of the Norilsk Nickel Mining and Metallurgical Company (MMC) (Norilsk District), the Kola MMC (Norilsk and Pechenga districts), South African companies Western Platinum, Impala Platinum, Anglo Platinum, Lonmin Platinum (Bushveld and Zimbabwe regions).

Measurement results of Cu and Ni isotopic composition are given as δ parameters in Fig. 5. Noteworthy is the classification of objects under two groups on the basis of δCu: the I group (−3.1…−1.2) is represented by samples of the Norilsk nickel and the Kola MMC; the II group (−1.1…−0.0) comprises samples of the SA companies. Within the first group, samples of the above two objects almost completely overlap in δCu and δNi values, although as regards δNi, samples of the Norilsk nickel have a smaller variation range (−0.1…−1.5) as compared with samples of the Kola MMC. The similarity of objects of the first group in respect of Cu and Ni isotopic composition is due to similar and partly overlapping sources of raw materials.

Within the second group, samples of different SA companies are also completely overlapping in δCu and partly in δNi. In the Cu–Ni isotope systematics, the products from two different regions of the raw material—Bushveld and Zimbabwe—completely overlap unlike Rb–Sr isotopic systematics based on radiogenic isotopes. Obviously, isotopic variations of nickel, unlike copper, are characterized by a

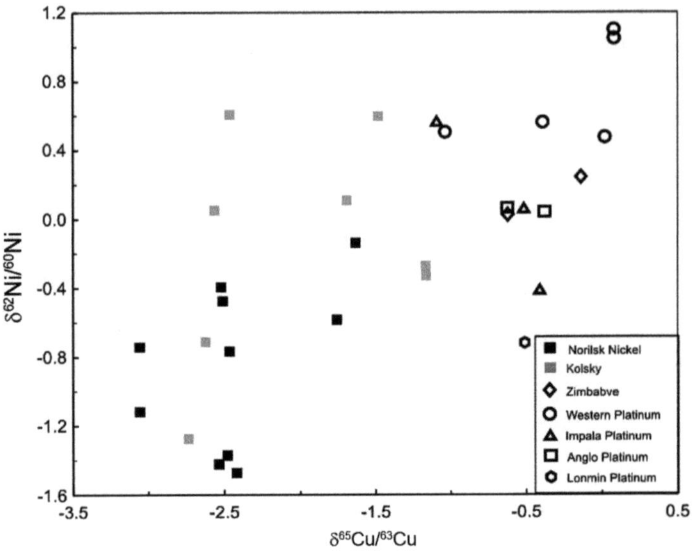

Fig. 5 $\delta^{65}Cu/^{63}Cu$ and $\delta^{62}Ni/^{60}Ni$ diagram for Ni–Cu–PGE metallurgical products

more narrow variation range of δ parameters, which might be caused by two valence states, Ni^0 and Ni^{2+}, of nickel and three valence states of copper, Cu^0, Cu^{1+} and Cu^{2+}. Copper isotopic composition variations are, to a greater extent, associated with different isotopic composition of raw material sources, whereas for nickel, presumably, the isotopic composition variations are, in a greater degree, accounted for by technological processes.

Figure 6 shows δ parameters for two isotopic ratios of nickel. Both parameters show good correlation, which is due to the vectors in accords of the natural, technogenic and instrumented mass-fractionation for this type of objects unlike some low-temperature natural processes.

The data suggest that isotopic variations of stable copper and nickel isotopes can be successfully used to identify the products and semi-finished products of Ni–Cu–PGE metallurgy, especially in combination with isotopic data on other elements (Sr, Pb, S).

Variations of nickel isotope ratios during carbonylation: a probable model of kinetic fractionation in nature. Due to appearance of a new generation of analytical instruments, such as MC ICPMS, it became possible to study a number of unconventional isotope systems, including Ni. The materials of the technological process of producing metallic nickel were studied in the so-called mond process. This process consists in extraction of nickel as tetracarbonyl $Ni(CO)_4$ from Ni-containing materials (commercial ores, ore concentrates, matte, etc.). The samples were taken at the operating process plant of the OAO “Norilsk Nickel”.

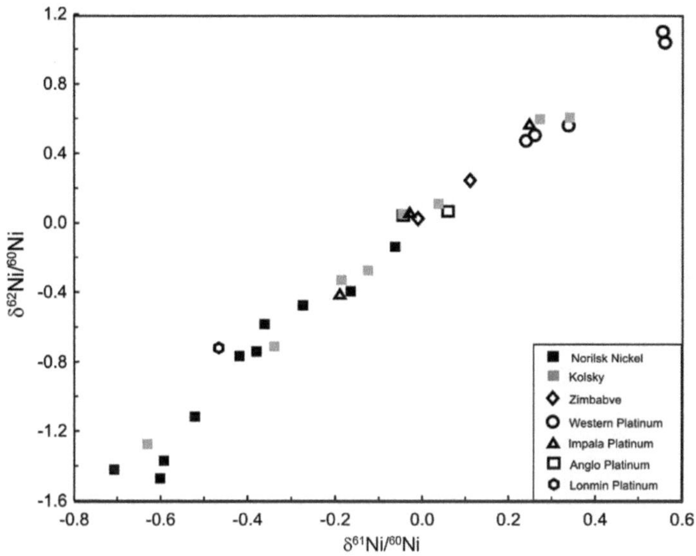

Fig. 6 $\delta^{61}Ni/^{60}Ni$ and $\delta^{62}Ni/^{60}Ni$ diagram for Ni–Cu–PGE metallurgical products

Source material are the remains of Ni-anodes of electrochemical production. Before entering the mond process, it is subjected to high-temperature melting and subsequent crushing, which ensures its isotopic homogeneity.

Sample 1—material in the reaction chamber is exposed to a long-term flow impact of the synthesis gas at 200 °C and P 200 atm. Synthesis gas is produced by burning coke and contains 95–97% CO and H_2 as well as small amounts of SO_2 and CO_2. Such conditions ensure a vigorous material carbonylation: $x\text{Me} + y\text{CO} \Leftrightarrow \square\text{Me}_x(\text{CO})_y$.

Sample 2—carbonylation feedstock residues after removal of a substantial part of Ni and other metals forming volatile carbonyls (Fe, Co, MPG, etc.). An admixture in the sample is amorphous carbon forming as a result of Bell-Boudoir reaction $2CO \rightarrow CO_2 + C$.

Sample 3—carbonyl condensate residues solidified by burning after rectification stripping of most of tetracarbonyl Ni (T_{boiling} 43 °C at atm. pressure) from other, less volatile, carbonyls.

Sample 4—powder-like metallic nickel obtained by thermal decomposition of Ni carbonyl: $Ni(CO)_4 \Longrightarrow Ni_{(\text{metal})} + 4CO$.

Results and interpretation. Figure 7 shows isotopic values in $\delta^{62}Ni/^{60}Ni$ and $^{61}Ni/^{60}Ni$ versus the NIST986 standard in the studied samples. Measurements on ^{58}Ni and ^{64}Ni masses were not used because of isobaric interference of ^{58}Fe and ^{64}Zn. The comparison shows that the final product—carbonyl Ni (sample 4) is most enriched in the light isotope ^{60}Ni as compared with both the source (sample 1), and the intermediate materials of the carbonylation process.

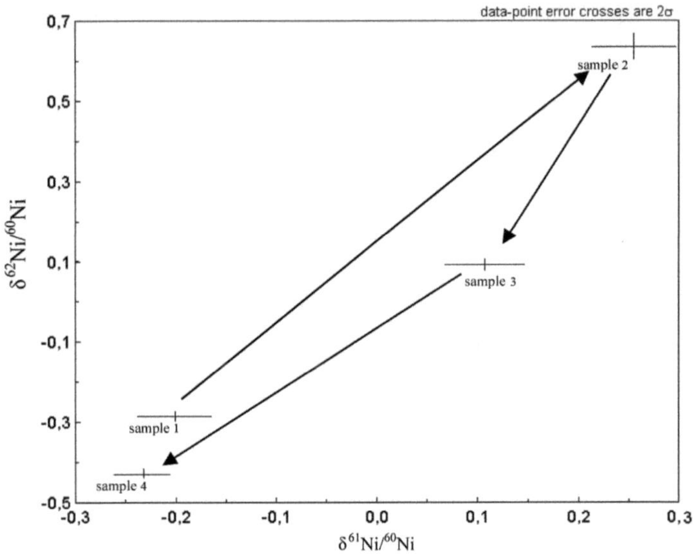

Fig. 7 Isotopic composition of Ni in the studied samples

Gradual transition of Ni from solid metal-bearing compounds to the gas phase, $Ni(CO)_4$, entrained by the gas flow, subsequent condensation and distillation create the conditions for kinetic fractionation.

The established regular change in isotopic composition of Ni at a relatively low-temperature and low-pressure gas extraction as volatile carbonyl can be regarded as a model of natural kinetic fractionation of Ni isotopes and, possibly, of some other transition metals prone to form carbonyls. Reducing gases containing CO and H_2 in natural processes can both have a juvenile origin, and form at thermal heating (contacts of intrusions, metamorphism) due to carbon-containing compounds and water vapour in the geological environments.

Strontium and Neodymium Isotopes

Yevgeny Bogomolov, Boris Belyatsky, Robert Krymsky and Yury Pushkarev

Abstract The chapter presents data on Rb–Sr and Sm–Nd TIMS analysis of pyroxenes, plagioclases, gross samples of igneous rocks and sulphide ores. In most Sm–Nd analyses of minerals, the isotope system (pyroxene–plagioclase) does not reflect the age of intrusions and shows a rejuvenated or aged isochronous age. Only for the Kharaelakh intrusion, one 241 ± 32 Ma isochron was obtained, close to the U–Pb age of zircons. The primary isotope ratio $^{87}Sr/^{88}Sr$ and the ε_{Nd} parameter of samples are calculated for the age of 250 Ma based on the average U–Pb age of zircons. The primary ratio of plagioclases strontium from magmatic rocks of intrusions (0.7032–0.7090) is lower than that of ore sulphides (0.7081–0.7116). The heterogeneity of Sr and Nd isotope composition of minerals and rocks within a single intrusion is caused by different degrees of impact of ore and host-rock contamination. In the gross samples of igneous rocks from the studied massifs, the primary ratio of strontium varies from 0.7049 to 0.7128, and ε_{Nd} varies from −7.7 to +7.0. These data indicate a significant effect of the crustal component in the formation of igneous rocks during the injection of mantle material with the parameters $^{87}Sr/^{88}Sr = 0.7037$ and $\varepsilon_{Nd} = +9.5$. Based on Nd and Sr isotopic data, it is impossible to clearly distinguish groups of intrusions with different ore content.

Rb–Sr, U–Pb, Pb–Pb and other [1] methods of isotopic studies of rocks were repeatedly used to study sulphide platinoid–nickel–copper deposits. Along with the Rb–Sr method, the Sm–Nd isotopic method was also applied. These methods complement each other well due to varying degrees of closure of Rb–Sr and Sm–Nd isotopic systems in minerals in relation to the superimposed metamorphic processes as well as due to the use of different sets of minerals in them.

Rb–Sr method is based on radioactive decay of ^{87}Rb isotope and its transformation into a stable radiogenic isotope ^{87}Sr by emitting a beta particle with a half-life $T_{1/2} = 4.89 \times 10^{10}$ years [2].

Y. Bogomolov (✉) · B. Belyatsky · R. Krymsky · Y. Pushkarev
Russian Geological Research Institute (VSEGEI), St. Petersberg, Russia
e-mail: evgeniy_bogomolov@vsegei.ru

O. Petrov (ed.), *Isotope Geology of the Norilsk Deposits*, Springer Geology,
https://doi.org/10.1007/978-3-030-05216-4_4

Sm–Nd method is based on radioactive decay of ^{147}Sm isotope and its transformation by α-decay into a stable radiogenic isotope ^{143}Nd with a half-life $T_{1/2} = 1.06 \times 10^{11}$ years [2].

1 Methodology, Samples

Analysis of Rb–Sr and Sm–Nd systems of rocks and minerals was performed using isotope dilution method for determining concentrations of rubidium, strontium, samarium and neodymium. For this purpose, weighed amounts of solutions of mixed indicators ^{87}Rb–^{84}Sr and ^{149}Sm–^{150}Nd were added to the pre-ground weighed samples. Then, the prepared samples were decomposed in a mixture of nitric and hydrofluoric acids. The extraction of rubidium and strontium was performed by cation exchange chromatography on resin of AG50W-X8 with following purification of strontium on microcolumn with Eichrom Sr specific resin. The extraction of samarium and neodymium was performed in two stages. The first one is cation exchange chromatography on AG50W-X8 resin to separate rare earth elements from the total mass of the rock and mineral matter. The second stage is extraction chromatography using liquid cation exchange extractant HDEHP on a Teflon carrier.

Isotopic analysis of Rb, Sr, Sm and Nd was made on the nine-collector TRITON mass spectrometer in a static mode. Correction for strontium isotope fractionation was carried out by means of normalizing the measured values using ^{88}Sr/^{86}Sr = 8.37521. Normalized ratios were adjusted to the value ^{87}Sr/^{86}Sr = 0.71025 in the international isotope standard NBS-987. A correction for isotopic fractionation of neodymium was carried out by normalizing the measured values using ratio ^{148}Nd/^{144}Nd = 0.241578. Normalized ratios were adjusted to the value ^{143}Nd/^{144}Nd = 0.511860 in the international isotope standard La Jolla.

The error of determining the contents of Rb, Sr, Sm, Nd is 0.5%. The level of the blank experiment (pg) is: 30 for Rb, 30 for Sr, 20 for Sm and 40 for Nd. This makes it possible to analyse at a rather high level the strontium and neodymium isotope ratios even in samples with low contents of these elements. Plotting of isochronous dependencies and calculation of the age of the investigated rocks as well as of the initial ratio $(^{87}\mathrm{Sr}/^{86}\mathrm{Sr})_0$ and ε_{Nd} parameter were carried out by ISOPLOT software [3] using the following constants: $\lambda_{87Rb} = 1.42 \times 10^{-11}$ year^{-1}, $\lambda_{147Sm} = 6.54 \times 10^{-12}$ year^{-1}, $(^{143}\mathrm{Nd}/^{144}\mathrm{Nd}_{CHUR})^{\circ} = 0.512636$, $(^{147}\mathrm{Sm}/^{144}\mathrm{Nd}_{CHUR})^{\circ} = 0.1967$ [4–6].

In calculations, the following values of relative errors were introduced for determining the Rb–Sr and Sm–Nd data: 0.5% for ^{87}Rb/^{86}Sr ratio, 0.5% for ^{147}Sm/^{144}Nd, 0.01% for ^{87}Sr/^{86}Sr, 0.003% for ^{143}Nd/^{144}Nd. ε_{Nd} parameter was determined with the precision of ±0.5.

Model of neodymium isotope evolution in the Earth uses the concept of CHUR, i.e. of a homogeneous chondrite reservoir; modern values of its parameters are given in [2, 5]. Initial ^{143}Nd/^{144}Nd ratios in igneous and metamorphic crustal rocks are compared with the corresponding ratios in CHUR during rock crystallization. Since differences in these ratios are rather small, an “epsilon” parameter is defined as 10^4-fold increased value of relative difference in the initial $^{143}Nd/^{144}Nd$ ratio in rock found by plotting an isochron for rock samples with the corresponding ratio in CHUR, calculated for the time which is determined by the slope of this isochron.

In the studied samples of plagioclases, samarium and neodymium contents ranged from 0.126 to 0.669 ppm Sm and from 0.735 to 4.951 ppm Nd; and in pyroxene samples, from 1.404 to 3.100 ppm Sm and from 3.307 to 9.677 ppm Nd. This indicates the presence of a small number or absence of older inclusions in these minerals. Therefore, in 2006–2008, an attempt was made to determine the age by Sm–Nd method. A large part of the results of such estimates within the determination error is consistent with zircon age of 250 Ma in case of positive values of ε_{Nd} parameter (from 0.3 to 1.0) for commercially mineralized (Kharaelakh, Norilsk-1) and mineralized (Chernogorsky) intrusions. In most cases, when dating on the basis of plagioclase-pyroxene pairs there are significant deviations from this value both towards increasing (to 945 ± 49 Ma for plagioolivinite from the commercially mineralized Kharaelakh intrusion), and towards decreasing (to 38 ± 45 Ma for melanotroctolite from the commercially mineralized Talnakh intrusion), which can be due either to the lack of the original isotopic equilibrium in rock minerals, or to the disturbance of the closure of Sm–Nd system in them caused by secondary processes. Elevated age values are associated with a possible presence of a certain share of the older crustal component in the studied rocks. However, in one case, it was possible to obtain a three-point isochron (plagioclase-bulk-pyroxene) for the commercially mineralized Kharaelakh intrusion (sample 844-1.): $t = 241 \pm 32$ Ma, $\varepsilon_{Nd} = +0.9$ MSWD = 0.0061 (Fig. 1).

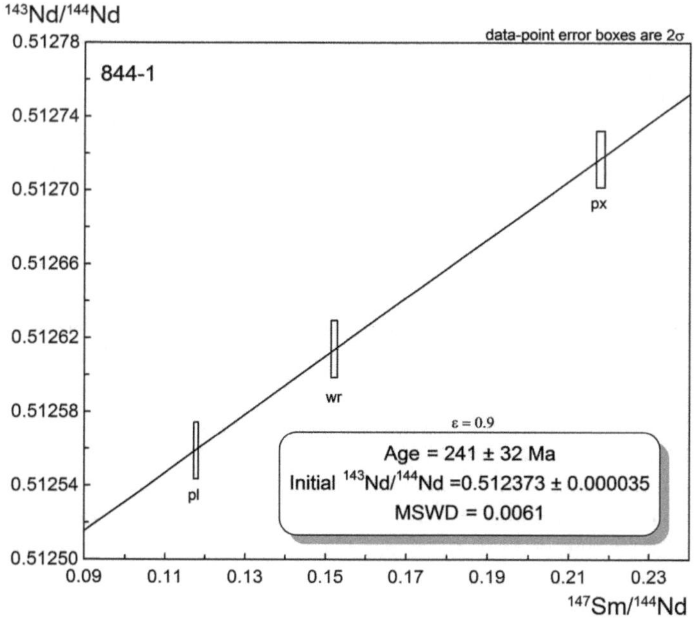

Fig. 1 Sm–Nd isochron after rock sample as a whole and minerals from Kharaelakh intrusive, Bh. KZ-844

However, this is rather the exception than the rule, because, in most cases, when plotting isochrones from three or more points the calculations give significant dating errors and unacceptable values of MSWD parameter. For example, for plagioolivinite from the commercially mineralized Kharaelakh intrusion (844-2), the system plagioclase-bulk-pyroxene gives the result $t = 882 \pm 6400$ Ma, $\varepsilon_{Nd} = -0.3$ MSWD = 416; and for melanotroctolite from the commercially mineralized Talnakh intrusion (T-16), the system plagioclase-olivine-pyroxene, $t = 42 \pm 1200$ Ma, $\varepsilon_{Nd} = -2.7$, MSWD = 17.

2 Results and Discussion

In 2006–2008, CIR VSEGEI performed 391 Rb–Sr and 370 Sm–Nd analyses including 215 Rb–Sr analyses and 198 Sm–Nd analyses in bulk samples as well as 73 Rb–Sr and 73 Sm–Nd analyses in plagioclases, 76 Rb–Sr and 73 Sm–Nd in pyroxenes, 27 Rb–Sr in sulphides, 23 Sm–Nd in olivines, 1 Rb–Sr in calcite and 2 Rb–Sr in anhydrites (Tables 1, 2 and 3).

When interpreting the data, the assessments of the initial isotopic ratios for the mantle material [4] dated at 250 Ma were used. For strontium, the value $(^{87}Sr/^{86}Sr)_0 = 0.703723$ was used. Numerical value of ε_{Nd} parameter is +9.5 [7].

Evaluation of the initial ratio of strontium isotopes in the studied samples could be made rather strictly for $^{87}Sr/^{86}Sr$ ratio in plagioclases with the minimum Rb–Sr ratios. The range of variations in them is 0.7032–0.7080. One can also assess the initial strontium ratio for bulk samples by making a correction for the geological age about 250 Ma obtained from U–Pb zircon data for the majority of zircons. Such an assessment gives the variation range 0.7032–0.7090 for the samples investigated in 2006–2008. Strontium isotope analysis of sulphides from these samples gives higher $^{87}Sr/^{86}Sr$ values (0.7081–0.7116). This is, apparently, due to the influence of the crustal component on the mantle component of the fluid, the intrusion of which resulted in ore formation. Rb–Sr dating is not possible due to a very small variation range of rubidium–strontium ratio.

There is also a significant variation of ε_{Nd} parameter values calculated when plotting isochrones (from −4.8 to 4.4). This points to a significant influence of the crustal component on the intruded mantle fluid. The above data also indicate a lack of complete isotope homogenization of neodymium in the studied rocks.

The results confirm a multi-stage metasomatic impact on the mantle fluid, which affected the rubidium–strontium and samarium–neodymium systems in the studied rocks and the composing minerals.

For the most conservative samarium–neodymium isotopic system (not to mention the rubidium–strontium system!), model age values cannot be assigned to any real geological event, even if a two-stage model is used, taking into account only a single interaction of the mantle protolith with crustal matter.

Table 1 Results of Sm–Nd analysis in rocks of Norilsk-1 intrusive

Sample	Rock	Sm, ppm	Nd, ppm	$^{147}Sm/^{144}Nd$	$^{143}Nd/^{144}Nd$	2σ	ε_{Nd} (250)
Bh. MN-2							
51	Gabbrodiorite	3.763	19.850	0.1146	0.512450	6	−1.0
52		4.101	18.270	0.1357	0.512445	4	−1.8
53	Olivine-free gabbro	1.456	5.809	0.1516	0.512635	7	1.4
55		2.764	9.960	0.1678	0.512644	7	1.0
56		2.448	8.749	0.1692	0.512669	9	1.5
57	Olivine-bearing gabbro	2.062	7.132	0.1748	0.512646	6	0.9
59		2.355	8.376	0.1700	0.512657	8	1.2
61		2.284	7.975	0.1731	0.512704	10	2.0
63	Plagiowehrlite	2.174	7.057	0.1863	0.512660	14	0.8
64		1.143	4.095	0.1687	0.512653	8	1.2
65		1.082	3.394	0.1662	0.512682	9	1.8
66		1.769	6.615	0.1617	0.512646	8	1.3
67		1.079	4.084	0.1597	0.512611	8	0.7
68	Troctolite	1.060	4.091	0.1566	0.512655	14	1.6
69a	Olivine-bearing gabbro	1.255	4.454	0.1704	0.512642	7	0.9
69b	Olivine-bearing leucogabbro	1.274	4.607	0.1672	0.512947	7	7.0
69b		1.288	4.707	0.1655	0.512630	14	0.8
71	Plagiowehrlite	1.380	5.029	0.1659	0.512651	7	1.2
72	Gabbro-troctolite	1.083	3.786	0.1729	0.512674	9	1.5
73	Olivine-free gabbro	2.601	9.011	0.1745	0.512645	8	0.8
74	Pyroxene leucogabbro	3.088	10.640	0.1755	0.512641	4	0.7
75	Gabbro olivine	1.047	3.732	0.1697	0.512312	45	−5.5
76	Olivine-free gabbro	3.562	14.910	0.1445	0.512485	6	−1.3
77	Troctolite	2.753	8.878	0.1875	0.512849	24	4.4
77-1		2.737	9.669	0.1711	0.512643	9	0.9
78	Olivine-free leucogabbro	4.439	17.560	0.1529	0.512511	4	−4.1
N1-11	Gabbrodiorite	3.334	14.980	0.1346	0.512429	6	−2.1
N1-12	Leucogabbro	1.074	4.063	0.1598	0.512666	6	1.7
Bh. MS-18							
18-1	Leucogabbro	2.464	9.631	0.1547	0.512609	10	0.8
18-2		1.301	4.899	0.1605	0.512655	9	1.5
18-3	Olivine-bearing leucogabbro	2.622	9.652	0.1643	0.512627	9	0.8
18-4	Olivine-bearing gabbro	2.813	10.230	0.1665	0.512602	8	0.3
Bh. KN-97							
97-1	Olivine-bearing gabbro	2.485	9.488	0.1583	0.512660	7	1.7
97-2	Plagiowehrlite	2.423	9.075	0.1615	0.512630	12	1.0
97-3	Troctolite	1.683	6.450	0.1578	0.512668	4	1.8
97-6		1.657	6.291	0.1593	0.512694	6	2.3

Table 2 Results of Sm–Nd analysis in rock-forming minerals of Norilsk-1 intrusive

Sample	Rock, mineral	Sm, ppm	Nd, ppm	$^{147}Sm/^{144}Nd$	$^{143}Nd/^{144}Nd$	2σ
Bh. Kn-97						
Kn-97-1	Plagioclase	0.391	1.887	0.1252	0.512551	5
Kn-97-1	Pyroxene	1.938	4.717	0.2484	0.512762	6
Kn-97-2	Plagioclase	0.508	2.408	0.1276	0.512440	9
Kn-97-2	Pyroxene	2.225	5.433	0.2476	0.512766	10
Kn-97-3	Plagioclase	0.578	2.481	0.1408	0.512596	10
Kn-97-3	Orthopyroxene	3.555	8.432	0.2549	0.512794	7
Kn-97-3	Clinopyroxene	1.159	2.704	0.2592	0.512908	8
Kn-97-4	Plagioclase	0.737	3.438	0.1296	0.512605	12
Kn-97-4	Pyroxene	4.029	9.708	0.2510	0.512769	8
Kn-97-5	Plagioclase	0.480	2.299	0.1261	0.512446	11
Kn-97-5	Pyroxene	4.053	10.100	0.2426	0.512776	9
Kn-97-6	Plagioclase	0.565	2.388	0.1431	0.512613	14
Kn-97-6	Pyroxene	3.215	7.816	0.2487	0.512848	8
Bh. MN-2						
N1-1	Plagioclase	0.690	3.704	0.1126	0.512390	11
N1-1	Pyroxene	3.759	11.530	0.1971	0.512549	5
N1-2	Plagioclase	1.153	5.025	0.1387	0.512576	10
N1-2	Pyroxene	8.861	26.070	0.2055	0.512706	7
N1-3	Plagioclase	0.232	1.331	0.1054	0.512358	18
N1-3	Pyroxene	3.133	8.751	0.2164	0.512712	8
N1-4	Plagioclase	0.218	1.071	0.1233	0.512547	18
N1-4	Pyroxene	2.067	5.817	0.2149	0.512746	9
N1-5	Plagioclase	0.448	2.113	0.1281	0.512590	9
N1-5	Pyroxene	1.731	4.192	0.2497	0.512769	9
N1-6	Plagioclase	0.553	2.578	0.1297	0.512598	12
N1-6	Pyroxene	1.778	4.418	0.2434	0.512770	5
N1-7	Plagioclase	0.259	1.241	0.1261	0.512606	11
N1-7	Pyroxene	3.212	8.447	0.2299	0.512731	7
N1-8	Plagioclase	0.236	1.094	0.1302	0.512601	13
N1-8	Pyroxene	3.728	9.007	0.2503	0.512771	13
N1-9	Plagioclase	0.262	1.276	0.1239	0.512554	8
N1-9	Pyroxene	1.766	4.329	0.2467	0.512794	7
N1-10	Plagioclase	0.456	2.409	0.1144	0.512524	8
N1-10	Pyroxene	5.187	15.030	0.2087	0.512612	6

Given below are the variations of the initial strontium isotope ratios and ε_{Nd} parameter (250 Ma) for three wells selected as representatives of the commercially mineralized, mineralized and weakly mineralized intrusions. Commercially mineralized Kharaelakh intrusion (well KZ-963, Fig. 2, Table 4) is characterized by

Table 3 Results of Sm–Nd analysis in rocks of Talnakh and Lower Talnakh intrusives

Sample	Rock, mineral	Sm, ppm	Nd, ppm	$^{147}Sm/^{144}Nd$	$^{143}Nd/^{144}Nd$	2σ	ε_{Nd} (250)
Talnakh intrusive (Bh. OUG-2)							
T-1	Gabbrodiorite	3.802	13.960	0.1647	0.512641	14	1.1
T-2	Diorite-pegmatite	6.139	25.680	0.1445	0.512595	5	0.8
T-3	Gabbrodiorite	7.490	29.020	0.1560	0.512603	4	0.6
T-5	Leucogabbro, olivine-bearing	3.514	12.960	0.1639	0.512627	6	0.8
T-6	Gabbro, olivine-bearing	2.863	10.310	0.1678	0.512643	9	1.0
T-8		1.977	7.105	0.1683	0.512634	6	0.8
T-10		1.898	6.950	0.1651	0.512627	6	0.8
T-12		2.187	7.954	0.1662	0.512649	15	1.2
T-13	Plagiowehrlite, miner.	1.470	5.337	0.1666	0.512645	7	1.1
T-14		1.221	4.593	0.1608	0.512561	12	−0.4
T-15		0.896	3.209	0.1688	0.512648	25	1.1
T-15-132	Massive Cu-Ni ore	0.012	0.079	0.0946	0.512800	223	6.4
T-16	Melanotroctolite, miner.	1.508	6.350	0.1436	0.512477	11	−1.4
T-17	Plagiopyroxenite, miner.	1.958	7.435	0.1592	0.512593	13	0.3
T-18	Gabbro, olivine-bearing, miner.	1.599	8.109	0.1193	0.512343	6	−3.3
T-19	Metasomatite	10.560	54.150	0.1180	0.512500	8	−0.1
T-26	Hornfels	5.924	28.940	0.1238	0.512140	8	−7.3
T-30	Alkali metasomatite	3.560	18.920	0.1138	0.512320	4	−3.6
Lower Talnakh intrusive (Bh. TG-31)							
31-1	Olivine-free gabbro	2.831	11.680	0.1466	0.512309	4	−4.8
31-2	Olivinic gabbro	2.016	8.518	0.1431	0.512291	3	−5.1
31-3	Melanotroctolite	3.621	11.020	0.1987	0.512551	5	−1.8
31-4	Troctolitic gabbro	1.711	7.277	0.1421	0.512299	6	−4.9
31-5	Plagiowehrlite	1.590	6.631	0.1450	0.512297	7	−5.0
31-6	Olivinic gabbro	1.669	6.940	0.1454	0.512295	5	−5.1
31-7	Plagiowehrlite	1.373	5.445	0.1525	0.512328	7	−4.6
31-8		1.160	4.607	0.1523	0.512315	5	−4.9
31-9		1.360	5.763	0.1427	0.512305	7	−4.8
31-10		1.430	6.106	0.1416	0.512295	9	−4.9
31-11	Melanotroctolite	3.006	12.700	0.1431	0.512488	5	−1.2
31-12	Plagioolivinite	1.382	6.326	0.1321	0.512297	5	−4.6
31-13	Plagiowehrlite	1.369	5.520	0.1499	0.512299	8	−5.1
31-14		1.462	6.052	0.1461	0.512325	8	−4.5
31-15	Melanotroctolite	2.149	9.230	0.1408	0.512303	4	−4.8
31-16		1.974	8.520	0.1401	0.512278	4	−5.2
31-17		1.869	7.969	0.1418	0.512268	5	−5.5
31-18	Olivine-free gabbro	2.814	11.560	0.1472	0.512302	6	−5.0
31-19	Olivinic gabbro	1.662	6.883	0.1459	0.512297	7	−5.0
31-20	Plagioolivinite	1.200	5.032	0.1441	0.512327	4	−4.4

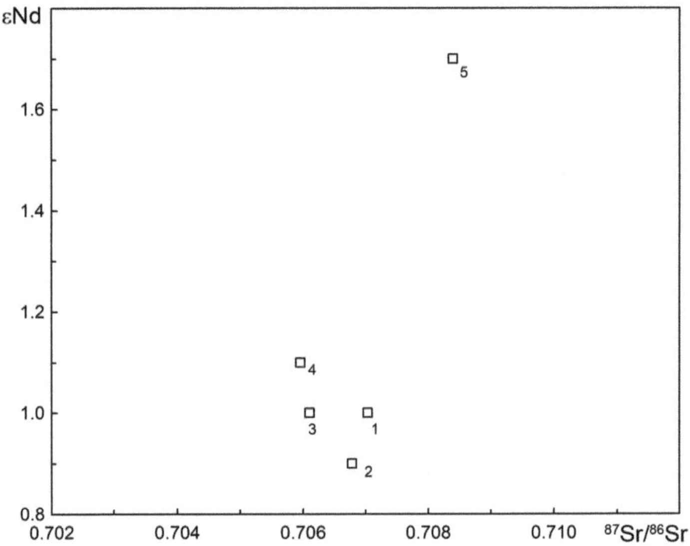

Fig. 2 Initial Nd and Sr isotopic composition for the age of 250 Ma from rocks of Kharaelakh intrusive, Bh. KZ-963. Numbers of imaging points correspond to Table 4

positive values of ε_{Nd} parameter (250 Ma) from 0.9 to 1.7 with its growth with increasing well depth. Apparently, this is due to a lower addition of the crustal component into the mantle fluid with increasing depth. The initial strontium ratio in sulphides is around 0.709 throughout the investigated well depth. Ratios 0.707 and below traced in ultramafic rocks are, most likely, accounted for by the secondary processes and low rubidium and strontium contents in ultramafic rocks.

Zub-Marksheidersky mineralized intrusion (well MP-27, Fig. 3, Table 6) is characterized by positive values of ε_{Nd} parameter (250 Ma) about 2.0 through the entire well depth studied. The initial strontium ratio gradually decreases from 0.709 to 0.706 with increasing well depth. This is, most likely, accounted for by a contribution of the crustal component to the mantle fluid with increasing depth.

Weakly mineralized Lower Talnakh intrusion (well TG-31, Fig. 4, Table 8) is characterized by negative values of the ε_{Nd} parameter (250 Ma), −5.0, throughout the well depth studied. This indicates a much stronger influence of the crust on the mantle fluid as compared with mineralized and commercially mineralized intrusions. The initial Sr ratio varies fairly smoothly from 0.7085 to 0.7076 with increasing well depth. The initial strontium ratio in plagioclases from 0.7087 to 0.7077 supports the trend of bulk samples. In all three cases, variations of the ε_{Nd} parameter (250 Ma) make it possible to confidently enough diagnose a greater or a lesser influence of the crust on the mantle fluid. This is also supported by variations of the initial isotopic ratio of strontium in the same objects.

The studied rocks lie within a broad range of variations, both as regards isotopic composition of the initial strontium ratio, and the value of the ε_{Nd} parameter defined

Table 4 Results of Sm–Nd and Rb–Sr analysis of samples (Kharaelakh intrusive, Bh. KZ-963)

Sample	Rock, mineral	$^{147}Sm/^{144}Nd$	$^{143}Nd/^{144}Nd$	2σ	ε_{Nd} (250)	$^{87}Rb/^{86}Sr$	$^{87}Sr/^{86}Sr$	2σ	IR (250)
963-21	Olivine-free gabbro	0.1678	0.512641	4	1.0	0.1343	0.707513	11	0.707036
963-23	Olivinic gabbro	0.1648	0.512634	5	0.9	0.1822	0.707428	12	0.706780
963-29	Melanotroctolite	0.1702	0.512648	4	1.0	0.1469	0.706625	18	0.706103
963-31		0.1622	0.512639	7	1.1	0.1855	0.706716	10	0.705956
963-65	Olivinic gabbro	0.1429	0.512636	7	1.7	0.2241	0.709194	7	0.708397

Table 5 Results of Sm–Nd analysis in rock-forming minerals of Talnakh and Lower Talnakh intrusives

Sample	Rock, mineral	Sm, ppm	Nd, ppm	$^{147}Sm/^{144}Nd$	$^{143}Nd/^{144}Nd$	2σ
Talnakh intrusive (Bh. OUG-2)						
T-1	Plagioclase	0.669	3.360	0.1203	0.512540	4
T-1	Pyroxene	1.556	3.718	0.2530	0.512777	7
T-2	Plagioclase	0.524	2.061	0.1536	0.512594	3
T-2	Pyroxene	3.100	9.677	0.1937	0.512670	6
T-5	Plagioclase	0.640	2.807	0.1377	0.512580	4
T-5	Pyroxene	2.042	5.892	0.2095	0.512701	8
T-6	Plagioclase	1.185	4.951	0.1447	0.512584	4
T-6	Pyroxene	2.175	4.831	0.2722	0.512730	8
T-8	Plagioclase	0.269	1.363	0.1194	0.512548	5
T-8	Pyroxene	1.855	4.858	0.2309	0.512750	12
T-10	Plagioclase	0.375	1.756	0.1291	0.512531	5
T-10	Pyroxene	1.454	3.458	0.2542	0.512765	12
T-12	Plagioclase	0.248	1.274	0.1178	0.512520	6
T-12	Pyroxene	1.404	3.307	0.2566	0.512767	12
T-13	Plagioclase	0.337	1.645	0.1238	0.512559	9
T-13	Pyroxene	2.231	5.374	0.2510	0.512707	8
T-14	Plagioclase	0.217	1.141	0.1151	0.512519	10
T-14	Pyroxene	2.908	7.503	0.2347	0.512718	7
T-15	Plagioclase	0.143	0.735	0.1179	0.512581	13
T-15	Pyroxene	2.343	6.300	0.2248	0.512635	8
T-16	Plagioclase	0.189	1.101	0.1038	0.512453	8
T-16	Pyroxene	2.620	7.096	0.2232	0.512483	26
T-17	Plagioclase	0.394	1.913	0.1247	0.512551	9
T-17	Pyroxene	2.470	6.488	0.2302	0.512739	9
T-18	Plagioclase	0.395	2.479	0.0963	0.512307	14
T-18	Pyroxene	1.725	6.855	0.1521	0.512387	8
T-19	Plagioclase	0.368	1.724	0.1289	0.512490	16
T-19	Pyroxene	1.432	3.390	0.2555	0.512780	8
Lower Talnakh intrusive (Bh. TG-31)						
31-1	Plagioclase	0.267	1.627	0.0992	0.512265	12
31-1	Pyroxene	1.427	4.026	0.2143	0.512399	7
31-3	Plagioclase	0.184	1.005	0.1107	0.512247	12
31-3	Pyroxene	1.482	4.104	0.2187	0.512487	10
31-7	Plagioclase	0.126	0.746	0.1020	0.512542	25
31-7	Pyroxene	1.707	4.711	0.2191	0.512406	6
31-9	Plagioclase	0.221	1.211	0.1103	0.512191	18
31-9	Pyroxene	1.508	4.141	0.2202	0.512443	12
31-10	Plagioclase	0.215	1.352	0.0964	0.512287	21

(continued)

Table 5 (continued)

Sample	Rock, mineral	Sm, ppm	Nd, ppm	$^{147}Sm/^{144}Nd$	$^{143}Nd/^{144}Nd$	2σ
31-10	Pyroxene	2.247	6.216	0.2185	0.512437	5
31-11	Plagioclase	0.317	1.826	0.1048	0.512345	10
31-11	Pyroxene	1.909	5.317	0.2171	0.512544	10
31-13	Plagioclase	0.324	1.745	0.1122	0.512329	9
31-13	Pyroxene	1.722	5.099	0.2042	0.512402	7
31-16	Plagioclase	0.224	1.321	0.1027	0.512293	17
31-16	Pyroxene	1.578	4.572	0.2086	0.512663	18

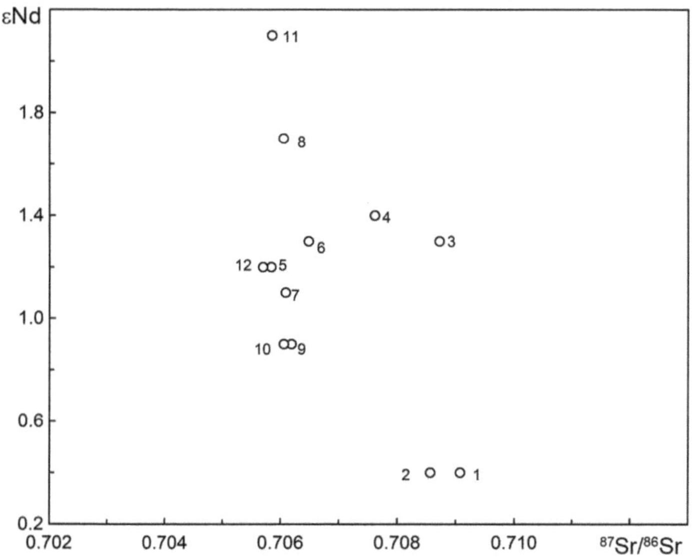

Fig. 3 Initial Nd and Sr isotopic composition for the age of 250 Ma from rocks of Zub-Marksheidersky intrusive, Bh. MP-27. Numbers of imaging points correspond to Table 6

from the initial isotopic composition of neodymium. It should be noted that the imaging points of rocks from different intrusions form certain segregated sets of points, which leads to the conclusion on their origin from multiple sources differing in neodymium and strontium isotopic composition.

At least, three such sets of imaging points or combinations of groups can be distinguished. One cloud is characterized by the variation range of $^{87}Sr/^{86}Sr$ ratio from 0.7075 to 0.7088 and of the ε_{Nd} parameter from −4.2 to −6.0. It is composed of the results of Nd–Sr analysis of rocks in the Lower Talnakh, Lower Norilsk and Zelenaya Griva intrusions. All of them belong to the weakly mineralized type of intrusions (Tables 10, 11, 12, 13, 14, 15, 16, 17, 18, 19, 20, 21, 22, 23 and 24).

Another group is characterized by $^{87}Sr/^{86}Sr$ variation range from 0.7055 to 0.7063 and the ε_{Nd} parameter from −4.8 to −0.8. It is made up of the results of

Table 6 Results of Sm–Nd and Rb–Sr analysis of samples (Zub-Marksheidersky intrusive, Bh. MP-27)

Sample	Rock, mineral	$^{147}Sm/^{144}Nd$	$^{143}Nd/^{144}Nd$	2σ	ε_{Nd} (250)	$^{87}Rb/^{86}Sr$	$^{87}Sr/^{86}Sr$	2σ	IR (250)
27-1	Olivine-free gabbro	0.1371	0.512563	6	0.4	0.8898	0.712241	8	0.709076
27-3		0.1412	0.512567	6	0.4	0.7802	0.711338	8	0.708563
27-4	Gabbrodiorite	0.1377	0.512607	5	1.3	0.9439	0.712076	7	0.708719
27-5	Olivinic gabbro	0.1758	0.512676	6	1.4	0.3065	0.708709	8	0.707619
27-6		0.1549	0.512630	4	1.2	0.2514	0.706734	8	0.705840
27-7		0.1701	0.512662	7	1.3	0.2448	0.707353	8	0.706482
27-8	Gabbro-troctolite	0.1682	0.512647	3	1.1	0.3918	0.707481	9	0.706088
27-10	Ultramafite	0.1734	0.512688	8	1.7	0.2727	0.707024	8	0.706054
27-11	Olivinic gabbro	0.1616	0.512629	6	0.9	0.2030	0.706910	8	0.706188
27-12		0.1640	0.512629	3	0.9	0.5685	0.708078	8	0.706056
27-13	Gabbro-troctolite	0.1615	0.512688	6	2.1	0.4662	0.707511	9	0.705853
27-14	Olivinic gabbro	0.1781	0.512670	5	1.2	0.2837	0.706707	6	0.705698

Table 7 Results of Sm–Nd analysis in rocks of Kharaelakh, Zelenaya Griva, Lower Norilsk and Kruglogorsky intrusives

Sample	Rock	Sm, ppm	Nd, ppm	$^{147}Sm/^{144}Nd$	$^{143}Nd/^{144}Nd$	2σ	ε_{Nd} (250)
Kharaelakh intrusive (boreholes KZ-844 and KZ-963)							
844-1	Gabbro, olivine-bearing	2.440	9.696	0.1521	0.512614	12	1.0
844-2	Plagioolivinite	1.275	4.846	0.1590	0.512618	3	1.1
844-3,4		1.220	3.856	0.1913	0.512670	7	0.8
844-6	Gabbro, olivine-bearing	1.316	4.906	0.1622	0.512621	5	0.8
844-7	Melanotroctolite	0.919	3.487	0.1593	0.512616	5	0.8
844-10		2.000	7.297	0.1657	0.512635	7	0.9
844-15	Leucogabbro, olivine-bearing	1.688	5.895	0.1732	0.512672	5	1.4
844-10, 11	Impreg. of sulphides	0.004	0.021	0.1269	0.512020	250	−9.8
844-18	Massive sulphides	0.031	0.207	0.0901	0.512210	20	−5.0
963-21	Gabbro, olivine-free, alt.	2.410	8.686	0.1678	0.512641	4	1.0
963-23	Gabbro, olivine-free.	3.631	13.320	0.1648	0.512634	5	0.9
963-29	Melanotroctolite, miner.	0.121	4.306	0.1702	0.512648	4	1.0
963-31	Gabbro-troctolite, miner.	2.200	8.202	0.1622	0.512639	7	1.1
963-65	Leucogabbro, olivine-bearing	3.060	12.950	0.1429	0.512636	7	1.7
Zelenaya Griva intrusive (Bh. F-233)							
F-233-2	Gabbro, olivine-bearing, alt.	2.284	9.888	0.1396	0.512297	7	−4.8
F-233-4	Metasomatite	2.568	9.670	0.1605	0.512343	4	−4.6
F-233-5	Gabbro-troctolite	1.687	7.245	0.1408	0.512302	7	−4.8
F-233-6	Plagiowehrlite	1.321	5.443	0.1467	0.512300	5	−5.0
F-233-7	Gabbro-troctolite	1.437	5.932	0.1465	0.512289	9	−5.2
F-233-10	Melanotroctolite, miner.	1.215	5.202	0.1412	0.512256	8	−5.7
F-233-11	Wehrlite, carbonatized	1.815	6.588	0.1666	0.512330	7	−5.1
F-233-12	Plagiowehrlite	1.485	6.136	0.1463	0.512308	9	−4.8
F-233-15	Melanotroctolite	1.907	7.804	0.1477	0.512258	6	−5.9
F-233-16	Gabbro, olivine-free.	2.295	10.130	0.1370	0.512251	8	−5.6

(continued)

Table 7 (continued)

Sample	Rock	Sm, ppm	Nd, ppm	$^{147}Sm/^{144}Nd$	$^{143}Nd/^{144}Nd$	2σ	ε_{Nd} (250)
Lower Norilsk intrusive (Bh. NP-37)							
37-4	Gabbro, olivine-bearing	1.680	7.023	0.1446	0.512299	5	−5.0
37-5		1.837	7.605	0.1460	0.512260	6	−5.8
37-7		2.035	8.615	0.1428	0.512291	6	−5.1
37-8b	Melanotroctolite	1.783	7.474	0.1442	0.512287	8	−5.2
37-9a	Metasomatite	1.463	3.791	0.2333	0.512620	9	−1.5
Kruglogorsky intrusive (Bh. MP-2)							
K-1	Ferrogabbro, altered	3.332	11.480	0.1756	0.512687	4	1.6
K-2	Leucogabbro	2.471	8.824	0.1693	0.512668	6	1.5
K-3		2.714	9.602	0.1709	0.512692	4	1.9
K-4	Leucogabbro, altered	2.456	8.869	0.1675	0.512655	5	1.3
K-6	Gabbro-troctolite, altered	2.829	10.240	0.1671	0.512697	3	2.1
K-8	Gabbro-troctolite, miner.	1.916	6.744	0.1717	0.512686	4	1.7
K-9	Olivinic gabbro, altered	2.852	10.130	0.1702	0.512675	4	1.6
K-10	Olivinic gabbro	3.352	11.930	0.1699	0.512701	5	2.1

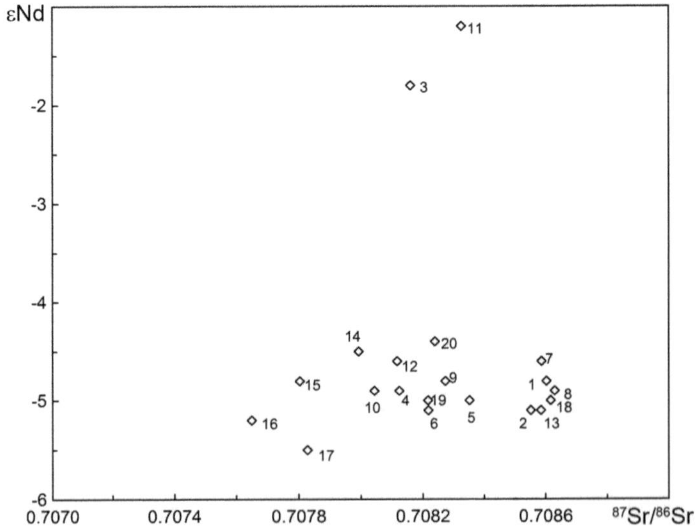

Fig. 4 Initial Nd and Sr isotopic composition for the age of 250 Ma from rocks of Lower Talnakh intrusive, Bh. TG-31. Numbers of imaging points correspond to Table 8

Table 8 Results of Sm–Nd and Rb–Sr analysis of samples (Lower Talnakh intrusive, Bh. TG-31)

Sample	Rock, mineral	$^{147}Sm/^{144}Nd$	$^{143}Nd/^{144}Nd$	2σ	ε_{Nd} (250)	$^{87}Rb/^{86}Sr$	$^{87}Sr/^{86}Sr$	2σ	IR (250)
31-1	Olivine-free gabbro	0.1466	0.512309	4	−4.8	0.4283	0.710126	7	0.708603
31-2	Olivinic gabbro	0.1431	0.512291	3	−5.1	0.5787	0.710610	7	0.708552
31-3	Melanotroctolite	0.1987	0.512551	5	−1.8	0.1825	0.708665	7	0.708016
31-4	Troctolitic gabbro	0.1421	0.512299	6	−4.9	0.1404	0.708623	25	0.708124
31-5	Plagiowehrlite	0.1450	0.512297	7	−5.0	0.2086	0.709094	10	0.708352
31-6	Olivinic gabbro	0.1454	0.512295	5	−5.1	0.1403	0.708717	8	0.708218
31-7	Plagiowehrlite	0.1525	0.512328	7	−4.6	0.0482	0.708759	10	0.708587
31-8		0.1523	0.512315	5	−4.9	0.2331	0.709459	23	0.708630
31-9		0.1427	0.512305	7	−4.8	0.2894	0.709302	8	0.708273
31-10		0.1416	0.512295	9	−4.9	0.2205	0.708828	6	0.708044
31-11	Melanotroctolite	0.1431	0.512488	5	−1.2	0.1943	0.709016	8	0.708325
31-12	Plagioolivinite	0.1321	0.512297	5	−4.6	0.3695	0.709431	6	0.708117
31-13	Plagiowehrlite	0.1499	0.512299	8	−5.1	0.0810	0.708873	9	0.708585
31-14		0.1461	0.512325	8	−4.5	0.3301	0.709166	8	0.707992
31-15	Gabbro-troctolite	0.1408	0.512303	4	−4.8	0.3628	0.709093	7	0.707803
31-16	Melanotroctolite	0.1401	0.512278	4	−5.2	0.3662	0.708951	27	0.707649
31-17		0.1418	0.512268	5	−5.5	0.3648	0.709125	8	0.707828
31-18	Olivine-free gabbro	0.1472	0.512302	6	−5.0	0.4262	0.710133	9	0.708617
31-19	Olivinic gabbro	0.1459	0.512297	7	−5.0	0.1382	0.708709	9	0.708217
31-20	Plagioolivinite	0.1441	0.512327	4	−4.4	0.2944	0.709286	7	0.708239

Table 9 Results of Sm–Nd analysis in rock-forming minerals of Kharaelakh, Mikchangda and Binyuda intrusives

Sample	Rock, mineral	Sm, ppm	Nd, ppm	$^{147}Sm/^{144}Nd$	$^{143}Nd/^{144}Nd$	2σ
Kharaelakh intrusive (boreholes KZ-844 and KZ-963)						
844-1	Plagioclase	0.249	1.278	0.1178	0.512559	19
844-1	Pyroxene	2.828	7.854	0.2177	0.512717	16
844-2	Plagioclase	0.464	2.438	0.1151	0.512018	21
844-2	Pyroxene	2.932	7.801	0.2272	0.512713	13
844-3,4	Plagioclase	0.147	0.703	0.1263	0.512442	33
844-3,4	Pyroxene	1.725	4.667	0.2235	0.512743	9
844-6	Plagioclase	0.172	0.936	0.1110	0.512526	32
844-6	Pyroxene	2.341	6.543	0.2163	0.512699	4
844-7	Plagioclase	0.234	1.246	0.1138	0.512469	47
844-7	Pyroxene	1.919	4.978	0.2331	0.512732	13
844-7	Olivine	0.089	0.369	0.1505	0.512656	46
844-10,11	Plagioclase	0.240	1.338	0.1085	0.512500	16
844-10,11	Pyroxene	4.970	13.650	0.2201	0.512727	10
844-15	Plagioclase	0.163	0.899	0.1093	0.512527	20
844-15	Pyroxene	3.684	10.150	0.2195	0.512721	10
963-30	Pyroxene	2.868	6.798	0.2550	0.512856	2
963-30	Olivine	1.431	5.942	0.1451	0.512048	5
Mikchangda intrusive (Bh. MD-48)						
48-9	Plagioclase	0.107	0.578	0.1116	0.512599	16
48-9	Pyroxene	1.771	4.499	0.2380	0.513012	7
48-16	Plagioclase	0.246	1.138	0.1306	0.512633	5
48-16	Pyroxene	1.716	4.165	0.2491	0.513832	6
48-18	Plagioclase	0.478	2.123	0.1361	0.512621	5
48-18	Pyroxene	1.642	3.876	0.2562	0.513167	9
48-25	Plagioclase	0.192	0.937	0.1240	0.512636	9
48-25	Pyroxene	2.211	5.506	0.2428	0.512952	7
48-27,28	Plagioclase	0.128	0.578	0.1341	0.512658	17
48-27,28	Pyroxene	2.044	4.939	0.2503	0.513030	5
48-30	Plagioclase	0.207	0.939	0.1332	0.512702	10
48-30	Pyroxene	1.612	3.756	0.2594	0.513047	7
48-30	Pyroxene 2	2.088	4.839	0.2609	0.512894	11
48-32,33,34	Plagioclase	0.218	1.052	0.1256	0.512594	18
48-32,33,34	Pyroxene	1.942	4.921	0.2386	0.513062	10
Binyuda intrusive (boreholes S-1 and S-3)						
S1-1	Plagioclase	0.422	2.706	0.0944	0.512277	16
S1-1	Pyroxene	3.858	11.300	0.2064	0.512467	4
S1-2	Plagioclase	0.375	2.352	0.0964	0.512290	6

(continued)

Table 9 (continued)

Sample	Rock, mineral	Sm, ppm	Nd, ppm	$^{147}Sm/^{144}Nd$	$^{143}Nd/^{144}Nd$	2σ
S1-2	Pyroxene	3.809	11.280	0.2042	0.512451	4
S1-3	Plagioclase	0.345	2.127	0.0979	0.512298	12
S1-3	Pyroxene	3.537	10.280	0.2079	0.512459	4
S1-5	Plagioclase	0.421	2.241	0.1137	0.512333	6
S1-5	Pyroxene	2.657	7.777	0.2066	0.512484	4
S3-2	Plagioclase	0.302	1.940	0.0942	0.512321	6
S3-2	Pyroxene	3.322	9.820	0.2045	0.512510	5
Guli Massif						
G17/24	Olivine	0.276	1.362	0.1225	0.512400	8
G17/48		0.256	0.942	0.1641	0.512770	14

analyzed rocks from the Binyuda and, partly, Talnakh intrusions. They belong to potentially mineralized and commercially mineralized types of intrusions.

The third set is characterized by the variation range of $^{87}Sr/^{86}Sr$ ratio from 0.7050 to 0.7092 and the ε_{Nd} parameter from −1.8 to +4.2. It is constituted based on results of rock analysis from the Talnakh (partially), Norilsk-1, Kharaelakh, Imangda, Chernogorsky, Zub-Marksheidersky, Vologochan, South Pyasino, Mikchangda, Oganer intrusions and picrites from volcanics of the Gudchikhin Formation. They belong to commercially mineralized, mineralized and potentially mineralized intrusions, and to non-mineralized formations. A more careful analysis allows distinguishing the available Nd–Sr data only for three non-mineralized picrites from the volcanics of the Gudchikhin Formation as a separate cloud characterized by the variation range of $^{87}Sr/^{86}Sr$ ratio from 0.7050 to 0.7061 and the ε_{Nd} parameter from 3.4 to 4.2.

Currently available six Nd–Sr analyses of rocks from the non-mineralized Oganer intrusion form a combination characterized by the variation range of $^{87}Sr/^{86}Sr$ ratio from 0.7055 to 0.7071 and the ε_{Nd} parameter from 1.4 to 3.4. It is only partly overlapping with the cloud of the potentially mineralized Mikchangda and the mineralized Imangda intrusions. This does not allow on the basis of the Nd–Sr isotopic systematics to strictly grade the mineralized, potentially mineralized and non-mineralized intrusions. Unfortunately, a relatively small number of Nd–Sr analyses in general, and from the rocks of non-mineralized formations, in particular, do not allow within the framework of this work a clear delineation of the set of imaging points that make up the groups assigned to the sources of commercially mineralized, mineralized, potentially, weakly and non-mineralized intrusions and formations.

In 2013 and 2014, 105 additional analyses of the Rb–Sr and Sm–Nd systems were made in bulk samples (Tables 25, 26, 27, 28, 29 and 30).

Evaluation of the initial strontium ratio based on bulk samples in this case was also carried out by introducing into the results of isotope studies a correction for the geological age of about 250 Ma obtained from the U–Pb data for the majority of

Table 10 Results of Sm–Nd analysis in rocks of Chernogorsk, Vologochan, South Pyasina and Imangda intrusives

Sample	Rock, mineral	Sm, ppm	Nd, ppm	$^{147}Sm/^{144}Nd$	$^{143}Nd/^{144}Nd$	2σ	ε_{Nd} (250)
Chernogorsk intrusive (Bh. MP-2bis)							
CH-2	Olivine-bearing gabbro	2.578	8.904	0.1750	0.512649	5	0.9
CH-6		2.414	8.634	0.1690	0.512637	6	0.9
CH-9		2.076	7.293	0.1721	0.512654	5	1.1
CH-11	Gabbro-troctolite, miner.	1.226	4.461	0.1661	0.512668	7	1.6
CH-13	Gabbro-troctolite, altered	1.688	6.206	0.1644	0.512720	8	2.6
CH-15	Olivine-bearing leucogabbro	2.368	9.263	0.1545	0.512636	6	1.3
Vologochan intrusive (Bh. OV-29)							
29-1	Metasomatite	7.760	29.500	0.1632	0.512638	3	1.1
29-2	Gabbrodiorite	3.852	13.940	0.1670	0.512650	4	1.2
29-3		2.404	8.432	0.1724	0.512676	7	1.5
29-4	Gabbro, olivine-free, altered	3.282	11.880	0.1671	0.512657	4	1.3
29-5		2.547	9.080	0.1696	0.512647	7	1.0
29-6		2.517	9.086	0.1675	0.512637	5	0.9
29-7		2.181	7.587	0.1738	0.512646	5	0.9
29-8	Olivine-bearing gabbro	2.357	8.861	0.1608	0.512637	7	1.1
29-9	Gabbro-troctolite	1.783	6.442	0.1673	0.512627	14	0.7
29-10	Melanotroctolite	1.606	6.150	0.1578	0.512657	6	1.6
29-11	Gabbro-troctolite	2.099	7.601	0.1670	0.512642	7	1.0
29-12		1.742	6.285	0.1675	0.512659	5	1.3
29-13	Melanotroctolite	1.347	4.822	0.1689	0.512669	3	1.5
29-14		2.063	7.377	0.1691	0.512670	4	1.5
29-15	Olivinic gabbro	2.379	8.357	0.1722	0.512673	5	1.5
29-16	Troctolite, miner.	1.632	5.788	0.1705	0.512710	5	2.2
29-17	Gabbro-troctolite with sulph.	2.205	8.113	0.1643	0.512707	4	2.4
29-18		1.797	7.248	0.1499	0.512475	3	−1.7
29-20	Metasomatite	3.572	13.350	0.1618	0.512667	6	1.7
29-24	Melanotroctolite, miner.	1.373	5.012	0.1657	0.512647	11	1.2

(continued)

Table 10 (continued)

Sample	Rock, mineral	Sm, ppm	Nd, ppm	$^{147}Sm/^{144}Nd$	$^{143}Nd/^{144}Nd$	2σ	ε_{Nd} (250)
South Pyasina intrusive (Bh. OV-25)							
25-4	Olivine-bearing gabbro, altered	2.390	8.592	0.1682	0.512657	4	1.3
25-9,10	Gabbro, olivinic	2.980	10.670	0.1688	0.512634	7	0.8
25-20	Leucogabbro, olivinic	1.819	6.752	0.1629	0.512635	7	1.0
25-31	Melanotroctolite, miner.	1.674	6.567	0.1542	0.512572	5	0.1
25-35	Gabbro-troctolite	2.189	8.737	0.1515	0.512571	4	0.1
25-36	Gabbro-troctolite, miner.	1.920	7.494	0.1549	0.512581	5	0.2
25-41	Gabbro-troctolite	1.878	6.633	0.1712	0.512661	6	1.3
25-44	Gabbro, olivinic	2.834	9.978	0.1717	0.512670	8	1.4
Imangda intrusive (Bh. KP-4)							
4-1	Olivine-bearing gabbro	1.822	6.307	0.1747	0.512680	6	1.5
4-2	Olivine-bearing leucogabbro	0.856	3.118	0.1659	0.512640	6	1.0
4-3	Gabbro-troctolite	2.060	7.443	0.1674	0.512644	6	1.1
4-4	Plagiowehrlite	1.980	7.154	0.1673	0.512633	6	0.8
4-5		1.665	5.967	0.1687	0.512687	6	1.9
4-6		1.551	5.477	0.1712	0.512646	6	1.0
4-7	Melanotroctolite	1.661	5.943	0.1690	0.512654	7	1.2
4-8		1.682	5.963	0.1705	0.512667	6	1.4
4-9	Olivinic leucogabbro	1.788	6.562	0.1648	0.512660	7	1.4
4-10	Gabbro, olivinic	2.320	8.068	0.1739	0.512686	5	1.7

zircons. This estimate gave a variation range of 0.7049–0.7128. At the same time, for the Oktyabrsky deposit, it is 0.7060–0.7128; for the Koevsky area, 0.7053–0.7112; for the Tangarylakh area, 0.7050–0.7060, for the section along the Mokulay Creek, 0.7047–0.7053. The results of Rb–Sr analysis suggest a twofold origin of the intrusions composed of mantle material contaminated to a variable degree by crustal matter. The assessment of the initial strontium isotopic ratio for the **mantle material aged 250 Ma** gave the value **$(^{87}Sr/^{86}Sr)_0 = 0.703723$**.

Data of Sm–Nd analysis of the studied rocks also suggest a mantle origin of the intrusion, which underwent varying degrees of the impact of the crustal component, since the assessment of the value of the ε_{Nd} parameter from bulk samples adjusted to the geological age of 250 Ma gives a variation range from −7.2 to 6.3. This is consistent with the results of testing samples from the collection of 2006–2008; the variation range in them is from −7.7 to +7.0. The numerical value of **the ε_{Nd} parameter for the mantle matter of this age is +9.5**, which clearly indicates a

Table 11 Results of Sm–Nd analysis in rock-forming minerals in ore-bearing intrusives

Sample	Rock, mineral	Sm, ppm	Nd, ppm	$^{147}Sm/^{144}Nd$	$^{143}Nd/^{144}Nd$	2σ
Zub-Marksheidersky intrusive (Bh. MP-27)						
27-3	Plagioclase	1.301	7.961	0.0998	0.512505	7
27-4		0.224	1.184	0.1143	0.512459	9
27-4	Pyroxene	15.750	68.940	0.1381	0.512529	4
27-5	Plagioclase	0.313	1.659	0.1140	0.512546	9
27-5	Pyroxene	2.542	7.802	0.1970	0.512703	8
27-7	Plagioclase	0.594	3.295	0.1090	0.512481	8
27-7	Pyroxene	1.756	4.301	0.2469	0.512767	7
27-10	Plagioclase	0.227	1.175	0.1169	0.512558	6
27-10	Pyroxene	2.372	5.601	0.2561	0.512808	6
27-10	Olivine	0.030	0.090	0.2099	0.512932	19
27-13		0.031	0.110	0.1699	0.512747	11
27-13	Plagioclase	0.174	0.931	0.1130	0.512534	7
27-13	Pyroxene	1.749	4.216	0.2509	0.512791	8
Chernogorsk intrusive (Bh. MP-2bis)						
CH-11	Plagioclase	0.131	0.666	0.1187	0.512597	12
CH-11	Pyroxene	2.089	4.969	0.2542	0.512802	5
CH-13	Plagioclase	0.187	0.959	0.1179	0.512597	8
CH-13	Pyroxene	2.334	7.140	0.1976	0.512847	7
Vologochan intrusive (Bh. OV-29)						
29-2,3	Plagioclase	0.426	1.976	0.1304	0.512580	7
29-2,3	Pyroxene	1.244	2.911	0.2584	0.512816	8
29-5,6	Plagioclase	0.415	2.029	0.1237	0.512585	7
29-5,6	Pyroxene	2.345	7.679	0.1846	0.512699	6
29-8	Plagioclase	0.173	0.879	0.1191	0.512621	7
29-8	Pyroxene	1.675	3.895	0.2600	0.512793	4
29-9	Plagioclase	0.183	0.998	0.1109	0.512704	18
29-9	Pyroxene	1.805	4.792	0.2277	0.512749	7
29-16	Plagioclase	0.185	1.008	0.1111	0.512600	19
29-16	Pyroxene	2.479	6.254	0.2397	0.512794	4
29-17	Plagioclase	0.190	0.968	0.1186	0.512707	15
29-17	Pyroxene	2.657	6.518	0.2465	0.512793	6
29-17	Olivine	0.121	0.351	0.2046	0.512619	11
South Pyasina intrusive (Bh. OV-25)						
25-35	Plagioclase	0.158	0.925	0.1033	0.512489	16
25-35	Pyroxene	2.387	6.644	0.2172	0.512720	2
25-35	Olivine	0.111	0.441	0.1490	0.512569	9
Imangda intrusive (Bh. KP-4)						
4-5	Pyroxene 1	1.790	4.973	0.2175	0.512511	7
4-5	Pyroxene 2	1.926	5.349	0.2176	0.512767	3

(continued)

Table 11 (continued)

Sample	Rock, mineral	Sm, ppm	Nd, ppm	$^{147}Sm/^{144}Nd$	$^{143}Nd/^{144}Nd$	2σ
4-5	Plagioclase	0.449	2.180	0.1246	0.512565	9
4-5	Olivine	0.091	0.321	0.1753	0.512682	12
Talnakh intrusive (Bh. OUG-2)						
T-10	Olivine	0.440	1.531	0.1745	0.512764	4
T-13		1.101	3.221	0.2060	0.512687	7
T-14		0.170	0.631	0.1638	0.512687	18
T-15		0.320	1.161	0.1672	0.512775	5
T-16		0.271	0.961	0.1671	0.512534	4
Lower Talnakh intrusive (Bh. TG-31)						
31-3	Olivine	0.151	0.581	0.1596	0.512715	11
31-7		0.530	2.341	0.1364	0.512735	7
31-9		0.070	0.270	0.1631	0.512761	5
31-10		0.050	0.250	0.1341	0.512092	7
31-11		0.161	0.630	0.1590	0.512794	8
31-13		0.171	0.631	0.1616	0.512789	10
31-16		0.110	0.391	0.1630	0.512805	15

significant impact of the crustal component on the intruded mantle fluid. We are forced to resort to qualitative evaluations of the contribution of each source, because we do not have isotopic coordinates of the second term of the two-component mixture—crustal matter.

Therefore, from the data on the Rb–Sr and Sm–Nd systematics it follows:

- for the first time, such a large number of high-precision isotopic Rb–Sr and Sm–Nd analyses of rocks in the Norilsk and Taimyr province intrusions were received. This enabled, by means of the joint use of the two isotopic systems in rock and mineral samples from 15 different intrusions, to perform a more complete analysis of possible sources of the studied rocks;
- heterogeneity of Sr and Nd isotopic composition in the same intrusion is caused by the fact that Rb–Sr and Sm–Nd systems in different minerals respond in different ways to the occurring superimposed processes. This is due to different mobility of the considered elements, which vary markedly in their physico-chemical properties;
- variations in Sr and Nd isotopic data in different rocks (minerals) of the same intrusion are accounted for by different preservation degree of Rb–Sr and Sm–Nd systems in rocks and their constituent minerals;
- a set of data on the strontium and neodymium isotopic composition clearly indicates a significant influence of the crustal component on the intruded mantle fluid;

Table 12 Results of Sm–Nd analysis in rocks of barren intrusives

Sample	Rock, mineral	Sm, ppm	Nd, ppm	$^{147}Sm/^{144}Nd$	$^{143}Nd/^{144}Nd$	2σ	ε_{Nd} (250)
Agatsky intrusive (Bh. OV-25)							
25-25	Olivine-bearing dolerite	2.674	10.230	0.1581	0.512240	5	−6.5
25-26		2.728	10.430	0.1581	0.512242	4	−6.5
25-27		2.746	10.520	0.1578	0.512243	5	−6.5
Daldykan intrusive (Bh. NP-37)							
37-34	Ferrogabbropegmatite	5.650	20.000	0.1708	0.512756	6	3.1
37-39	Dolerite, olivine-bearing	3.829	13.220	0.1752	0.512771	4	3.3
37-43	Dolerite, olivinic	4.337	15.150	0.1730	0.512779	3	3.5
Ergalakh intrusive (Bh. NP-37)							
37-48	Dolerite, titanium-augitic	8.120	37.450	0.1311	0.512320	6	−4.1
37-52	Trachydolerite	8.028	36.890	0.1316	0.512338	4	−3.8
Oganer intrusive (boreholes NP-37 and MD-48)							
37-58	Dolerite, olivine-free.	4.178	14.810	0.1706	0.512692	5	1.9
37-60	Leucodolerite, olivine-bearing	2.654	9.358	0.1715	0.512671	8	1.4
37-61		3.292	11.720	0.1699	0.512664	4	1.4
48-2	Olivinic diabase	2.856	10.080	0.1714	0.512728	5	2.6
48-4	Leucodolerite, olivine-bearing	2.990	10.440	0.1731	0.512775	8	3.4
48-7	Olivine-bearing dolerite	2.771	9.668	0.1733	0.512755	4	3.0
Lower Vologochan intrusive (Bh. OV-38)							
38-3	Olivine-bearing gabbro	3.785	13.590	0.1684	0.512710	5	2.3
38-6		4.684	17.220	0.1644	0.512672	8	1.7
38-12		3.180	11.390	0.1688	0.512708	4	2.3
Morongo intrusive							
M-4	Olivine-bearing gabbro	3.089	10.730	0.1470	0.512742	5	2.8
M-6		1.971	6.820	0.1747	0.512757	6	3.0
M-9		2.076	7.180	0.1748	0.512756	5	3.0
Ruinny intrusive							
Ru-2	Troctolite	1.099	3.863	0.1719	0.512727	10	2.5
Ru-5		0.646	2.295	0.1702	0.512734	14	2.7
Pegmatoid intrusive							
P-11	Pyroxenite	3.057	9.609	0.1923	0.512776	5	2.8
Gudchikha Formation (Bh. KhS51)							
KhS51/1-1	Picrite	2.366	8.710	0.1642	0.512759	4	3.4
KhS51/1-2		1.887	6.842	0.1668	0.512795	4	4.0
KhS51/1-3		1.851	6.640	0.1685	0.512808	8	4.2

(continued)

Table 12 (continued)

Sample	Rock, mineral	Sm, ppm	Nd, ppm	$^{147}Sm/^{144}Nd$	$^{143}Nd/^{144}Nd$	2σ	ε_{Nd} (250)
Guli Massif							
G17/27	Dunite	0.041	0.133	0.1839	0.512960	48	6.7
G17/36		0.036	0.138	0.1580	0.512790	48	4.2
G17/46		0.054	0.209	0.1568	0.512987	48	8.1
G17/48		0.178	0.684	0.1571	0.512758	51	3.6
GU-03	Chromitite	0.031	0.120	0.1559	0.513313	46	14.5
K-17		0.049	0.186	0.1594	0.513500	20	18.0

Table 13 Results of Sm–Nd analysis in olivines of intrusives in Norilsk Province

Sample	Intrusive, borehole number	Sm, ppm	Nd, ppm	$^{147}Sm/^{144}Nd$	$^{143}Nd/^{144}Nd$	2σ
T-10	Talnakh, OUG-2	0.440	1.531	0.1745	0.512764	4
T-13		1.101	3.221	0.2060	0.512687	7
T-14		0.170	0.631	0.1638	0.512687	18
T-15		0.319	1.159	0.1672	0.512775	5
T-16		0.271	0.961	0.1671	0.512534	4
844-7	Kharaelakh, KZ-844	0.089	0.369	0.1505	0.512656	6
963-30	Kharaelakh, KZ-963	1.431	5.942	0.1451	0.512048	5
29-17	Vologochan, OV-29	0.121	0.351	0.2046	0.512619	11
27-10	Zub-Marksheidersky, MP-27	0.030	0.090	0.2099	0.512932	19
27-13		0.031	0.110	0.1699	0.512747	11
25-35	South Pyasina, OV-25	0.111	0.441	0.1490	0.512569	9
4-5	Imangda, KP-4	0.091	0.321	0.1753	0.512682	5
31-3	Lower Talnakh, TG-31	0.151	0.581	0.1596	0.512715	11
31-7		0.530	2.341	0.1364	0.512735	7
31-9		0.070	0.270	0.1631	0.512761	5
31-10		0.050	0.250	0.1341	0.512092	7
31-11		0.161	0.630	0.1590	0.512794	4
31-13		0.170	0.629	0.1616	0.512789	10
31-16		0.111	0.391	0.1630	0.512805	8
G 17/24	Guli Massif	0.276	1.362	0.1225	0.512400	8
G 17/48		0.256	0.942	0.1641	0.512770	14
30-33		0.023	0.136	0.1028	0.512448	24
31-35		0.029	0.102	0.1750	0.512468	43

Table 14 Results of Rb–Sr analysis in rocks of Norilsk-1 intrusive

Sample	Rock, mineral	Rb, ppm	Sr, ppm	$^{87}Rb/^{86}Sr$	$^{87}Sr/^{86}Sr$	2σ	IR (250)
Bh. MN-2							
51	Gabbrodiorite	21.09	439.2	0.1388	0.706489	10	0.705995
52		22.00	380.5	0.1671	0.706567	14	0.705973
53	Gabbro	16.79	277.4	0.1749	0.706752	7	0.706130
55		8.00	278.8	0.0827	0.706528	10	0.706234
56		7.07	311.7	0.0656	0.706435	11	0.706202
57	Olivinic gabbro	12.82	234.2	0.1581	0.706561	11	0.705999
59		6.58	219.5	0.0867	0.706041	10	0.705733
61		13.20	220.6	0.1729	0.706476	11	0.705861
63	Plagiowehrlite	17.33	80.6	0.6217	0.708457	10	0.706246
64		8.01	111.9	0.2069	0.706734	16	0.705998
65		8.35	109.3	0.2208	0.706777	16	0.705992
66	Melanotroctolite	26.25	120.5	0.6297	0.708476	10	0.706236
67	Taxitic gabbro	13.18	153.6	0.2479	0.706809	19	0.705927
68	Troctolite	16.66	127.7	0.3768	0.707484	10	0.706144
69a	Olivinic leucogabbro	12.12	130.0	0.2694	0.706926	15	0.705968
69b	Leucogabbro	15.81	141.7	0.3225	0.707123	10	0.705976
71	Plagiowehrlite	11.48	122.8	0.2700	0.706835	10	0.705875
72	Gabbrotroctolite	9.85	140.3	0.2028	0.706466	10	0.705745
73	Mineralized gabbro	3.85	179.3	0.0620	0.705933	10	0.705713
74	Leucogabbro	6.37	193.0	0.0954	0.706093	10	0.705754
75	Olivinic gabbro	13.86	238.4	0.1680	0.706157	10	0.705559
76	Gabbro	8.41	263.4	0.0922	0.706083	12	0.705755
77	Troctolite	9.96	257.7	0.1117	0.706051	10	0.705654
78	Gabbro	12.61	298.7	0.1220	0.706167	10	0.705733
N1-11	Gabbrodiorite	27.26	441.8	0.1783	0.706577	8	0.705943
N1-12	Leucogabbro	16.15	175.0	0.2667	0.706785	6	0.705836
Bh. MS-18							
18-1	Leucogabbro	29.10	1156.0	0.0727	0.706975	9	0.706716
18-2		38.47	584.9	0.0.1901	0.707469	10	0.706793
18-3	Olivinic leucogabbro	8.72	237.9	0.1059	0.706223	8	0.705847
18-4	Olivinic gabbro	30.00	348.8	0.2485	0.707335	9	0.706451
Bh. KN-97							
97-1	Olivinic gabbro	10.44	154.0	0.1960	0.706556	8	0.705859
97-2	Plagiowehrlite	11.78	162.3	0.2098	0.706821	11	0.706075
97-3	Troctolite	20.34	185.4	0.3170	0.707071	7	0.705944
97-6		20.38	186.2	0.3162	0.707111	15	0.705987

Table 15 Results of Rb–Sr analysis in rock-forming minerals of Norilsk-1 intrusive

Sample	Rock, mineral	Rb, ppm	Sr, ppm	$^{87}Rb/^{86}Sr$	$^{87}Sr/^{86}Sr$	2σ
Bh. Kn-97						
Kn-97-1	Plagioclase	8.930	504.90	0.0511	0.705887	9
Kn-97-1	Pyroxene	0.440	19.20	0.0658	0.707517	16
Kn-97-2	Plagioclase	8.921	371.40	0.0694	0.706289	12
Kn-97-2	Pyroxene	0.582	20.14	0.0835	0.706079	14
Kn-97-3	Plagioclase	32.840	386.40	0.2456	0.706755	10
Kn-97-3	Orthopyroxene	0.651	16.18	0.1166	0.700030	11
Kn-97-3	Clinopyroxene	0.361	16.06	0.0644	0.705921	16
Kn-97-4	Plagioclase	8.621	432.90	0.0575	0.705877	11
Kn-97-4	Pyroxene	0.542	20.41	0.0763	0.707224	10
Kn-97-5	Plagioclase	16.560	404.60	0.1183	0.706167	12
Kn-97-5	Pyroxene	0.623	20.85	0.0855	0.706252	12
Kn-97-6	Plagioclase	42.780	529.50	0.2335	0.706727	10
Kn-97-6	Pyroxene	0.592	16.85	0.1007	0.706394	16
Bh. MN-2						
N1-1	Plagioclase	23.550	709.90	0.0959	0.706350	10
N 1-1	Pyroxene	2.048	39.51	0.1498	0.706703	14
N 1-2	Plagioclase	23.460	668.70	0.1014	0.706377	12
N 1-2	Pyroxene	0.777	26.16	0.0859	0.707252	18
N 1-3	Plagioclase	8.499	400.90	0.0613	0.706020	7
N 1-3	Pyroxene	1.013	27.11	0.1080	0.706620	19
N 1-4	Plagioclase	12.920	512.80	0.0728	0.706466	12
N 1-4	Pyroxene	1.155	38.81	0.0860	0.706435	16
N 1-5	Plagioclase	19.070	419.20	0.1315	0.706523	14
N 1-5	Pyroxene	0.509	25.98	0.0566	0.706149	16
N 1-6	Plagioclase	23.720	397.80	0.1723	0.706149	12
N 1-6	Pyroxene	0.653	26.09	0.0723	0.705955	27
N 1-7	Plagioclase	14.220	394.10	0.1043	0.705957	10
N 1-7	Pyroxene	0.899	30.77	0.0845	0.706228	12
N 1-8	Plagioclase	31.670	455.40	0.2009	0.706593	9
N 1-8	Pyroxene	0.561	23.50	0.0691	0.706362	12
N 1-9	Plagioclase	16.780	406.60	0.1192	0.705935	9
N 1-9	Pyroxene	0.536	19.37	0.0800	0.705930	33
N 1-10	Plagioclase	11.050	552.90	0.0577	0.705825	8
N 1-10	Pyroxene	2.142	30.16	0.2052	0.706469	10

Table 16 Results of Rb–Sr analysis in rocks of Talnakh and Lower Talnakh intrusives

Sample	Rock, mineral	Rb, ppm	Sr, ppm	$^{87}Rb/^{86}Sr$	$^{87}Sr/^{86}Sr$	2σ	IR (250)
Talnakh intrusive (Bh. OUG-2)							
T-1	Alkali metasomatite, miner.	31.80	466.6	0.1970	0.708398	6	0.707697
T-2	Diorite-pegmatite	17.99	171.1	0.3038	0.708018	46	0.706937
T-3	Gabbrodiorite	26.84	200.7	0.3865	0.708041	31	0.706666
T-5	Leucogabbro, olivine-bearing	16.29	272.2	0.1730	0.706735	36	0.706119
T-6	Gabbro, olivine-bearing	18.84	229.8	0.2369	0.706913	9	0.706070
T-8		12.57	226.0	0.1607	0.706779	111	0.706207
T-10	Gabbro, olivinic	8.61	210.2	0.1183	0.706257	7	0.705836
T-12		7.48	222.0	0.0973	0.706329	57	0.705983
T-13	Plagiowehrlite, miner.	6.06	145.3	0.1206	0.706357	12	0.705928
T-14		11.81	123.3	0.2769	0.707523	56	0.706538
T-15		8.23	117.2	0.2028	0.707153	93	0.706432
T-16	Melanotroctolite, miner.	18.53	256.3	0.2089	0.707303	9	0.706560
T-17	Plagiopyroxenite, miner.	25.12	263.4	0.2757	0.707533	141	0.706552
T-18	Gabbro, olivine-bearing, miner.	24.51	452.4	0.1566	0.707648	11	0.707091
T-19	Metasomatite	32.50	443.7	0.2117	0.708488	7	0.707735
T-20		19.69	225.4	0.2524	0.709311	7	0.708413
T-26	Hornfelsed mudstone	107.00	177.7	1.7406	0.715789	10	0.709599
T-27		47.29	78.63	1.7392	0.714759	10	0.708574
T-28	Mudstone	139.80	136.8	2.9563	0.719479	8	0.708965
T-29	Metasomatite, mineralized	45.12	527.1	0.2474	0.708635	7	0.707755
T-30	Alkali metasomatite, miner.	25.94	546.4	0.1372	0.708068	6	0.707580
Lower Talnakh intrusive (Bh. TG-31)							
31-1	Olivine-free gabbro	33.62	226.9	0.4283	0.710126	7	0.708603
31-2	Olivinic gabbro	26.50	132.3	0.5787	0.710610	7	0.708552
31-3	Melanotroctolite	13.06	206.8	0.1825	0.708665	7	0.708016
31-4	Troctolitic gabbro	12.42	255.7	0.1404	0.708623	25	0.708124
31-5	Plagiowehrlite	14.90	206.4	0.2086	0.709094	10	0.708352
31-6	Olivinic gabbro	13.53	278.8	0.1403	0.708717	8	0.708218
31-7	Plagiowehrlite	2.50	149.7	0.0482	0.708759	10	0.708587
31-8		9.49	117.7	0.2331	0.709459	23	0.708630
31-9		11.26	112.5	0.2894	0.709302	8	0.708273
31-10		12.10	158.6	0.2205	0.708828	6	0.708044
31-11	Melanotroctolite	13.53	201.2	0.1943	0.709016	8	0.708325
31-12	Plagioolivinite	17.14	134.0	0.3695	0.709431	6	0.708117
31-13	Plagiowehrlite	4.445	158.6	0.0810	0.708873	9	0.708585
31-14		15.66	137.1	0.3301	0.709166	8	0.707992

(continued)

Table 16 (continued)

Sample	Rock, mineral	Rb, ppm	Sr, ppm	$^{87}Rb/^{86}Sr$	$^{87}Sr/^{86}Sr$	2σ	IR (250)
31-15	Melanotroctolite	20.22	161.1	0.3628	0.709093	7	0.707803
31-16		20.53	162.1	0.3662	0.708951	27	0.707649
31-17		19.91	157.7	0.3648	0.709125	8	0.707828
31-18	Olivine-free gabbro	33.47	227.0	0.4262	0.710133	9	0.708617
31-19	Olivinic gabbro	13.37	279.5	0.1382	0.708709	9	0.708217
31-20	Plagioolivinite	10.93	107.3	0.2944	0.709286	7	0.708239

Table 17 Results of Rb–Sr analysis in rock-forming minerals of Talnakh and Lower Talnakh intrusives

Sample	Rock, mineral	Rb, ppm	Sr, ppm	$^{87}Rb/^{86}Sr$	$^{87}Sr/^{86}Sr$	2σ
Talnakh intrusive (Bh. OUG-2)						
T-1	Plagioclase	11.060	410.50	0.0779	0.706463	6
T-1	Pyroxene	0.699	20.32	0.0994	0.706567	25
T-2	Plagioclase	1.870	53.99	0.1001	0.707350	10
T-2	Pyroxene	0.632	17.85	0.1013	0.706439	27
T-5	Plagioclase	9.487	424.00	0.0647	0.706180	6
T-5	Pyroxene	0.969	19.96	0.1404	0.706565	11
T-6	Plagioclase	28.900	355.60	0.2349	0.706865	7
T-6	Pyroxene	3.101	24.07	0.3726	0.707840	26
T-8	Plagioclase	6.374	424.90	0.0433	0.706047	7
T-8	Pyroxene	1.671	32.49	0.1488	0.706546	19
T-10	Plagioclase	7.631	415.10	0.0531	0.705957	12
T-10	Pyroxene	0.209	23.39	0.0258	0.706096	10
T-12	Plagioclase	4.183	425.00	0.0284	0.705907	7
T-12	Pyroxene	0.191	24.06	0.0321	0.706028	12
T-13	Plagioclase	3.581	443.90	0.0233	0.705882	10
T-13	Pyroxene	0.621	19.70	0.0915	0.706018	12
T-14	Plagioclase	21.250	419.50	0.1464	0.706611	16
T-14	Pyroxene	0.689	19.10	0.1047	0.706774	18
T-15	Plagioclase	9.309	424.30	0.0634	0.706107	21
T-15	Pyroxene	0.369	19.88	0.0539	0.706782	21
T-16	Plagioclase	15.360	485.90	0.0914	0.706387	5
T-16	Pyroxene	2.240	22.76	0.2838	0.707653	18
T-17	Plagioclase	21.360	449.60	0.1373	0.706380	15
T-17	Pyroxene	2.841	24.04	0.3415	0.707575	10
T-18	Plagioclase	17.230	853.90	0.0583	0.707090	14
T-18	Pyroxene	4.059	24.56	0.4783	0.708783	26
T-19	Plagioclase	7.930	424.50	0.0540	0.705956	13
T-19	Pyroxene	0.261	24.36	0.0307	0.706095	22

(continued)

Table 17 (continued)

Sample	Rock, mineral	Rb, ppm	Sr, ppm	$^{87}Rb/^{86}Sr$	$^{87}Sr/^{86}Sr$	2σ
Lower Talnakh intrusive (Bh. TG-31)						
31-1	Plagioclase	94.890	602.90	0.4549	0.710317	10
31-1	Pyroxene	0.532	20.92	0.0735	0.708195	12
31-3	Plagioclase	21.110	510.80	0.1194	0.708175	6
31-3	Pyroxene	0.129	18.91	0.0198	0.707829	9
31-7	Plagioclase	3.530	568.20	0.0180	0.708058	13
31-7	Pyroxene	0.281	18.85	0.0431	0.707876	20
31-9	Plagioclase	3.421	511.80	0.0193	0.708074	7
31-9	Pyroxene	0.089	17.83	0.0146	0.707859	19
31-10	Plagioclase	4.226	581.00	0.0210	0.708006	6
31-10	Pyroxene	0.149	19.70	0.0218	0.707964	16
31-11	Plagioclase	7.691	500.10	0.0444	0.707930	7
31-11	Pyroxene	2.231	18.97	0.3400	0.707944	18
31-13	Plagioclase	6.209	491.90	0.0365	0.707792	8
31-13	Pyroxene	0.149	19.55	0.0218	0.708513	16
31-16	Plagioclase	3.759	501.70	0.0217	0.707792	8
31-16	Pyroxene	0.189	18.94	0.0293	0.708044	14

Table 18 Results of Rb–Sr analysis in rocks of Kharaelakh, Zub-Marksheidersky, Kruglogorsky and Zelenaya Griva intrusives

Sample	Rock, mineral	Rb, ppm	Sr, ppm	$^{87}Rb/^{86}Sr$	$^{87}Sr/^{86}Sr$	2σ	IR (250)
Kharaelakh intrusive (boreholes KZ-844 u KZ-963)							
844-1	Gabbro, olivinic	15.51	271.0	0.1654	0.707016	27	0.706428
844-1a		15.46	270.8	0.1650	0.707028	40	0.706441
844-2	Plagioolivinite	5.13	056.6	0.2624	0.706973	15	0.706040
844-3,4		6.24	127.1	0.1419	0.708196	18	0.707691
844-6	Gabbro, olivinic	4.44	184.0	0.0697	0.706967	13	0.706719
844-7	Melanotroctolite	14.88	153.4	0.2802	0.706544	16	0.705547
844-7b		3.69	150.9	0.0707	0.706576	113	0.706324
844-10		8.07	159.7	0.1460	0.707305	15	0.706786
844-15	Leucogabbro, olivine-bearing	19.56	289.0	0.1956	0.708677	11	0.707981
963-5	Hornfels	7.63	376.5	0.0585	0.708827	93	0.708618
963-21	Gabbro, olivine-free, altered	13.45	289.5	0.1343	0.707513	11	0.707036
963-23	Gabbro olivine-free.	19.94	316.2	0.1822	0.707428	12	0.706780
963-29	Melanotroctolite, miner.	5.16	101.6	0.1469	0.706625	18	0.706103
963-31	Gabbro-troctolite, miner.	13.53	210.8	0.1855	0.706616	10	0.705956
963-65	Gabbro, leucocrat., olivine-bearing	31.44	405.4	0.2241	0.709194	7	0.708397

(continued)

Table 18 (continued)

Sample	Rock, mineral	Rb, ppm	Sr, ppm	$^{87}Rb/^{86}Sr$	$^{87}Sr/^{86}Sr$	2σ	IR (250)
Zub-Marksheidersky intrusive (Bh. MP-27)							
27-1	Metasomatite	79.75	259.1	0.8898	0.712241	8	0.70906
27-3		76.60	283.8	0.7802	0.711338	8	0.708563
27-4	Amph.-biotite-alb. rock	77.28	236.7	0.9439	0.712076	7	0.708719
27-5	Diorite	34.29	323.3	0.3065	0.708709	8	0.707619
27-6	Olivine-bearing gabbro	19.39	222.9	0.2514	0.706734	8	0.705840
27-7		18.19	214.7	0.2448	0.707353	8	0.706482
27-8	Gabbro-troctolite	24.01	155.2	0.3918	0.707481	9	0.706088
27-10	Plagiowehrlite, miner.	9.07	096.1	0.2727	0.707024	8	0.706054
27-11	Gabbro-troctolite	12.46	177.4	0.2030	0.706910	8	0.706188
27-12		34.84	177.1	0.5685	0.708078	8	0.706056
27-13		28.06	173.9	0.4662	0.707511	9	0.705853
27-14	Gabbro, olivine-bearing, altered	20.16	205.3	0.2837	0.706707	6	0.705698
Kruglogorsky intrusive (Bh. MP-2)							
K-1	Ferrogabbro	10.65	244.8	0.1257	0.706572	9	0.706125
K-2	Leucogabbro (2)	34.16	451.4	0.2187	0.708328	7	0.707550
K-3	Leucogabbro	32.27	644.0	0.1448	0.708310	8	0.707795
K-4	Leucogabbro, altered	38.45	986.1	0.1127	0.708407	7	0.708006
K-6	Melanotroctolite, mineralized	11.43	379.8	0.0869	0.707212	8	0.706903
K-8	Gabbro-troctolite	7.53	276.1	0.0789	0.706252	9	0.705972
K-9	Olivinic gabbro	18.93	363.5	0.1505	0.707525	9	0.706990
K-10		19.65	358.0	0.1586	0.707447	8	0.706883
Zelenaya Griva intrusive (Bh. F-233)							
F-233-2	Olivine-bearing gabbro	32.39	257.5	0.3637	0.709296	8	0.708003
F-233-4	Gabbro-troctolite, miner.	61.91	467.8	0.3826	0.709832	9	0.708471
F-233-5	Gabbro-troctolite	20.80	247.8	0.2425	0.709068	7	0.708205
F-233-6	Plagiowehrlite	15.82	140.9	0.3245	0.709546	25	0.708392
F-233-7	Gabbro-troctolite	18.65	233.5	0.2309	0.708931	9	0.708110
F-233-10	Miner. troctolite	24.40	183.0	0.3855	0.709451	10	0.708080
F-233-11	Troctolite (1)	39.81	58.6	1.9639	0.715612	9	0.708628
F-233-12	Plagiowehrlite	19.45	89.7	0.6268	0.710899	22	0.708670
F-233-15	Melanotroctolite	39.56	137.7	0.8307	0.711370	11	0.708416
F-233-16	Olivine-free gabbro	53.99	434.8	0.3589	0.709864	7	0.708588

Table 19 Results of Rb–Sr analysis in rocks of Chernogorsk, Vologochan, South Pyasina and Imangda intrusives

Sample	Rock, mineral	Rb, ppm	Sr, ppm	$^{87}Rb/^{86}Sr$	$^{87}Sr/^{86}Sr$	2σ	IR (250)
Chernogorsk intrusive (Bh. MP-2bis)							
CH-2	Olivine-bearing gabbro	29.18	359.6	0.2346	0.708113	12	0.707279
CH-6		7.59	262.8	0.0835	0.706695	8	0.706398
CH-9	Olivinic gabbro	11.53	235.8	0.1412	0.706902	15	0.706400
CH-11	Gabbro-troctolite, miner.	14.65	222.1	0.1907	0.706962	8	0.706763
CH-13	Gabbro-troctolite, altered	21.44	206.2	0.3006	0.708105	7	0.707036
CH-15	Olivine-bearing leucogabbro	34.83	299.3	0.3364	0.710175	7	0.706931
Vologochan intrusive (Bh. OV-29)							
29-1	Metasomatite	13.26	184.8	0.2073	0.708222	26	0.707129
29-2	Gabbrodiorite	35.48	348.7	0.2940	0.708393	10	0.707374
29-3		31.09	242.8	0.3701	0.708247	12	0.706931
29-4	Gabbro, olivine-free, altered	41.91	312.4	0.3877	0.708506	9	0.707127
29-5	Gabbro, olivine-bearing, altered	30.73	299.6	0.2965	0.707826	10	0.706772
29-6		20.95	249.0	0.2431	0.706955	10	0.706091
29-7		18.43	234.0	0.2276	0.706887	12	0.706078
29-8	Olivine-bearing gabbro	12.95	218.9	0.1708	0.706413	8	0.705806
29-9	Gabbro-troctolite	9.74	152.8	0.1843	0.706677	16	0.706022
29-10	Melanotroctolite	9.35	118.9	0.2273	0.706637	5	0.705829
29-11	Gabbro-troctolite	8.29	152.8	0.1569	0.706336	9	0.705778
29-12		9.42	160.3	0.1698	0.706472	6	0.705868
29-13	Melanotroctolite	5.63	124.6	0.1305	0.706191	7	0.705727
29-14		8.29	152.5	0.1571	0.706158	7	0.705600
29-15	Olivinic gabbro	15.02	179.4	0.2420	0.706930	9	0.706069
29-16	Troctolite, miner.	11.45	161.9	0.2043	0.706683	6	0.705956
29-17	Gabbro-troctolite with sulph.	9.71	211.1	0.1329	0.706517	6	0.706045
29-18		11.26	215.5	0.1510	0.707144	9	0.706607
29-20	Metasomatite	76.26	234.2	0.9414	0.711430	9	0.708080
29-24	Melanotroctolite, miner.	7.26	126.5	0.1659	0.706431	5	0.705841
South Pyasina intrusive (Bh. OV-25)							
25-4	Olivine-bearing gabbro, altered	19.98	311.7	0.1853	0.707861	9	0.707157
25-9,10	Gabbro, olivinic	28.04	255.6	0.3170	0.707425	9	0.706298
25-20	Leucogabbro, olivinic	7.94	211.4	0.1085	0.707447	8	0.707061
25-31	Melanotroctolite, miner.	7.24	147.5	0.1419	0.707240	8	0.706735
25-35	Gabbro-troctolite	8.04	189.9	0.1223	0.706948	7	0.706513
25-36	Gabbro-troctolite, miner.	8.64	178.8	0.1395	0.706801	9	0.706305
25-41	Gabbro-troctolite	7.09	150.9	0.1359	0.706058	8	0.705575

(continued)

Table 19 (continued)

Sample	Rock, mineral	Rb, ppm	Sr, ppm	$^{87}Rb/^{86}Sr$	$^{87}Sr/^{86}Sr$	2σ	IR (250)
25-44	Gabbro, olivinic	15.79	227.2	0.2008	0.706552	10	0.705838
Imangda intrusive (Bh. KP-4)							
4-1	Olivine-bearing gabbro	6.17	431.8	0.0413	0.706921	16	0.706774
4-2	Olivine-bearing leucogabbro	6.03	468.2	0.0372	0.707036	14	0.706904
4-3	Gabbro-troctolite	16.03	591.9	0.0783	0.708067	12	0.707788
4-4	Plagiowehrlite	7.90	144.5	0.1580	0.707581	28	0.707019
4-5		5.81	137.6	0.1220	0.706601	17	0.706167
4-6		5.80	148.8	0.1127	0.706546	17	0.706145
4-7	Melanotroctolite	6.15	115.6	0.1536	0.706031	15	0.705485
4-8		6.01	119.9	0.1449	0.706064	29	0.705549
4-9	Olivinic leucogabbro	7.61	404.7	0.0543	0.707043	10	0.706850
4-10	Gabbro, olivinic	9.61	270.6	0.1026	0.706757	18	0.706392

Table 20 Results of Rb–Sr analysis in rocks of Mikchangda, Binyuda, Dyumptalej, Lower Norilsk, Morongo, Ruinny and Byrranga intrusives and picrites of Gudchikha Formation

Sample	Rock, mineral	Rb, ppm	Sr, ppm	$^{87}Rb/^{86}Sr$	$^{87}Sr/^{86}Sr$	2σ	IR (250)
Mikchangda intrusive (boreholes MD-48 and MD-50)							
48-9	Melanotroctolite, altered	5.13	202.1	0.0733	0.706929	8	0.706668
48-16	Olivine-bearing gabbro, altered	9.80	214.1	0.1323	0.705988	9	0.705517
48-18	Olivinic gabbro	8.90	204.7	0.1256	0.706187	9	0.705740
48-25	Melanotroctolite	7.16	301.6	0.0686	0.707175	7	0.706931
48-27, 28	Plagiowehrlite, altered	6.62	157.9	0.1212	0.706842	8	0.706411
48-30	Melanotroctolite, altered	8.71	224.6	0.1120	0.706161	7	0.705763
48-32, 33, 34	Olivine-bearing gabbro, altered	28.42	475.1	0.1729	0.708259	9	0.707644
50-1	Microgabbro	46.17	562.6	0.2372	0.708357	8	0.707514
50-3	Olivine-bearing gabbro	42.79	823.3	0.1502	0.708863	8	0.708329
50-4	Melanotroctolite	13.42	242.2	0.1601	0.707740	8	0.707170
50-5	Plagiowehrlite	14.69	164.9	0.2575	0.706350	8	0.705434

(continued)

Table 20 (continued)

Sample	Rock, mineral	Rb, ppm	Sr, ppm	$^{87}Rb/^{86}Sr$	$^{87}Sr/^{86}Sr$	2σ	IR (250)
Binyuda intrusive (boreholes S-1, S-3)							
S1-1.2	Olivinite	2.54	62.6	0.1173	0.706077	9	0.705660
S1-3-3.5		0.94	18.6	0.1452	0.706353	15	0.705837
S1-7-7.5	Plagioolivinite	7.01	1225.0	0.1653	0.706675	8	0.706087
S1-33.0		2.31	53.4	0.1247	0.706257	9	0.705814
S1-38.0		2.71	51.0	0.1538	0.706299	9	0.705752
S1-53.2	Plagiowehrlite	3.99	52.0	0.2218	0.706593	9	0.705804
S1-58.0		4.31	65.0	0.1915	0.706472	10	0.705791
S1-123.0	Melanotroctolite	8.65	127.9	0.1954	0.706702	10	0.706007
S1-128.0		11.73	156.9	0.2162	0.706762	8	0.705993
V-52 (S1-5)	Melanotroctolite, miner.	7.36	141.4	0.1503	0.706311	8	0.705776
S-18.0	Olivinite	7.73	60.5	0.3693	0.707673	8	0.706360
S-65.0		2.51	58.7	0.1237	0.706607	13	0.706167
Dyumptalej intrusive (Bh. TP-43)							
43-3	Metasomatite	9.14	601.6	0.0439	0.704852	8	0.704696
43-9	Leucocrat. ferrogabbro	6.73	642.4	0.0303	0.704756	9	0.704648
43-13	Olivine-bearing ferrogabbro	13.81	969.3	0.0412	0.704786	9	0.704639
43-20	Ferrogabbro, troctolitic	1.21	387.1	0.0090	0.704537	8	0.704505
43-27	Ferrogabbro, olivine-free	25.13	376.8	0.1928	0.706753	10	0.706067
43-28	Gabbro-troctolite	11.96	399.9	0.0864	0.705230	8	0.704922
43-29	Melanotroctolite	21.30	133.0	0.4627	0.706686	10	0.705041
Lower Norilsk intrusive (Bh. NP-37)							
37-5	Olivine-bearing gabbro	9.97	193.8	0.1486	0.708066	7	0.707537
37-6		11.04	190.8	0.1672	0.708235	8	0.707640
37-8b	Melanotroctolite	11.24	181.5	0.1789	0.708250	7	0.707614
37-9a	Metasomatite	57.87	43.4	3.8634	0.722070	126	0.708330
37-13	Gabbro	147.90	461.3	1.0967	0.714960	27	0.711060
Morongo intrusive							
M-4	Olivine-bearing gabbro	8.78	179.3	0.1415	0.704988	7	0.704484
M-6		5.70	144.4	0.1140	0.705021	15	0.704616
M-9	Gabbro, olivinic	5.99	152.3	0.1136	0.704960	9	0.704556

(continued)

Table 20 (continued)

Sample	Rock, mineral	Rb, ppm	Sr, ppm	$^{87}Rb/^{86}Sr$	$^{87}Sr/^{86}Sr$	2σ	IR (250)
Gudchikha Formation							
KhS51/ 1-1	Picrite	2.40	111.0	0.0625	0.705258	7	0.705036
KhS51/ 1-2		3.47	108.9	0.0920	0.706419	8	0.706092
KhS51/ 1-3		3.44	108.0	0.0921	0.706416	6	0.706088
Ruinny intrusive							
Ru-2	Troctolite	3.36	161.0	0.0603	0.704684	5	0.704469
Ru-5		1.77	150.2	0.0341	0.704669	6	0.704548
Byrranga intrusive							
74b	Dolerite	19.54	163.8	0.3448	0.708088	9	0.706862

Table 21 Results of Rb–Sr analysis in rocks of non-ore-bearing intrusives

Sample	Rock, mineral	Rb, ppm	Sr, ppm	$^{87}Rb/^{86}Sr$	$^{87}Sr/^{86}Sr$	2σ	IR (250)
Ergalakh intrusive (Bh. NP-37)							
37-48	Dolerite, tatanium-augitic	26.73	488.6	0.1581	0.706454	30	0.705892
37-52	Trachydolerite	33.20	478.8	0.2003	0.706463	8	0.705751
Oganer intrusive (boreholes NP-37, MD-48)							
37-58	Dolerite, olivine-free.	5.64	244.8	0.0666	0.706056	19	0.705819
37-60	Leucodolerite, olivine-bearing	13.73	308.8	0.1285	0.706961	6	0.706504
37-61		16.62	260.0	0.1841	0.706191	11	0.705536
48-2	Olivinic diabase	9.91	201.2	0.1423	0.705919	7	0.705485
48-4	Leucodolerite, olivine-bearing	12.51	237.4	0.1522	0.706498	12	0.705956
48-7	Dolerite, olivine-bearing	15.64	435.3	0.1038	0.707485	8	0.707115
Lower Vologochan intrusive (Bh. OV-38)							
38-3	Gabbro, olivine-bearing	21.40	357.0	0.1730	0.707672	8	0.707057
38-6		22.40	269.0	0.2410	0.707453	24	0.706595
38-12		16.70	303.0	0.1601	0.706623	10	0.706056
Daldykan intrusive (Bh. NP-37)							
37-34	Dolerite, olivine-bearing	15.83	318.7	0.1435	0.706738	8	0.706228
37-39		10.22	272.8	0.1082	0.705768	12	0.705383
37-43	Dolerite, olivinic	11.59	199.7	0.1678	0.705980	52	0.704384
Agatsky intrusive (Bh. OV-25)							
25-25	Dolerite, olivine-bearing	9.31	247.6	0.1086	0.707082	8	0.706696
25-26		10.18	252.7	0.1164	0.707094	9	0.706680
25-27		9.05	261.9	0.0998	0.706959	8	0.706604
Pegmatoid intrusive							
P-11	Pyroxenite	8.23	351.1	0.0677	0.706920	9	0.706679

Table 22 Results of Rb–Sr analysis in rock-forming minerals of Kharaelakh, Mikchangda and Binyuda intrusives

Sample	Rock, mineral	Rb, ppm	Sr, ppm	$^{87}Rb/^{86}Sr$	$^{87}Sr/^{86}Sr$	2σ
Kharaelakh intrusive (boreholes KZ-844 u KZ-963)						
844-1	Plagioclase	8.16	431.30	0.0547	0.706386	12
844-1	Pyroxene	1.55	28.62	0.1567	0.706560	15
844-2	Plagioclase	4.03	433.60	0.0269	0.706322	9
844-2	Pyroxene	0.90	28.83	0.0907	0.707352	15
844-3,4	Plagioclase	26.56	422.20	0.1818	0.708341	9
844-3,4	Pyroxene	0.26	23.78	0.0320	0.706594	8
844-6	Plagioclase	4.27	410.00	0.0301	0.706324	8
844-6	Pyroxene	0.60	31.35	0.0552	0.706304	27
844-7	Plagioclase	1.74	408.40	0.0123	0.706277	12
844-7	Pyroxene	0.26	129.10	0.0058	0.704572	9
844-10,11	Plagioclase	13.78	534.40	0.0745	0.706801	10
844-10,11	Pyroxene	1.19	24.67	0.1397	0.706887	12
844-15	Plagioclase	26.74	427.70	0.1807	0.708359	9
844-15	Pyroxene	3.21	35.51	0.2611	0.707978	10
963-30	Plagioclase	1.33	392.10	0.0098	0.705705	8
Oganer intrusive (Bh. MD-48)						
48-2	Plagioclase	0.32	81.24	0.0115	0.705788	15
48-2	Pyroxene	0.52	18.52	0.0818	0.705709	22
48-7	Plagioclase	14.56	947.10	0.0444	0.707277	8
48-7	Pyroxene	0.37	24.73	0.0435	0.706826	12
Mikchangda intrusive (Bh. MD-48)						
48-9	Plagioclase	4.01	443.10	0.0261	0.706010	7
48-9	Pyroxene	0.25	22.26	0.0326	0.706318	19
48-16	Plagioclase	1.28	362.00	0.0102	0.705246	8
48-16	Pyroxene	0.61	26.06	0.0677	0.704603	17
48-18	Plagioclase	3.00	404.90	0.0214	0.705476	5
48-18	Pyroxene	0.29	17.59	0.0476	0.705897	12
48-25	Plagioclase	2.77	436.70	0.0183	0.705695	7
48-25	Pyroxene	1.18	38.68	0.0884	0.706615	10
48-27,28	Plagioclase	12.79	449.70	0.0822	0.706121	5
48-27,28	Pyroxene	0.80	35.73	0.0650	0.706313	11
48-30	Plagioclase	1.88	407.40	0.0133	0.705752	4
48-30	Pyroxene	0.34	28.33	0.0349	0.706118	12
48-30	Pyroxene 2	0.57	19.57	0.0843	0.706015	14
48-32,33,34	Plagioclase	16.04	605.50	0.0765	0.706869	3
48-32,33,34	Pyroxene	0.32	28.42	0.0325	0.706894	14

(continued)

Table 22 (continued)

Sample	Rock, mineral	Rb, ppm	Sr, ppm	$^{87}Rb/^{86}Sr$	$^{87}Sr/^{86}Sr$	2σ
Zub-Marksheidersky intrusive (Bh. MP-27)						
27-3	Plagioclase	46.01	581.70	0.2286	0.709637	7
27-4	Plagioclase	4.23	410.40	0.0298	0.705914	8
27-4	Pyroxene	9.72	21.98	1.2793	0.713901	10
27-5	Plagioclase	20.98	498.50	0.1216	0.707550	8
27-5	Pyroxene	1.41	26.38	0.1548	0.707407	12
27-7	Plagioclase	58.09	606.90	0.2767	0.709531	6
27-7	Pyroxene	0.48	16.57	0.0836	0.706267	10
27-10	Plagioclase	1.80	412.20	0.0126	0.705738	7
27-10	Pyroxene	0.44	16.01	0.0801	0.706231	16
27-13	Plagioclase	1.21	396.30	0.0088	0.705623	8
27-13	Pyroxene	0.32	17.05	0.0548	0.706139	10

Table 23 Results of Rb–Sr analysis in rock-forming minerals of Chernogorsk, Vologochan, South Pyasina, Imangda intrusives and Gudchikha Formation

Sample	Rock, mineral	Rb, ppm	Sr, ppm	$^{87}Rb/^{86}Sr$	$^{87}Sr/^{86}Sr$	2σ
Chernogorsk intrusive (Bh. MP-2bis)						
CH-11	Plagioclase	12.85	460.40	0.0807	0.706314	8
CH-11	Pyroxene	0.63	18.28	0.0990	0.706391	17
CH-13	Plagioclase	2.39	417.60	0.0166	0.706064	13
CH-13	Pyroxene	0.90	20.34	0.1277	0.706857	15
Vologochan intrusive (Bh. OV-29)						
29-2,3	Plagioclase	33.76	538.80	0.1811	0.707563	5
29-2,3	Pyroxene	1.48	23.46	0.1825	0.707124	17
29-5,6	Plagioclase	13.63	410.60	0.0960	0.706435	7
29-5,6	Pyroxene	0.43	16.13	0.0768	0.706080	12
29-8	Plagioclase	1.23	374.30	0.0095	0.705655	6
29-8	Pyroxene	0.33	20.11	0.0467	0.706033	22
29-9	Plagioclase	1.49	414.50	0.0104	0.705807	5
29-9	Pyroxene	0.32	19.71	0.0466	0.705953	12
29-16	Plagioclase	2.44	419.80	0.0168	0.705683	4
29-16	Pyroxene	0.77	25.16	0.0887	0.706165	8
29-17	Plagioclase	2.96	393.00	0.0218	0.705803	6
29-17	Pyroxene	0.76	17.96	0.1221	0.706376	15
South Pyasina intrusive (Bh. OV-25)						
25-35	Plagioclase	0.98	456.60	0.0062	0.706252	8
25-35	Pyroxene	0.27	22.04	0.0360	0.706348	11
Imangda intrusive (Bh. KP-4)						
4-5	Plagioclase	0.62	429.00	0.0042	0.705567	8
4-5	Pyroxene I	0.62	10.38	0.1716	0.705890	12
4-5	Pyroxene II	0.20	16.71	0.0340	0.705419	12

(continued)

Table 23 (continued)

Sample	Rock, mineral	Rb, ppm	Sr, ppm	$^{87}Rb/^{86}Sr$	$^{87}Sr/^{86}Sr$	2σ
Picrites of Gudchikha Formation						
KhS51/2	Plagioclase	3.86	548.00	0.0203	0.705991	12
KhS51/2	Pyroxene	1.02	31.61	0.0934	0.706492	12
KhS51/3	Plagioclase	3.76	537.60	0.0202	0.705993	10
KhS51/3	Pyroxene	0.79	259.90	0.0088	0.708006	13
3b/S-08	Calcite	0.01	939.60	0.0001	0.708339	8
4b/S-08	Anhydrite	0.04	1042.00	0.0001	0.708207	8
5/S-08-16		0.01	1284.00	0.0000	0.708786	8

Table 24 Results of Rb–Sr analysis in massive disseminated Cu–Ni ores

Sample	Ore type	Rb, ppm	Sr, ppm	$^{87}Rb/^{86}Sr$	$^{87}Sr/^{86}Sr$	2σ	IR (250)
Talnakh intrusive (Bh. OUG-2)							
OUG-2, 79 (1239.6)	Mass. sulphide	0.258	2.645	0.2817	0.710365	93	0.709363
OUG-2, 87 (1248.20		0.075	14.130	0.0154	0.709337	74	0.709282
OUG-2, 96 (1257.7)		0.069	4.878	0.0411	0.709295	47	0.709149
OUG-2, T-13 (1208.9)	Impreg. sulphide	0.314	9.737	0.0933	0.706439	25	0.706107
T-14 (1213.8)		0.233	2.428	0.2781	0.709910	53	0.708921
T-15 (1217.8–1218.3)		0.305	3.583	0.2465	0.708357	27	0.707480
T-16 (1223.0–1223.3)		0.685	9.516	0.2081	0.707607	34	0.706867
T-17 (1226.3–1226.6)		0.604	14.780	0.1181	0.708297	22	0.707877
T-15-132	Massive ore	0.158	11.240	0.0405	0.708399	24	0.708255
Kharaelakh intrusive (boreholes KZ-844, KZ-963)							
844-3,4	Impreg. sulphide	0.921	3.146	0.8469	0.707451	87	0.704439
844-10,11		0.193	26.950	0.0207	0.708229	12	0.708155
844-18	Mass. sulphide	0.027	3.556	0.0218	0.711727	82	0.711649
844-19		12.210	14.530	2.4320	0.719279	49	0.710630
844-20		0.102	32.480	0.0091	0.709142	9	0.709110
963-5		0.092	17.150	0.0154	0.709078	28	0.709023
963-17		0.811	5.372	0.4365	0.709529	11	0.707976

(continued)

Table 24 (continued)

Sample	Ore type	Rb, ppm	Sr, ppm	$^{87}Rb/^{86}Sr$	$^{87}Sr/^{86}Sr$	2σ	IR (250)
963-30	Impreg. sulphide	0.221	4.390	0.1457	0.707995	26	0.707477
963-31		0.231	5.490	0.1215	0.707428	25	0.706996
963-38		1.059	4.147	0.7387	0.711278	30	0.708651
963-78	Mass. sulphide	1.242	4.894	0.7341	0.712121	27	0.709510
963-86		1.642	9.227	0.5146	0.710614	25	0.708784
963-95		0.292	4.969	0.1702	0.710092	29	0.709487
Chernogorsk intrusive (Bh. MP-2bis)							
CH-11	Impreg. sulphide	0.254	2.126	0.3461	0.710881	68	0.709650
Zub-Marksheidersky intrusive (Bh. MP-27)							
27-13	Impreg. sulphide	0.446	2.303	0.5599	0.710116	66	0.708124
Lower Norilsk intrusive (Bh. NP-37)							
37-12	Impreg. sulphide	4.329	39.600	0.3159	0.709659	12	0.708535
Mikchangda intrusive (Bh. MD-48)							
48-32,33	Impreg. sulphide	0.366	130.50	0.0081	0.707912	10	0.707883
South Pyasina intrusive (Bh. OV-25)							
25-35	Impreg. sulphide	0.944	32.960	0.0828	0.707561	10	0.707266

Table 25 Results of Sm–Nd analysis in rocks of Norilsk Province

Sample	Sm, ppm	Nd, ppm	$^{147}Sm/^{144}Nd$	$^{143}Nd/^{144}Nd$	2σ	ε_{Nd} (250)
ZF-13-429,9	4.066	15.430	0.1593	0.512504	4	−1.4
ZF-13-441,9	1.096	3.891	0.1702	0.512667	7	1.4
ZF-13-452,3	2.286	8.389	0.1648	0.512663	6	1.5
ZF-13-462,3	2.924	11.490	0.1538	0.512499	6	−1.3
ZF-13-483,2	2.150	8.392	0.1549	0.512609	6	0.8
ZF-13-521,8	34.040	149.400	0.1378	0.512642	3	2.0
ZF-21-446,0	4.442	11.290	0.2379	0.512739	15	0.7
ZF-30-346,1	2.163	8.662	0.1510	0.512616	6	1.0
ZF-30-357,1	2.926	10.710	0.1652	0.512907	18	6.3
ZF-30-375,9	2.596	8.989	0.1746	0.512656	15	1.1
ZF-30-458,7	2.298	8.239	0.1686	0.512672	11	1.6
ZF-30-387,7	4.132	15.110	0.1653	0.512643	12	1.1
ZF-30-388,7	3.886	15.540	0.1512	0.512643	5	1.6
ZF-31-539,8	2.325	10.480	0.1341	0.512642	6	2.1
ZF-43-625,5	3.499	12.270	0.1724	0.512677	8	1.5

(continued)

Table 25 (continued)

Sample	Sm, ppm	Nd, ppm	$^{147}Sm/^{144}Nd$	$^{143}Nd/^{144}Nd$	2σ	ε_{Nd} (250)
RT-2-1364,1	5.737	27.330	0.1269	0.512620	9	1.9
RT-2-1371,2	2.896	10.650	0.1645	0.512650	16	1.3
RT-2-1425,0	1.533	5.673	0.1634	0.512694	14	2.2
RT-2-1436,2	1.221	4.402	0.1677	0.512689	7	1.9
RT-2-1453,7	4.992	22.440	0.1345	0.512628	8	1.8
OM-32-1057,3	1.898	6.830	0.1681	0.512673	5	1.6
OM-32-1084,8	1.864	6.688	0.1685	0.512676	16	1.6
OM-32-1090,0	2.412	8.736	0.1669	0.512643	6	1.0
OM-32-1114,4	1.853	6.548	0.1711	0.512676	6	1.6
OM-32-1149,8	2.008	10.040	0.1209	0.512332	8	−3.6
OM-123-952,7	1.520	5.462	0.1682	0.512652	20	1.2
OM-123-980,0	2.345	8.535	0.1661	0.512645	6	1.1
OM-123-1005,6	3.102	6.619	0.2833	0.512674	5	−2.1
OM-123-1009,6	1.854	6.756	0.1662	0.512652	6	1.2
OM-123-1089,1	10.260	47.080	0.1317	0.512347	3	−3.6
OM-123-1112,0	11.560	55.120	0.1268	0.512270	6	−5.0
OV-28-690,1	4.832	24.310	0.1202	0.512479	5	−0.7
OV-28-703,2	3.928	14.370	0.1653	0.512663	6	1.5
OV-28-745,4	2.110	7.522	0.1696	0.512611	18	0.3
OV-28-813,8	2.375	8.933	0.1607	0.512656	7	1.5
OV-28-840,1	7.582	35.630	0.1287	0.512627	4	2.0
OV-36-1403,1	5.908	26.850	0.1330	0.512455	3	−1.5
OV-36-1424,0	3.359	13.980	0.1453	0.512627	12	1.4
OV-36-1438,0	4.341	16.660	0.1575	0.512579	4	0.1
OV-36-1463,7	2.523	9.085	0.1679	0.512670	4	1.5
OV-36-1539,2	4.655	16.670	0.1689	0.512706	5	2.2
MP-2f/g	1.436	4.9770	0.1745	0.512685	6	1.6
MP-25c/g	2.823	10.110	0.1688	0.512634	7	0.8
657/2 dolerite	3.701	13.250	0.1688	0.512668	7	1.5
N-2 basalt	7.574	34.240	0.1337	0.512413	5	−2.4
N-3 basalt	4.342	19.820	0.1324	0.512163	6	−7.2
12N05 basalt	11.460	54.750	0.1265	0.512357	4	−3.2
12N07 basalt	2.668	9.510	0.1696	0.512690	5	1.9
12N08 basalt	1.858	6.796	0.1653	0.512799	5	4.1
12N15	2.755	9.971	0.1670	0.512677	7	1.7
12N17	2.634	9.091	0.1752	0.512645	8	0.8
12N18 dolerite	1.857	16.670	0.06735	0.512669	4	4.7
12N19a	5.906	22.250	0.1605	0.512615	8	0.7
12N19b	5.351	19.740	0.1639	0.512614	9	0.4
12N21 hornfels	6.683	32.750	0.1233	0.512284	11	−4.6

Table 26 Results of Sm–Nd analysis in rocks of Norilsk Province

Sample	Sm, ppm	Nd, ppm	$^{147}Sm/^{144}Nd$	$^{143}Nd/^{144}Nd$	2σ	ε_{Nd} (250)
Oktyabrskoe deposit (boreholes KZ 361bis, KZ 931, KZ 952, KZ 1084, KZ 1089, KZ 1112, KZ 1319, KZ 1535)						
KZ 361bis-1062,8	1.444	6.832	0.1278	0.512619	5	1.8
KZ 361bis-1063,4	0.548	2.691	0.1230	0.512603	9	1.7
KZ 361bis-1074,4	3.056	13.980	0.1321	0.512584	5	1.0
KZ 361bis-1075,6	0.229	1.204	0.1148	0.512484	10	−0.4
KZ 931-643,9	1.063	5.535	0.1161	0.512619	6	2.2
KZ 931-645	0.177	1.006	0.1064	0.512621	8	2.6
KZ 952-971,1	1.403	5.524	0.1536	0.512516	7	−1.0
KZ 952-1010,4	2.414	10.880	0.1342	0.512505	6	−0.6
KZ 952-1013,5	1.717	7.829	0.1326	0.512492	8	−0.8
KZ 1084-1146,9	0.473	2.142	0.1334	0.512548	6	0.3
KZ 1089-1154,4	0.004	0.019	0.1309	0.512542	25	0.2
KZ 1089-1155,6	0.039	0.265	0.0901	0.512540	19	1.5
KZ 1089-1156,7	0.351	1.416	0.1499	0.512626	6	1.3
KZ 1112-1092,4	0.551	2.698	0.1234	0.512600	8	1.6
KZ 1112-1094,8	0.356	1.729	0.1246	0.512611	17	1.8
KZ 1112-1098,4	0.225	1.072	0.1269	0.512637	10	2.2
KZ 1112-1100,4	0.164	0.724	0.1367	0.512645	19	2.1
KZ 1112-1102,4	0.245	1.122	0.1319	0.512610	8	1.5
KZ 1112-1103,8	0.136	0.543	0.1511	0.512668	16	2.0
KZ 1319-598,4	1.955	7.471	0.1582	0.512526	9	−1.0
KZ 1319-626,8	2.496	8.641	0.1747	0.512680	6	1.5
KZ 1319-638,4	1.054	3.991	0.1597	0.512642	8	1.3
KZ 1319-641,4	0.565	2.305	0.1482	0.512644	8	1.7
KZ 1535-1489,2	1.980	7.039	0.1701	0.512672	5	1.5
KZ 1535-1491,0	1.237	4.307	0.1737	0.512662	9	1.2
KZ 1535-1495,9	1.398	5.004	0.1689	0.512671	6	1.5
Koevo area (Bh. PK-11)						
PK-11-320,1	2.750	9.959	0.1669	0.512797	9	4.1
PK-11-353,3	3.855	14.620	0.1594	0.512783	6	4.0
PK-11-416,0	4.242	14.720	0.1743	0.512797	8	3.8
PK-11-449,8	4.799	20.590	0.1409	0.512534	6	−0.2
PK-11-477,3	4.609	17.690	0.1575	0.512682	9	2.1
PK-11-503,5	5.805	24.770	0.1417	0.512397	5	−2.9
PK-11-529,3	5.455	23.620	0.1396	0.512527	7	−0.3

(continued)

Table 26 (continued)

Sample	Sm, ppm	Nd, ppm	$^{147}Sm/^{144}Nd$	$^{143}Nd/^{144}Nd$	2σ	ε_{Nd} (250)
Tangnarylakh area						
13096a	4.052	13.960	0.1756	0.512765	7	3.2
13096b	3.860	13.180	0.1771	0.512775	7	3.3
13097a	3.258	10.810	0.1822	0.512771	6	3.1
13097b	3.432	11.980	0.1732	0.512784	6	3.6
Section along Mokulaj Creek						
13005	2.918	10.640	0.1658	0.512700	7	2.2
13016	3.063	10.610	0.1746	0.512666	7	1.3
13020	3.366	11.380	0.1788	0.512715	5	2.1
13033	3.401	12.050	0.1706	0.512655	7	1.2
13049	3.404	11.520	0.1786	0.512737	6	2.5

Table 27 Rubidium and strontium in rocks of Norilsk Province

Sample	Rb, ppm	Sr, ppm	$^{87}Rb/^{86}Sr$	$^{87}Sr/^{86}Sr$	2σ	IR (250)
ZF-13-429,9	1.23	612.8	0.0058	0.708509	9	0.708488
ZF-13-441,9	17.49	256.8	0.1969	0.708287	23	0.707587
ZF-13-452,3	14.80	275.0	0.1556	0.708566	9	0.708013
ZF-13-462,3	0.29	295.7	0.0029	0.708409	10	0.708399
ZF-13-483,2	3.62	336.1	0.0311	0.708772	12	0.708661
ZF-13-521,8	41.89	153.9	0.7870	0.712754	24	0.709955
ZF-21-446,0	36.02	489.2	0.2128	0.708813	21	0.708056
ZF-30-346,1	10.20	244.6	0.1205	0.708950	8	0.708521
ZF-30-357,1	0.92	165.1	0.0162	0.708735	9	0.708677
ZF-30-375,9	16.81	228.8	0.2124	0.709334	7	0.708579
ZF-30-458,7	4.76	140.3	0.0979	0.707597	19	0.707249
ZF-30-387,7	14.52	310.7	0.1351	0.708289	26	0.707809
ZF-30-388,7	3.51	27.0	0.3752	0.709094	16	0.707760
ZF-31-539,8	118.40	57.3	5.9920	0.732585	27	0.711276
ZF-43-625,5	14.36	642.7	0.0646	0.708727	9	0.708497
RT-2-1364,1	145.10	54.8	7.6790	0.738073	12	0.710764
RT-2-1371,2	24.82	262.7	0.2730	0.708109	11	0.707138
RT-2-1425,0	6.60	89.5	0.2132	0.706682	10	0.705924
RT-2-1436,2	4.13	74.1	0.1611	0.706106	9	0.705533
RT-2-1453,7	125.10	118.0	3.0666	0.721705	10	0.710799
OM-32-1057,3	15.82	210.3	0.2174	0.706319	39	0.705546
OM-32-1084,8	9.57	153.6	0.1800	0.706282	11	0.705642
OM-32-1090,0	11.54	199.8	0.1669	0.706183	12	0.705589
OM-32-1114,4	18.67	349.8	0.2160	0.706478	6	0.705710

(continued)

Table 27 (continued)

Sample	Rb, ppm	Sr, ppm	$^{87}Rb/^{86}Sr$	$^{87}Sr/^{86}Sr$	2σ	IR (250)
OM-32-1149,8	0.63	131.2	0.0139	0.707230	8	0.707181
OM-123-952,7	12.95	272.5	0.1373	0.706095	32	0.705607
OM-123-980,0	11.67	231.4	0.1457	0.706238	15	0.705720
OM-123-1005,6	10.10	232.0	0.1258	0.706722	8	0.706275
OM-123-1009,6	6.56	222.0	0.0856	0.705838	35	0.705534
OM-123-1089,1	3.99	395.0	0.0292	0.706739	33	0.706635
OM-123-1112,0	71.10	342.5	0.3198	0.707478	11	0.706341
OV-28-690,1	20.33	511.6	0.1149	0.709135	10	0.708726
OV-28-703,2	16.01	275.1	0.1682	0.707776	15	0.707178
OV-28-745,4	13.45	221.7	0.1753	0.707115	14	0.706492
OV-28-813,8	18.82	206.9	0.2628	0.707422	11	0.706487
OV-28-840,1	111.30	479.6	0.6710	0.712478	25	0.710092
OV-36-1403,1	37.10	240.2	0.4464	0.710560	21	0.708972
OV-36-1424,0	1.01	608.4	0.0048	0.708631	12	0.708614
OV-36-1438,0	4.74	579.1	0.0237	0.708754	12	0.706670
OV-36-1463,7	37.65	639.5	0.1702	0.709002	9	0.708397
OV-36-1539,2	19.21	228.4	0.2430	0.706870	24	0.706006
MP-2f/g	6.42	239.8	0.0773	0.706544	12	0.706269
MP-25c/g	18.83	362.9	0.1022	0.707630	8	0.707267
657/2 dolerite	12.79	211.9	0.1744	0.706151	15	0.705531
N-2 basalt	24.36	433.9	0.1622	0.706062	7	0.705485
N-3 basalt	28.25	222.1	0.3677	0.710067	23	0.708759
12N05 basalt	50.28	375.1	0.3874	0.707458	9	0.706170
12N07 basalt	9.46	227.9	0.1199	0.705751	12	0.705325
12N08 basalt	4.54	95.9	0.1369	0.706216	10	0.705729
12N15	12.64	251.3	0.1454	0.706703	9	0.706186
12N17	24.78	510.6	0.1403	0.708227	8	0.707728
12N18 dolerite	28.31	473.1	0.1730	0.708826	14	0.708211
12N19a	33.86	494.3	0.1980	0.708973	19	0.708269
12N19b	44.07	582.8	0.2186	0.708985	11	0.708208
12N21 hornfels	31.81	740.6	0.1241	0.709136	6	0.708695

Table 28 Results of Rb–Sr analysis in rocks of Norilsk Province

Sample	Rb, ppm	Sr, ppm	$^{87}Rb/^{86}Sr$	$^{87}Sr/^{86}Sr$	2?σ	IR (250)
Oktyabrskoe deposit (boreholes KZ 361bis, KZ 931, KZ 952, KZ 1084, KZ 1089, KZ 1112, KZ 1319, KZ 1535)						
KZ 361bis-1062,8	33.020	57.66	1.6564	0.716378	20	0.710487
KZ 361bis-1063,4	9.438	13.32	2.0489	0.714048	19	0.706761
KZ 361bis-1074,4	39.100	62.51	1.8098	0.718572	17	0.712136
KZ 361bis-1075,6	2.532	4.50	1.6301	0.716229	21	0.710432
KZ 931-643,9	11.100	53.91	0.5952	0.711566	16	0.709449
KZ 931-645	2.276	22.45	0.2932	0.709973	22	0.708930
KZ 952-971,1	0.084	13.88	0.0175	0.708558	82	0.708496
KZ 952-1010,4	0.322	19.38	0.0480	0.709812	67	0.709641
KZ 952-1013,5	0.578	22.65	0.0738	0.708270	15	0.708008
KZ 1084-1146,9	1.330	6.22	0.6190	0.712165	10	0.709964
KZ 1089-1154,4	0.070	0.72	0.2811	0.712660	29	0.711660
KZ 1089-1155,6	0.942	1.09	2.5070	0.717628	15	0.708712
KZ 1089-1156,7	4.779	9.04	1.5291	0.716044	14	0.710606
KZ 1112-1092,4	17.320	158.40	0.3161	0.711042	10	0.709918
KZ 1112-1094,8	5.242	17.57	0.8629	0.712674	12	0.709605
KZ 1112-1098,4	3.793	5.98	1.8374	0.718077	36	0.711543
KZ 1112-1100,4	2.804	6.04	1.3435	0.717066	64	0.712288
KZ 1112-1102,4	4.195	10.15	1.1951	0.713728	36	0.709478
KZ 1112-1103,8	0.961	4.47	0.6227	0.712217	89	0.710002
KZ 1319-598,4	0.259	50.73	0.0148	0.708570	32	0.708517
KZ 1319-626,8	6.685	205.90	0.0938	0.706332	14	0.705998
KZ 1319-638,4	3.186	62.19	0.1481	0.708077	26	0.707550
KZ 1319-641,4	2.532	77.45	0.0945	0.706701	63	0.706365
KZ 1535-1489,2	6.485	147.50	0.1271	0.713281	64	0.712829
KZ 1535-1491,0	5.018	40.28	0.3601	0.709298	23	0.708017
KZ 1535-1495,9	6.890	74.58	0.2670	0.707301	10	0.706351
Koevo area (Bh. PK-11)						
PK-11-320,1	0.878	124.80	0.0203	0.706876	51	0.706804
PK-11-353,3	12.290	101.70	0.3492	0.708889	16	0.707647
PK-11-416,0	12.620	205.90	0.1771	0.706426	17	0.705796
PK-11-449,8	18.380	389.30	0.1364	0.705779	16	0.705294
PK-11-477,3	4.217	295.20	0.0413	0.711387	43	0.711240
PK-11-503,5	5.734	353.40	0.0469	0.707242	20	0.707075
PK-11-529,3	3.984	365.50	0.0315	0.706579	17	0.706467
Tangnarylakh area						
13096a	10.170	194.4	0.1512	0.705596	28	0.705058
13096b	6.899	191.5	0.1041	0.705367	23	0.704997
13097a	12.240	299.6	0.1181	0.706414	10	0.705994

(continued)

Table 28 (continued)

Sample	Rb, ppm	Sr, ppm	$^{87}Rb/^{86}Sr$	$^{87}Sr/^{86}Sr$	2?σ	IR (250)
13097b	11.690	216.4	0.1560	0.705712	10	0.705157
Section along Mokulaj Creek						
13005	12.780	189.5	0.1949	0.705894	15	0.705201
13016	21.270	223.4	0.2751	0.706120	12	0.705142
13020	5.991	174.9	0.0990	0.705086	11	0.704734
13033	2.063	198.9	0.0300	0.705407	10	0.705300
13049	9.725	163.9	0.1714	0.705531	14	0.704921

Table 29 Results of Sm–Nd analysis in rocks of tuff lava sequence (Norilsk ore District)

Sample	Rock, mineral	Sm, ppm	Nd, ppm	$^{147}Sm/^{144}Nd$	$^{143}Nd/^{144}Nd$	2σ	ε_{Nd} (250)
1163805	Gabbro	1.481	6.336	0.1413	0.511750	8	−15.6
306		1.915	9.164	0.1263	0.511575	6	−18.5
1109896		1.595	7.600	0.1269	0.511590	9	−18.2
5801250	Norite	1.947	7.819	0.1505	0.511889	8	−13.1
5801180	Pyroxenite	1.290	5.042	0.1547	0.511918	9	−12.7
305	Monzodiorite	4.430	23.150	0.1157	0.511410	4	−21.4
308	Norite	6.142	32.340	0.1148	0.511346	4	−22.6
1T-730,5		1.117	5.901	0.1144	0.511331	6	−22.9

Table 30 Results of Rb–Sr analysis in rocks of tuff lava sequence (Norilsk ore District)

Sample	Rock, mineral	Rb, ppm	Sr, ppm	$^{87}Rb/^{86}Sr$	$^{87}Sr/^{86}Sr$	2σ	IR (250)
1163805	Gabbro	5.283	75.27	0.2029	0.711768	34	0.711046
306		13.690	287.50	0.1376	0.710921	29	0.710432
1109896		18.480	333.70	0.1601	0.711338	11	0.710769
5801250	Norite	6.144	50.66	0.3508	0.716182	21	0.714934
5801180	Pyroxenite	14.760	23.21	1.8450	0.750730	30	0.744169
305	Monzodiorite	57.430	301.80	0.5507	0.722458	32	0.720500
308	Norite	39.900	558.10	0.2067	0.710678	18	0.710033
1T-730,5		7.455	331.10	0.0651	0.706145	13	0.705913

– the cause of a relatively high radiogenity of strontium isotopic composition is, possibly, a more significant contribution of the crustal component into the mineralized fluids as compared with non-mineralized and weakly mineralized ones.

References

1. Arndt NT, Czamanske GK, Walker RJ et al (2003) Geochemistry and origin of the intrusion hosts of the Noril'sk-Talnakh Cu–Ni–PGE sulphide deposits. Econ Geology 98:495–515
2. Faure G (1989) Fundamentals of isotope geology. M.: Mir, 590 p
3. Liew TC, Hofmann AW (1988) Precambrian crustal components, plutonic associations, plate environment of the Hercynian Fold Belt of central Europe: Indications from a Nd and Sr isotopic study. Contrib Mineral Petrol 98:129–138
4. Goldstein SJ, Jacobsen SB (1988) Nd and Sr isotopic systematics of river water suspended material: implications for crustal evolution. Earth Planet Sci Lett 87:249–265
5. De Paolo DJ, Wasserburg GJ (1976) Nd Isotopic variations and petrogenetic models. Geophys Rev Lett 3:249–252
6. Ludwig KR (1999) User's manual for Isoplot/Ex. Vers. 2.05. Berkeley Geochronology Center, Berkeley, Special Publication N 1a. 48 p
7. Jacobsen SB, Wasserburg GJ (1980) Sm-Nd isotopic evolution of chondrites. Earth Planet Sci Lett 50:139–155

Lead Isotopes

Boris Belyatsky, Yury Pushkarev, Edward Prasolov, Igor Kapitonov, Robert Krymsky and Sergey Sergeev

Abstract The discussed below results on Pb isotopic compositions have been obtained by both TIMS and LA-ICP-MS-MC techniques on sulphide and plagioclase mg-weight specimens and single grains correspondingly. General Pb isotopic compositions imply derivation from both, the mantle and crustal sources with the latter being dominant. The obtained results from the Talnakh intrusion notably discrepant from that of the Norilsk and Kharaelakh being less radiogenic, while the latter two cluster together, demonstrating a minor divergence between massive and disseminated ores. Analysis of $^{207}Pb/^{204}Pb$ versus $\delta^{34}S$ along with Th/U assumes three sources of mantle, lower curst and upper crust, while Pb isotopes alone do not provide distinguishing of the massive ores from poor one. Disparity of Pb isotopes in sulphides and plagioclases assumes their chemical and isotopic disequilibrium, precluding their coexistence in a single batch of melt.

Lead isotopic composition in samples of rocks and ores from massifs of the Norilsk district was studied for a possible identification of sources of ore and silicate matter. It was also planned to find out, whether lead isotopic composition in disseminated ores can serve as a source of information on the potentially industrial mineralization in the corresponding massif.

Owing to the U/Pb ratio value close to zero in plagioclases and sulphides, it is assumed that lead isotopic composition in these minerals should correspond to the initial isotopic composition in the source of magmatic melts and can be used for genetic constructions.

Isotope-geochemical basis principals of the method. In nature, there are four stable lead isotopes—^{204}Pb, ^{206}Pb, ^{207}Pb, ^{208}Pb (^{204}Pb—non-radiogenic isotope; ^{206}Pb, ^{207}Pb and ^{208}Pb—radiogenic isotopes, decay products of uranium and thorium isotopes: ^{238}U => ^{206}Pb; ^{235}U => ^{207}Pb; ^{232}Th => ^{208}Pb).

Compilation of a large set of experimental data on isotopic composition of "ordinary" lead from geological objects of different age led to plotting of the

B. Belyatsky (✉) · Y. Pushkarev · E. Prasolov · I. Kapitonov · R. Krymsky · S. Sergeev
Russian Geological Research Institute (VSEGEI), St. Petersburg, Russia
e-mail: boris_belyatskiy@vsegei.ru

O. Petrov (ed.), *Isotope Geology of the Norilsk Deposits*, Springer Geology,
https://doi.org/10.1007/978-3-030-05216-4_5

evolution curves of lead isotopic composition ($^{206}Pb/^{204}Pb$, $^{207}Pb/^{204}Pb$, $^{208}Pb/^{204}Pb$ ratios) throughout the entire history of the Earth in general (Fig. 1) as well as in different model reservoirs of the planet. They can be illustrated by graphic curves in a diagram with marks corresponding to the model age. Thus, the diagram (Fig. 2) shows the evolution lines of isotopic composition of lead according to the notions of plumbotectonics—an evolution model of lead isotopic composition by Zartman and Doe, which assumes that various geochemical (geological) reservoirs have individual U/Pb and Th/U ratios, which over time cause subconcordant curves corresponding to the composition of the main reservoirs (upper and lower crust, mantle, orogen).

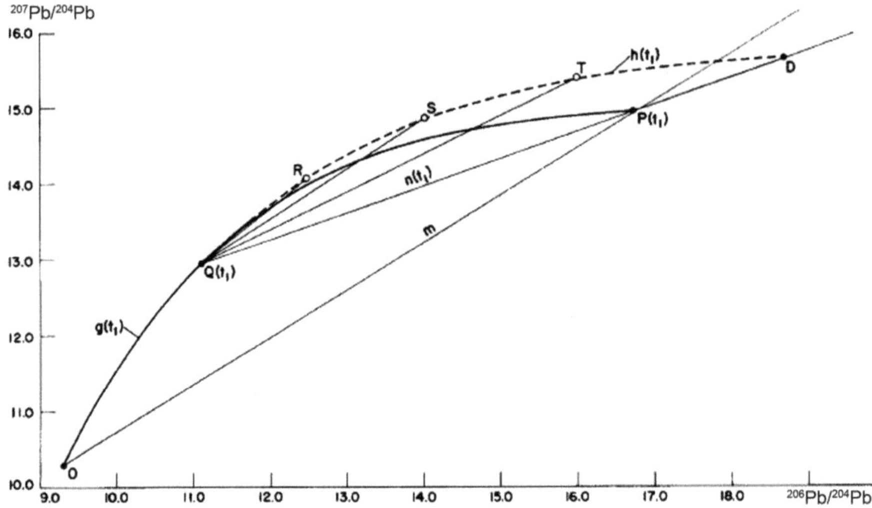

Fig. 1 Evolution of "ordinary" lead according to Stacey-Kramers model [1]

Fig. 2 Evolution lines of isotopic composition of lead in different geochemical reservoirs according to Doe and Zartman plumbotectonics model [2–4]. 1—upper crust; 2—lower crust; 3—orogen; 4—mantle

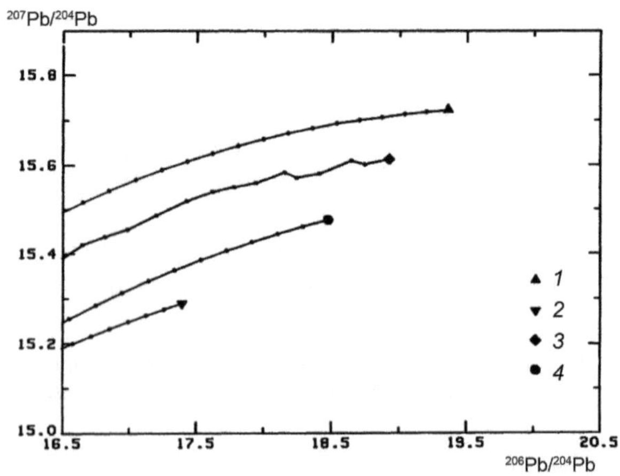

1 Methodology, Samples

Isotopic composition of lead was determined by isotope dilution method (TIMS) from milligram weighed portions of monomineral fractions with a purity of 90% or better for sulphides (192 samples) and plagioclases (117 samples). To determine possible causes of isotope ratio variation, local isotopic composition of lead from individual sulphide grains was also measured in 28 samples (of 20 individual grains) by means of laser sampling and ICP mass spectrometry.

Isotopic composition of lead was determined in sulphides and plagioclases from the intrusions with different extent and type of mineralization, including Norilsk-1, Talnakh and Kharaelakh massifs with abundant ore (Tables 1 and 2).

Lead separation from silicate and sulphide fractions was performed with a preliminary dissolution of samples in a mixture of hydrofluoric and nitric acids (plagioclase) or a mixture of hydrochloric and nitric acids (sulphides) and subsequent separation on 100 microcolumns with Eichrom Sr Spec resin in 1 N and 8 N HCl. Plagioclase fractions were grinded to powder and treated with nitric acid to remove the uranium-containing film minerals. For certain samples, a stepwise acid leaching with concentrated nitric and hydrochloric acids was applied for 4 h at 80 ° C. Further, the isotopic composition of leach residue was analysed. Samples (lead fractions in hydrochloric acid form) were applied on Re filament in a mixture with silica gel and 0,2 N H_3PO_4.

Isotopic analysis of lead by TIMS method was performed on the nine-collector mass spectrometer Triton TI (Thermo) in a single-filament version in the static mode of recording ionic currents of isotopes. $^{206}Pb/^{204}Pb$, $^{207}Pb/^{204}Pb$ and $^{208}Pb/^{204}Pb$ ratios were measured. Each measurement consisted of 50 blocks of 10 scans at 2.2–2.3 A current on the evaporator and temperature of 1300 °C. NIST 981 standard (50 ng) was measured before each sample batch. The average accuracy of the analyses was 0.05% for the $^{206}Pb/^{204}Pb$ ratio. Adjustment on the instrument mass fractionation bias was made on the basis of the average value of NIST 981 standard measurements ($^{206}Pb/^{204}Pb$ 16.9374, $^{207}Pb/^{204}Pb$ 15.4916, $^{208}Pb/^{204}Pb$ 36.7219) at the same temperature. The measured lead isotope ratios were adjusted for mass fractionation of 0.120% for $^{206}Pb/^{204}Pb$ and $^{207}Pb/^{204}Pb$; 0.135% amu for $^{208}Pb/^{204}Pb$. Blank (blank experiment) during the analysis did not exceed 0.2 ng for Pb; its composition is: $^{206}Pb/^{204}Pb$ 18.120, $^{207}Pb/^{204}Pb$ 15.542, $^{208}Pb/^{204}Pb$ 37.354. The ratio of blank lead to the sample did not exceed 1/200,000; thus, no adjustment for the blank lead content for the measured isotope ratios was applied.

Table 1 Isotopic composition of lead in sulphides from intrusives of Norilsk-Taimyr Region

Intrusion, borehole number	Sample number	Depth, m	Type of mineralization	Minerals, formula	$^{206}Pb/^{204}Pb$	2σ	$^{207}Pb/^{204}Pb$	2σ	$^{208}Pb/^{204}Pb$	2σ
Sulphides of industrial-ore-bearing intrusives										
Norilsk-1, Bh. MN-2	N1-1		Disseminated	Cu–Ni sulphides	17.9204	0.0093	15.5255	0.0070	37.9218	0.0213
	N1-2				18.1214	0.0091	15.5260	0.0069	38.0130	0.0209
	N1-3		MS horizon		18.0452	0.0093	15.5321	0.0070	37.9141	0.0212
	N1-6		Disseminated		18.0576	0.0092	15.5242	0.0069	37.9029	0.0207
	N1-7				18.0155	0.0093	15.5389	0.0070	37.9332	0.0212
	N1-8				18.0019	0.0093	15.5268	0.0070	37.8867	0.0211
	N1-9				18.0237	0.0093	15.5355	0.0070	37.9195	0.0212
	N1-12		MS horizon	Sulphides	18.0414	0.0007	15.5184	0.0005	37.8828	0.0013
Norilsk 1, Medvezhij open pit	Kn-97-1		Disseminated	Sn–pb	18.0621	0.0122	15.5811	0.0093	37.9849	0.0232
	Kn-97-1			Pd–pt–pb–bi	18.0551	0.0190	15.4259	0.0208	37.6437	0.0375
	Kn-97-1			Galena, pbs	18.2856	0.0194	15.4750	0.0197	37.9290	0.0383
	Kn-97-1			Cu–Ni sulphides	18.1040	0.0194	15.5500	0.0160	37.9836	0.0109
	Kn-97-2				18.1254	0.0189	15.5610	0.0161	38.0268	0.0118
	Kn-97-3			Clausthalite, pbse	18.0875	0.0180	15.4985	0.0165	37.8181	0.0292
	Kn-97-3			Galena, pbs	18.0878	0.0138	15.4488	0.0121	37.5038	0.0288
	Kn-97-3			Clausthalite, pbse	17.9935	0.0120	15.4295	0.0095	37.3963	0.0217
	Kn-97-3				17.9827	0.0201	15.4429	0.0203	37.7071	0.0415
	Kn-97-3			Galena, pbs	18.0710	0.0186	15.4473	0.0208	37.7036	0.0432
	Kn-97-3			Cu–Ni sulphides	18.0705	0.0202	15.5356	0.0190	37.9749	0.0383
Talnakh, Bh. OUG-2	26_V	1208.4			18.2409	0.0117	15.5926	0.0092	38.1166	0.0112
	21_V (d.27_V)	1208.9			18.1536	0.0007	15.5088	0.0006	37.8438	0.0015
	27_V	1208.9			18.2268	0.0118	15.5783	0.0092	38.0574	0.0067
	28_V	1209.4			18.2187	0.0115	15.5678	0.0092	38.0275	0.0101
	29_V	1210.0			18.2363	0.0118	15.5907	0.0092	38.1024	0.0065
	30_V	1210.5			18.2339	0.0118	15.5865	0.0092	38.0987	0.0087
	31_V	1211.0			18.1695	0.0005	15.5279	0.0005	37.9127	0.0010
	32_V	1211.5			18.1633	0.0004	15.5196	0.0004	37.8870	0.0009

(continued)

Table 1 (continued)

Intrusion, borehole number	Sample number	Depth, m	Type of mineralization	Minerals, formula	$^{206}Pb/^{204}Pb$	2σ	$^{207}Pb/^{204}Pb$	2σ	$^{208}Pb/^{204}Pb$	2σ
	33_V	1212.0			18.1616	0.0004	15.5209	0.0004	37.8919	0.0011
	34_V (d. 33_V)	1212.0			18.1631	0.0013	15.5240	0.0011	37.8994	0.0028
	35_V	1213.0			18.2068	0.0007	15.5391	0.0007	37.9488	0.0013
	37_V	1214.0			18.1831	0.0005	15.5414	0.0004	37.9541	0.0012
	38_V	1215.0			18.1741	0.0018	15.5257	0.0016	37.9040	0.0038
	39_V	1215.9			18.2909	0.0011	15.5351	0.0009	37.9288	0.0023
	43_V	1218.1			18.1747	0.0006	15.5365	0.0005	37.9329	0.0014
	44_V	1218.7			18.1548	0.0022	15.5090	0.0019	37.8484	0.0046
	45_V	1219.2			18.1695	0.0008	15.5330	0.0007	37.9216	0.0018
	6_V (d. 49 V)				18.1741	0.0010	15.5366	0.0009	37.9363	0.0022
	47_V	1220.2			18.1743	0.0006	15.5400	0.0005	37.9349	0.0014
	48_V	1220.7			18.1711	0.0005	15.5301	0.0005	37.9065	0.0012
	49_V	1221.4			18.1604	0.0025	15.5203	0.0022	37.8860	0.0052
	50_V	1221.9			18.1647	0.0010	15.5277	0.0009	37.9013	0.0020
	51_V	1222.1			18.1571	0.0026	15.5238	0.0023	37.8934	0.0056
	52_V	1222.6			18.1608	0.0037	15.5169	0.0031	37.8866	0.0077
	53_V	1223.2			18.2027	0.0009	15.5738	0.0009	38.0440	0.0027
	54_V	1223.7			18.1937	0.0011	15.5584	0.0010	38.0052	0.0023
	55_V	1224.3			18.1716	0.0004	15.5325	0.0004	37.9124	0.0012
	56_V	1224.8			18.1692	0.0009	15.5277	0.0007	37.9038	0.0019
	57_V	1225.3			18.1640	0.0014	15.5149	0.0012	37.8674	0.0029
	60_V	1227.3			18.2144	0.0005	15.5743	0.0005	38.0545	0.0012
	63_V	1229.5			18.1988	0.0012	15.5430	0.0011	37.9556	0.0027
	65_V	1230.5			18.1905	0.0009	15.5473	0.0008	37.9791	0.0020
	66_V	1231.1			18.1891	0.0011	15.5411	0.0009	37.9604	0.0024
	68_V	1231.8			18.2454	0.0052	15.5184	0.0043	37.9443	0.0106
	69_V				18.2340	0.0022	15.5457	0.0019	37.9719	0.0046

(continued)

Table 1 (continued)

Intrusion, borehole number	Sample number	Depth, m	Type of mineralization	Minerals, formula	$^{206}Pb/^{204}Pb$	2σ	$^{207}Pb/^{204}Pb$	2σ	$^{208}Pb/^{204}Pb$	2σ
	71S	1232.8	Massive		18.2457	0.0031	15.5259	0.0026	37.9351	0.0064
	72S	1233.3			18.4333	0.0036	15.5331	0.0029	38.0263	0.0070
	72r	1233.3		Pyrrhotite	18.3415	0.0091	15.5708	0.0080	38.1373	0.0184
	73S	1233.8		Cu–Ni sulphides	18.2550	0.0024	15.5437	0.0002	37.9824	0.0050
	74S	1234.4			18.2613	0.0008	15.5339	0.0007	37.9611	0.0017
	75S	1235.3			18.2677	0.0008	15.5606	0.0006	38.0503	0.0015
	76S	1236.4			18.2578	0.0024	15.5478	0.0021	37.9991	0.0051
	77S	1237.5			18.0692	0.0193	15.3960	0.0160	37.4087	0.0274
	78S	1238.6			18.2119	0.0061	15.5225	0.0053	37.8934	0.0125
	78p	1238.6		Pyrrhotite	18.2080	0.0087	15.5281	0.0082	38.0183	0.0181
Talnakh, Bh. OUG-2	79S	1239.6	Massive	Cu–Ni sulphides	18.2416	0.0055	15.5423	0.0048	38.0288	0.0116
	80S	1240.7			18.2388	0.0032	15.5249	0.0028	37.9299	0.0067
	81S	1241.8			18.2359	0.0032	15.5431	0.0028	37.9705	0.0066
	82S	1242.9			18.2157	0.0027	15.5282	0.0024	37.9140	0.0058
	82r	1242.9		Pyrrhotite	18.1366	0.0105	15.4626	0.0107	37.7791	0.0230
	83S	1244.0		Cu–Ni sulphides	18.2317	0.0022	15.5526	0.0020	37.9952	0.0047
	84S	1245.1			18.2940	0.0028	15.5528	0.0025	38.0575	0.0061
	85S	1246.2			18.2306	0.0005	15.5379	0.0005	37.9476	0.0012
	86S	1247.1			18.2275	0.0014	15.5495	0.0012	37.9798	0.0029
	87S	1248.2			18.2044	0.0013	15.5296	0.0011	37.9110	0.0026
	87sr	1248.2		Chalcopyrite	18.2625	0.0084	15.5822	0.0073	38.1159	0.0147
	87r	1248.2		Pyrrhotite	18.0905	0.0106	15.4662	0.0110	37.8339	0.0234
	88S	1249.3		Cu–Ni sulphides	18.2386	0.0016	15.5695	0.0014	38.0417	0.0033
	89S	1250.4			18.1982	0.0010	15.5270	0.0009	37.9003	0.0023
	89sr	1250.4		Chalcopyrite	18.2867	0.0089	15.5909	0.0071	38.1063	0.0126
	89r	1250.4		Pyrrhotite	18.1824	0.0088	15.5247	0.0084	37.9739	0.0195
	90S	1251.5		Cu–Ni sulphides	18.2530	0.0014	15.5702	0.0013	38.0516	0.0032
	91S	1252.6			18.2728	0.0013	15.5771	0.0012	38.1040	0.0028

(continued)

Table 1 (continued)

Intrusion, borehole number	Sample number	Depth, m	Type of mineralization	Minerals, formula	$^{206}Pb/^{204}Pb$	2σ	$^{207}Pb/^{204}Pb$	2σ	$^{208}Pb/^{204}Pb$	2σ
Talnakh, Bh. OUG-2	92S	1253.7	Massive		18.2154	0.0014	15.5245	0.0012	37.8938	0.0029
	93S	1254.5			18.2225	0.0014	15.5517	0.0013	37.9895	0.0031
	94S	1255.6			18.2174	0.0015	15.5580	0.0013	37.9932	0.0033
	95S	1256.7			18.1769	0.0044	15.5036	0.0038	37.8185	0.0089
	96S	1257.7			18.2039	0.0013	15.5353	0.0011	37.9243	0.0027
	96r	1257.7		Pyrrhotite	18.2420	0.0085	15.5670	0.0072	38.0508	0.0132
	97S	1258.3		Cu–Ni sulphides	18.1911	0.0015	15.5347	0.0012	37.9212	0.0030
	99S	1259.5			18.1971	0.0011	15.5183	0.0009	37.9873	0.0022
	100S	1261.0			18.2471	0.0024	15.5436	0.0021	38.0263	0.0048
	101S	1262.3			18.2069	0.0003	15.5142	0.0003	37.9133	0.0008
	102r (d. 89r)	1250.4		Pyrrhotite	18.2262	0.0084	15.5503	0.0071	37.9898	0.0153
	103sr (d. 89sr)	1250.4		Chalcopyrite	18.2629	0.0110	15.5835	0.0086	38.0391	0.0152
	104S (d. 89S)	1250.4		Cu–Ni sulphides	18.1941	0.0021	15.5164	0.0018	37.8652	0.0044
	T-13		Disseminated		18.1709	0.0009	15.5284	0.0011	37.9075	0.0037
	T-14				18.1945	0.0013	15.5520	0.0014	37.9923	0.0050
	T-15				18.1661	0.0012	15.5247	0.0009	37.9027	0.0028
	T-16				18.1461	0.0003	15.5085	0.0003	37.8326	0.0009
	T-17				18.1665	0.0009	15.5320	0.0011	37.9241	0.0037
	T-18				18.2112	0.0007	15.5501	0.0008	37.9879	0.0027
	T-19 (d. T-14)				18.1749	0.0008	15.5292	0.0009	37.9133	0.0030
	T-2				18.2447	0.0003	15.5220	0.0002	37.9126	0.0007
	T-3				18.1840	0.0004	15.5086	0.0004	37.9242	0.0014
	T-5 + T-6				18.1554	0.0002	15.5063	0.0001	37.8627	0.0004
	T-5-132		Massive		18.1722	0.0009	15.5340	0.0008	37.9052	0.0026
	T-20 (d.T-18)		Disseminated		18.1928	0.0007	15.5380	0.0008	37.9484	0.0028
Kharaelakh, Bh. KZ-844	844-2 V	949.5			18.1658	0.0003	15.5263	0.0003	37.8943	0.0008
	844-2 V				18.1753	0.0004	15.5322	0.0005	37.9253	0.0016
	844-3,4 V	955.0–956.0			18.1942	0.0008	15.5423	0.0008	37.9463	0.0019
	844-3.4 V				18.1998	0.0006	15.5534	0.0007	37.9892	0.0026

(continued)

Table 1 (continued)

Intrusion, borehole number	Sample number	Depth, m	Type of mineralization	Minerals, formula	$^{206}Pb/^{204}Pb$	2σ	$^{207}Pb/^{204}Pb$	2σ	$^{208}Pb/^{204}Pb$	2σ
Kharaelakh, Bh. KZ-844	844-6 V				18.1987	0.0004	15.5571	0.0004	37.9950	0.0011
	844-7			Galena, pbs	18.1885	0.0005	15.5508	0.0006	37.9645	0.0018
	844-7				18.1876	0.0011	15.5497	0.0011	37.9642	0.0031
	844-7				18.1529	0.0149	15.5185	0.0131	37.8575	0.0327
	844-7				18.1637	0.0021	15.5252	0.0017	37.8940	0.0043
	844-7				18.1832	0.0009	15.5453	0.0008	37.9524	0.0020
	844-7				18.1534	0.0028	15.5134	0.0023	37.8581	0.0056
	844-7				18.1768	0.0008	15.5374	0.0007	37.9267	0.0021
	844-7				18.1768	0.0007	15.5354	0.0007	37.9147	0.0020
	844-7 V			Cu–Ni sulphides	18.1710	0.0003	15.5274	0.0003	37.8927	0.0014
	844-10,11 V				18.1618	0.0004	15.5160	0.0004	37.8523	0.0014
	844-11 V	1021.0			18.1734	0.0003	15.5288	0.0003	37.8930	0.0008
	844-15			Galena, pbs	18.1660	0.0023	15.5225	0.0020	37.8796	0.0049
	844-15				18.1592	0.0025	15.5156	0.0021	37.8655	0.0053
	844-15 V			Cu–Ni sulphides	18.1905	0.0006	15.5336	0.0006	37.9250	0.0023
	844-18S	1046.0	Massive		18.1837	0.0003	15.5289	0.0002	37.8908	0.0007
	844-19S	1055.0			18.2055	0.0008	15.5431	0.0007	37.9504	0.0018
	844-19S				18.2014	0.0008	15.5413	0.0009	37.9364	0.0033
	844-20S	1063.0			18.1903	0.0004	15.5381	0.0004	37.9169	0.0009
Kharaelakh, Bh. KZ-963	963-5		Veinlet	Chalcopyrite	18.3415	0.0003	15.5573	0.0003	37.9809	0.0008
	963-12		Massive	Cu–Ni sulphides	18.2499	0.0009	15.5487	0.0011	37.9590	0.0036
	963-17				18.2317	0.0008	15.5500	0.0009	37.9705	0.0030
	963-17				18.2498	0.0005	15.5611	0.0005	38.0104	0.0018
	963-18				18.1909	0.0005	15.5304	0.0006	37.9042	0.0021
	963-25				18.2081	0.0004	15.5379	0.0004	37.9229	0.0017
	963-30		Disseminated		18.1812	0.0010	15.5409	0.0012	37.9533	0.0043
	963-30				18.1806	0.0007	15.5380	0.0008	37.9484	0.0030
	963-31				18.1916	0.0008	15.5243	0.0009	37.9024	0.0033

(continued)

Table 1 (continued)

Intrusion, borehole number	Sample number	Depth, m	Type of mineralization	Minerals, formula	$^{206}Pb/^{204}Pb$	2σ	$^{207}Pb/^{204}Pb$	2σ	$^{208}Pb/^{204}Pb$	2σ
Kharaelakh, Bh. KZ-963	963-37				18.2231	0.0006	15.5351	0.0007	37.9376	0.0026
	963-38				18.2558	0.0005	15.5458	0.0005	37.9748	0.0018
	963-54				18.3296	0.0025	15.5686	0.0021	38.0686	0.0050
	963-60				18.2680	0.0011	15.5376	0.0010	37.9305	0.0023
	963-71		Massive		18.3005	0.0018	15.5643	0.0014	38.0337	0.0037
	963-75				18.2438	0.0016	15.5380	0.0013	37.9333	0.0033
	963-75				18.2521	0.0005	15.5543	0.0006	37.9850	0.0023
	963-78				18.2753	0.0007	15.5555	0.0007	37.9921	0.0025
	963-86				18.1773	0.0003	15.5253	0.0004	37.8767	0.0013
	963-88				18.2333	0.0005	15.5324	0.0005	37.9024	0.0018
	963-88				18.2262	0.0006	15.5381	0.0007	37.9224	0.0025
	963-89				18.2244	0.0003	15.5285	0.0003	37.8783	0.0009
	963-95				18.2034	0.0002	15.5245	0.0002	37.8656	0.0008
Kharaelakh, Bh. TG-21	TG 21-3		Low-sulphide	Sulphides	18.4344	0.0027	15.6865	0.0024	38.4222	0.0062
Sulphides of ore-bearing intrusives										
Chernogorsk, MP-2 bis	CH-11		Disseminated	Sulphides	18.1901	0.0010	15.5390	0.0008	37.9510	0.0021
	CH-13			Cu–Ni sulphides	18.2212	0.0006	15.5543	0.0007	38.0052	0.0022
	CH-13 (d.)				18.2149	0.0006	15.5472	0.0006	37.9801	0.0022
Zub-Marksheidersky, MP-27	27-1			Sulphides	18.3776	0.0006	15.5617	0.0005	38.1201	0.0015
	27-3				18.5241	0.0005	15.6051	0.0006	38.3099	0.0019
	27-4				18.6329	0.0013	15.5993	0.0010	38.3203	0.0023
	27-10				18.2710	0.0011	15.5502	0.0010	37.9689	0.0026
	27-13				18.2996	0.0049	15.5488	0.0041	37.9748	0.0100
	27-13 (d.)			Cu–Ni sulphides	18.3066	0.0004	15.5573	0.0004	38.0176	0.0015
Vologochan, OV-29	29-16				18.2684	0.0009	15.5325	0.0008	37.8943	0.0020
	29-17				18.2103	0.0006	15.5222	0.0006	37.8554	0.0015
	29-17			Sulphides	18.1768	0.0005	15.5092	0.0006	37.8021	0.0017
	29-19			Cu–Ni sulphides	18.3240	0.0012	15.5510	0.0012	37.9567	0.0030

(continued)

Table 1 (continued)

Intrusion, borehole number	Sample number	Depth, m	Type of mineralization	Minerals, formula	$^{206}Pb/^{204}Pb$	2σ	$^{207}Pb/^{204}Pb$	2σ	$^{208}Pb/^{204}Pb$	2σ
Sulphides of potentially ore-bearing intrusives										
Mikchangda, MD-48	48-9		Disseminated	Sulphides	18.2657	0.0007	15.5579	0.0007	38.0510	0.0018
Binyuda, C-1	S1-1				17.9390	0.0082	15.4820	0.0067	37.7197	0.0168
	S-1-5			Cu–Ni sulphides	17.9853	0.0005	15.5208	0.0004	37.8351	0.0016
	S2-1		Veinlet		18.0438	0.0004	15.5469	0.0004	38.0525	0.0014
	S2-2				18.1687	0.0005	15.5648	0.0006	38.0320	0.0021
	S3-2		Disseminated	Sulphides	18.0425	0.0011	15.5234	0.0080	37.8778	0.0022
Dyumptalej, TP-43	43-11			Cu–Ni sulphides	17.8483	0.0004	15.4977	0.0004	37.7742	0.0015
Sulphides of weakly ore-bearing intrusives										
Lower Talnakh, TG-31	31 3		Disseminated	Cu–Ni sulphides	18.0657	0.0003	15.5218	0.0003	37.9223	0.0009
	31-9				18.2400	0.0003	15.5376	0.0003	37.9475	0.0011
	31-10				18.0917	0.0003	15.5298	0.0003	37.9385	0.0009
	31-11				18.1453	0.0002	15.5313	0.0002	37.9355	0.0007
	31-11 (d.)				18.1398	0.0003	15.5255	0.0003	37.9162	0.0008
	31-13				18.3171	0.0006	15.5471	0.0006	37.9741	0.0017
	31-16				18.1778	0.0004	15.5344	0.0004	37.9335	0.0012
	31-16 (d.)				18.1631	0.0006	15.5412	0.0008	37.9481	0.0028
Lower Norilsk, NP-37	37-1				18.2732	0.0005	15.5529	0.0006	37.9972	0.0020
	37-3				18.3003	0.0006	15.5758	0.0006	38.1835	0.0017
	37-4				18.1773	0.0003	15.5514	0.0004	38.0017	0.0012
	37-5				18.1837	0.0006	15.5723	0.0006	38.0752	0.0023
	37-6				18.1885	0.0007	15.5528	0.0008	38.0194	0.0028
	37-7				18.0829	0.0002	15.5254	0.0002	37.9211	0.0007
Lower Norilsk, NP-37	37-10		Disseminated	Cu–Ni sulphides	18.0722	0.0006	15.5332	0.0007	38.0625	0.0025
	37-11				18.0788	0.0008	15.5545	0.0010	38.1220	0.0061
	37-13				17.9688	0.0005	15.5836	0.0006	38.0249	0.0019
Zelenaya Griva, F-233	F-233-10				18.2485	0.0005	15.5751	0.0006	38.0926	0.0017
	(d.)				18.2531	0.0006	15.5749	0.0006	38.0923	0.0018

Table 2 Isotopic composition of lead in plagioclases of intrusives in Norilsk-Taimyr Region

Intrusive, borehole number	Sample number	Depth, m	$^{206}Pb/^{204}Pb$	2σ	$^{207}Pb/^{204}Pb$	2σ	$^{208}Pb/^{204}Pb$	2σ
Industrial-ore-bearing intrusives								
Norilsk 1, MN-2	N1-1		18.1322	0.0093	15.5437	0.0069	38.1115	0.0100
	N1-2		18.4708	0.0097	15.5099	0.0083	38.2075	0.0214
	N1-3		18.0334	0.0093	15.5607	0.0070	37.9806	0.0212
	N1-4		18.2838	0.0094	15.5719	0.0070	38.1961	0.0213
	N1-5		18.1521	0.0091	15.5518	0.0070	38.0356	0.0207
	N1-6		18.2387	0.0089	15.5810	0.0070	38.1418	0.0097
	N1-7		18.1053	0.0090	15.6012	0.0071	38.1368	0.0095
	N1-8		18.0964	0.0091	15.6129	0.0071	38.1720	0.0102
	N1-9		18.0925	0.0091	15.5759	0.0071	38.0697	0.0075
	N1-10		18.0237	0.0091	15.5567	0.0071	37.9792	0.0069
	N1-11		19.0254	0.0013	15.5467	0.0011	38.6753	0.0024
	N1-12		18.3295	0.0021	15.5474	0.0019	38.0184	0.0046
	18 1		18.2855	0.0221	15.5787	0.0168	38.2664	0.0442
Norilsk 1, Medvezhij open pit	Kn-97-1		18.0839	0.0194	15.5638	0.0160	38.0614	0.0117
	Kn-97-2		17.9920	0.0197	15.5747	0.0160	38.0344	0.0107
	Kn-97-3		18.0421	0.0183	15.5358	0.0164	37.9380	0.0214
Talnakh, OUG-2	T-1		18.5872	0.0041	15.5606	0.0033	38.2370	0.0084
	T-2		52.2556	0.0866	17.2751	0.0288	53.1118	0.0862
	T-4		18.1442	0.0075	15.5346	0.0063	37.8537	0.0154
	T-5		18.7791	0.0010	15.5454	0.0009	38.2276	0.0024
	T-6		19.1392	0.0008	15.5642	0.0008	38.4955	0.0019
	T-7		18.4102	0.3085	15.9458	0.3781	37.7498	0.8173
	T-8		18.2756	0.0009	15.5259	0.0007	37.9609	0.0019
	T-9		18.3368	0.0107	15.5779	0.0093	38.0376	0.0226
	T-10		18.3805	0.0016	15.5413	0.0013	38.0578	0.0030
	T-11		18.3581	0.0009	15.5566	0.0011	38.0231	0.0034
	T-12		18.0139	0.0015	15.5054	0.0012	37.7528	0.0032

(continued)

Table 2 (continued)

Intrusive, borehole number	Sample number	Depth, m	$^{206}Pb/^{204}Pb$	2σ	$^{207}Pb/^{204}Pb$	2σ	$^{208}Pb/^{204}Pb$	2σ
	T-13		18.3690	0.0028	15.5512	0.0023	38.0585	0.0055
	T-14		18.3459	0.0042	15.5375	0.0036	38.0088	0.0082
	T-15		18.2128	0.0053	15.5515	0.0044	37.9800	0.0105
	T-16		18.2192	0.0037	15.5393	0.0030	37.9454	0.0074
	T-17		18.5776	0.0048	15.5655	0.0038	38.1830	0.0095
	T-18		18.5295	0.0312	15.5491	0.0252	38.2575	0.0619
	T-19 (d.T-14)		18.3303	0.0071	15.5204	0.0061	37.9693	0.0142
Kharaelakh, KZ-844	844-1		18.5031	0.0046	15.5655	0.0038	38.2532	0.0096
	844-2		18.2497	0.0091	15.5866	0.0076	38.0170	0.0182
	844-3,4		19.1691	0.0225	15.6342	0.0187	38.7047	0.0448
	844-3,4 (d.)		19.2635	0.0137	15.6807	0.0113	38.9298	0.0279
	844-6		18.1993	0.0007	15.5211	0.0009	37.8730	0.0029
	844-6 (d.)		18.1725	0.0005	15.5048	0.0006	37.8142	0.0020
	844-7		18.2050	0.0024	15.5440	0.0019	37.9642	0.0048
	844-10,11		18.2281	0.0009	15.5595	0.0008	38.0169	0.0020
	844-15		18.2393	0.0008	15.5431	0.0008	37.9436	0.0022
	844-15 (d.)		18.2477	0.0007	15.5488	0.0007	37.9603	0.0020
Kharaelakh, TG-21	TG-21-3		18.4601	0.0020	15.5502	0.0017	38.0524	0.0040

(continued)

Table 2 (continued)

Intrusive, borehole number	Sample number	Depth, m	$^{206}Pb/^{204}Pb$	2σ	$^{207}Pb/^{204}Pb$	2σ	$^{208}Pb/^{204}Pb$	2σ
Ore-bearing intrusives								
Chernogorsk, MP-2bis	CH_11	216–1649	18.2569	0.0011	15.5110	0.0009	37.8769	0.0021
	CH-13	254–3235	18.3195	0.0017	15.5955	0.0015	38.1290	0.0042
Zub-Marksheidersky, MP-27	27 1	208–1641	19.4341	0.0007	15.5812	0.0050	38.7181	0.0015
	27 3	209–1642	18.4987	0.0006	15.5350	0.0005	38.0509	0.0001
	27 4	210–1643	18.1786	0.0018	15.5053	0.0015	37.8336	0.0037
	27-4 (d.)	275–3256	18.4230	0.0176	15.4422	0.0141	37.5657	0.0300
	27 5	211–1644	18.4288	0.0018	15.5618	0.0014	38.0847	0.0036
	27 7	212–1645	18.6807	0.0027	15.5778	0.0022	38.1369	0.0051
	27 10	213–1646	18.3113	0.0013	15.5054	0.0010	37.8748	0.0024
	27 13	214–1647	18.3281	0.0018	15.5290	0.0015	37.9465	0.0033
	27 14	215–1648	18.2127	0.0007	15.5086	0.0005	37.8404	0.0019
Vologochan, OV-29	29-2,3	227–1660	18.3677	0.0004	15.5518	0.0004	37.9830	0.0012
	29-2,3	182–1615	18.4820	0.0175	15.5870	0.0158	38.0200	0.0372
	29-5,6	228–1661	18.3308	0.0021	15.5252	0.0017	37.9259	0.0043
	29-5,6	183–1616	18.5550	0.0224	15.6310	0.0152	38.1130	0.0384
	29 8	229–1662	19.4997	0.0064	15.7440	0.0050	38.0514	0.0123
	29 8	184–1617	18.1770	0.0863	15.4010	0.0688	37.4270	0.0173
	29 9	230–1663	18.1964	0.0021	15.5190	0.0017	37.8655	0.0042
	29 9	185–1618	18.2220	0.0176	15.5590	0.0159	37.9210	0.0399
	29-16	231–1664	18.1880	0.0015	15.5152	0.0012	37.8581	0.0029
	29-16	186–1619	18.2520	0.0178	15.5890	0.0165	37.8870	0.0391
	29-17	232–1665	18.2646	0.0034	15.5165	0.0028	37.9009	0.0065
	29-17	187–1620	18.4530	0.0307	15.5410	0.0205	37.4760	0.0434

(continued)

Table 2 (continued)

Intrusive, borehole number	Sample number	Depth, m	$^{206}Pb/^{204}Pb$	2σ	$^{207}Pb/^{204}Pb$	2σ	$^{208}Pb/^{204}Pb$	2σ
Potentially ore-bearing intrusives								
Mikchangda, MD-48	48-9	262–3243	18.2171	0.0412	15.5032	0.0273	37.8294	0.0693
	48-9	217–1650	18.2506	0.0008	15.5817	0.0006	38.0772	0.0018
	48-16	263–3244	18.2021	0.0010	15.5289	0.0012	37.9256	0.0039
	48-18	264–3245	18.2720	0.0699	15.5169	0.0365	37.9602	0.0128
	48-25	265–3246	18.1783	0.0008	15.5150	0.0008	37.8749	0.0019
	48-27,28	266–3247	18.2257	0.0006	15.5135	0.0005	37.8745	0.0016
Binyuda, S1	S1-1	189–1622	18.1490	0.0171	15.5340	0.0153	38.0370	0.0340
	S1-1	234–1667	18.1239	0.0009	15.5053	0.0008	37.9101	0.0019
	S1-2	235–1668	17.8515	0.0006	15.4857	0.0005	37.7301	0.0013
	S1-2	190–1623	17.8020	0.0850	15.3510	0.0990	37.5280	0.0990
	S1-3	236–1669	17.8206	0.0006	15.4744	0.0005	37.6772	0.0016
	S1-3	191–1624	17.8520	0.0170	15.5360	0.0160	37.2280	0.0360
	S1-5	237–1670	18.0346	0.0005	15.5043	0.0005	37.8237	0.0019
	S1-5	192–1626	17.9510	0.0150	15.4930	0.0130	37.7190	0.0320
Binyuda, S3	S3-2	238–1671	18.0589	0.0009	15.4942	0.0008	37.8285	0.0019
	S3-2	193–1626	18.3420	0.0250	15.6110	0.0220	37.9220	0.0510

(continued)

Table 2 (continued)

Intrusive, borehole number	Sample number	Depth, m	$^{206}Pb/^{204}Pb$	2σ	$^{207}Pb/^{204}Pb$	2σ	$^{208}Pb/^{204}Pb$	2σ
Weakly ore-bearing intrusives								
Lower Talnakh, TG-31	31 1	172–1605	18.3275	0.0021	15.5565	0.0018	38.1118	0.0043
	31 1	194–1627	18.4010	0.0160	15.5720	0.0120	37.9000	0.0210
	31 1	239–1672	18.3193	0.0005	15.5308	0.0004	38.0472	0.0010
	313	177–1610	17.9877	0.0026	15.4913	0.0022	37.8575	0.0066
	313	195–1628	18.1650	0.0160	15.5770	0.0130	37.8750	0.0510
	313	240–1673	18.0045	0.0006	15.4926	0.0006	37.8613	0.0018
	317	173–1606	18.2433	0.0181	15.5656	0.0159	38.0881	0.0383
	317	196–1629	18.3720	0.0340	15.5720	0.0022	37.9140	0.0560
	317	241–1674	18.1740	0.0029	15.5214	0.0026	37.9862	0.0060
	31-9	178–1611	17.7569	0.0006	15.4973	0.0005	37.5485	0.0012
	31-9	197–1630	18.3070	0.0680	15.5870	0.0450	37.9010	0.1110
	31-9	242–1675	18.1559	0.0019	15.5151	0.0016	37.9756	0.0037
	31 10	174–1607	18.0150	0.0029	15.5309	0.0025	37.9750	0.0058
	31 10	198–1631	18.1790	0.0400	15.5700	0.0260	37.7710	0.0690
	31 10	243–1676	17.9569	0.0011	15.4850	0.0010	37.8304	0.0024
	31-11	179–1612	17.9717	0.0005	15.4640	0.0004	37.7716	0.0010
	31-11	199–1632	18.0240	0.0200	15.5160	0.0170	37.8860	0.0390
	31-11	244–1677	18.0002	0.0011	15.4885	0.0009	37.8436	0.0022
	31-13	180–1613	18.1184	0.0007	15.4919	0.0006	37.8985	0.0015
	31-13	200–1633	18.1860	0.0240	15.5470	0.0180	37.9200	0.0420
	31-13	245–1678	18.1835	0.0014	15.5262	0.0011	38.0640	0.0003
	31-16	181–1614	17.9753	0.0023	15.4671	0.0018	37.7552	0.0045
	31-16	201–1634	17.1760	0.0160	15.4530	0.0320	36.9990	0.0330
	31-16	246–1679	18.0522	0.0010	15.5050	0.0008	37.8711	0.0021

(continued)

Table 2 (continued)

Intrusive, borehole number	Sample number	Depth, m	$^{206}Pb/^{204}Pb$	2σ	$^{207}Pb/^{204}Pb$	2σ	$^{208}Pb/^{204}Pb$	2σ
Zelenaya Griva, F-233	F-233-2	256–3237	18.0230	0.0060	15.5286	0.0052	38.0150	0.0129
	F-233-7	257–3238	18.3846	0.0085	15.5362	0.0073	38.0173	0.0173
	7 (d.)	258–3239	18.3058	0.0204	15.4939	0.0178	37.9016	0.0393
	F-233-10	259–3240	18.3500	0.0016	15.5385	0.0015	38.4178	0.0429
Kruglogorsky, MP-2 bis	K-4	272–3253	18.4646	0.0265	15.5205	0.0229	37.9263	0.0544
	K-8	273–3254	18.2692	0.0030	15.5219	0.0026	37.8873	0.0064
	K-9	274–3255	18.3381	0.0403	15.4917	0.0293	37.9337	0.0968
Gudchikha picrites, KhS-51	KhS51/2	175–1608	18.2060	0.0023	15.5996	0.0019	37.8941	0.0046
	KhS51/3	176–1609	18.2156	0.0010	15.5776	0.0009	37.8331	0.0022

2 Results and Discussion

Summarized data isotopic composition of mineral fractions are plotted in coordinates of lead isotope ratios (Table 3 and Figs. 3 and 4). They clearly show the difference of lead isotopic composition in the Norilsk, Kharaelakh and Talnakh intrusive bodies. Ores of the Norilsk intrusion differ from the Talnakh and Kharaelakh ores by both lower $^{206}Pb/^{204}Pb$ ratios, and elevated $^{208}Pb/^{206}Pb$ ratios lying above the Stacey-Kramers model curve.

Figure 3 clearly shows that lead isotopic composition in the Talnakh massive ores differs from that in disseminated ores by a regular shift to the area of higher $^{206}Pb/^{204}Pb$ values. Specificity of the Norilsk ores from the low-sulphide horizon shows up in lower values of $^{207}Pb/^{204}Pb$ and $^{208}Pb/^{206}Pb$ ratios. All these differences in lead isotopic composition reflect the diversity of ore matter sources, which appear to be different not only for different intrusions, but sometimes also for different ore types within a single intrusive body, for example, in the case of three types of the Kharaelakh ores (disseminated, massive and ores of the low-sulphide horizon). Thus, lead isotopic composition in sulphide ores at three main commercial deposits in the Norilsk district is different, but always corresponds to the crustal-mantle mixtures dominated by the crustal component. This shows up in the fact that in the Pb–Pb diagram of isotopic systematics they lie under the Stacey-Kramers curve being shifted to the evolution curve of lead isotopic composition in the mantle (Fig. 5).

At the same time, lead isotopic composition in disseminated and massive ores of the same Kharaelakh deposit (Fig. 5) varies in the same limits; for the Talnakh deposit, it is also very close to it. This indicates that ore matter could come to the magmatic chambers of such ore bodies from the same source, but the isotopic composition of this source evolved over time (accumulation of radiogenic isotopes with a uniform U/Pb ratio). The similarity of lead isotopic composition in disseminated sulphides and massive ores at nigh-grade deposits can be used as an additional isotopic criterion of the productivity of intrusions with unknown mineralization.

Lead isotopic composition in both sulphides, and plagioclases (Tables 1 and 2) corresponds to the three-component crustal-mantle mixture dominated by the crustal component. The combined isotope systematics of lead and sulphur enables, presumably, to identify these three components. One of them is, possibly, mantle matter characterized by a slightly elevated μ value ($^{238}U/^{204}Pb$) in case of a "mantle" isotopic composition of sulphur. Two other components are crustal, and with identical isotopic compositions of sulphur ($\delta^{34}S$ 11–12‰), they differ significantly in μ value. At the same time, they have different values of Th/U ratio. One of the crustal components is characterized by an elevated Th/U ratio in rocks of the granulite metamorphic facies or in the lower crust matter; and the other component has a value of this ratio, which is virtually identical to that in the upper crust.

Difference between lead isotopic composition in sulphides and plagioclases clearly indicates that in commercially mineralized intrusions the ore matter and

Table 3 Results of microprobe analysis and determination of isotopic composition of lead in sulphides using LA-ICPI-MS

Borehole number/Depth, m	S	Fe	Co	Ni	Cu	Mineral, formula	$^{206}Pb/^{204}Pb$	2σ	$^{207}Pb/^{204}Pb$	2σ	$^{208}Pb/^{204}Pb$	2σ
Kharaelakh intrusive, Oktyabrskoe deposit												
Western flank												
ZF-13/441.9–442.5	54.36	45.64				Pyrite	18.104	0.019	15.458	0.016	37.645	0.044
	54.28	45.72					18.112	0.021	15.464	0.018	37.669	0.046
	54.72	45.28					18.108	0.025	15.451	0.021	37.626	0.052
	54.22	45.78					18.087	0.025	15.438	0.022	37.579	0.054
	54.87	45.13					18.101	0.026	15.455	0.022	37.623	0.056
	54.93	45.07					18.106	0.027	15.453	0.023	37.647	0.059
	54.94	45.06					18.116	0.027	15.455	0.023	37.647	0.059
	55.14	44.86					18.087	0.028	15.438	0.024	37.594	0.062
	54.32	45.68					18.095	0.030	15.432	0.025	37.589	0.067
	54.59	45.41					18.113	0.036	15.440	0.031	37.602	0.076
	54.76	45.24					18.090	0.037	15.447	0.031	37.584	0.080
	54.82	45.18					18.027	0.039	15.375	0.033	37.448	0.084
	55.00	45.00					18.097	0.041	15.452	0.034	37.627	0.105
	55.00	45.00					18.101	0.041	15.455	0.035	37.621	0.088
	55.03	44.97					18.142	0.043	15.490	0.036	37.717	0.090
	54.34	45.66					18.067	0.046	15.416	0.038	37.540	0.101
	54.39	45.61					18.117	0.050	15.464	0.043	37.667	0.110
	55.04	44.96					18.072	0.052	15.427	0.043	37.581	0.115
	55.20	44.80					18.168	0.059	15.511	0.050	37.795	0.123
	54.61	45.39					18.057	0.066	15.414	0.056	37.532	0.141
	35.61	31.29			33.10	Chalcopyrite	18.087	0.066	15.429	0.056	37.589	0.142
	54.44	45.56				Pyrite	18.024	0.069	15.366	0.059	37.391	0.154
	54.92	41.98	3.10				18.066	0.070	15.428	0.059	37.566	0.146
	55.05	44.95					18.222	0.073	15.549	0.061	37.881	0.154
	55.43	44.57					18.206	0.084	15.524	0.070	37.832	0.180
	54.91	45.09					18.165	0.084	15.509	0.070	37.775	0.178
	54.66	45.34					18.084	0.088	15.429	0.074	37.568	0.183
	54.43	45.57					18.079	0.093	15.430	0.079	37.563	0.194

(continued)

Table 3 (continued)

Borehole number/Depth, m	S	Fe	Co	Ni	Cu	Mineral, formula	$^{206}Pb/^{204}Pb$	2σ	$^{207}Pb/^{204}Pb$	2σ	$^{208}Pb/^{204}Pb$	2σ
	55.77	44.23					18.071	0.107	15.430	0.091	37.586	0.228
	55.00	45.00					17.922	0.134	15.299	0.112	37.029	0.306
	54.46	45.54					17.915	0.136	15.283	0.116	37.227	0.291
	54.46	45.54					18.308	0.158	15.627	0.134	38.082	0.342
	55.34	44.66					17.971	0.195	15.341	0.167	37.364	0.415
	54.61	45.39					18.187	0.216	15.505	0.181	37.846	0.457
ZF-13/448	52.40	47.60					18.093	0.019	15.450	0.017	37.629	0.045
	53.12	46.88					18.097	0.020	15.451	0.017	37.632	0.050
	52.40	47.60					18.089	0.021	15.445	0.018	37.603	0.045
	52.98	47.02					18.087	0.022	15.443	0.018	37.610	0.056
	52.45	47.55					18.090	0.022	15.448	0.019	37.605	0.048
	53.60	46.40					18.097	0.023	15.448	0.019	37.614	0.055
	52.35	47.65					18.094	0.023	15.451	0.020	37.619	0.050
	52.53	47.47					18.112	0.024	15.465	0.020	37.670	0.056
	53.60	46.40					18.070	0.025	15.423	0.021	37.552	0.054
	53.10	46.90					18.238	0.025	15.486	0.021	37.756	0.060
	52.58	47.42					18.088	0.025	15.438	0.022	37.577	0.057
	52.31	47.69					18.106	0.029	15.449	0.025	37.628	0.063
	53.02	46.98					18.068	0.036	15.436	0.031	37.594	0.078
	52.58	47.42					18.029	0.073	15.397	0.061	37.490	0.161
	52.59	47.41					18.260	0.446	15.449	0.381	37.663	0.934
ZF-13/478.4–479	52.02	47.98					18.276	0.021	15.470	0.018	37.712	0.048
	52.28	47.72					18.312	0.036	15.493	0.030	37.764	0.074
	52.46	47.54					18.322	0.045	15.470	0.038	37.728	0.091
	52.56	47.44					18.250	0.053	15.456	0.045	37.682	0.112
	53.14	46.86					18.293	0.069	15.531	0.058	37.858	0.148
	52.43	47.57					18.397	0.085	15.549	0.071	37.936	0.176
	52.38	47.62					18.333	0.093	15.514	0.082	37.911	0.215
	53.02	46.98					18.273	0.104	15.474	0.088	37.756	0.214
	53.10	46.90					18.120	0.119	15.336	0.098	37.403	0.246
	52.73	47.27					18.364	0.119	15.554	0.099	37.908	0.247

(continued)

Table 3 (continued)

Borehole number/Depth, m	S	Fe	Co	Ni	Cu	Mineral, formula	$^{206}Pb/^{204}Pb$	2σ	$^{207}Pb/^{204}Pb$	2σ	$^{208}Pb/^{204}Pb$	2σ
	52.88	47.12					18.307	0.124	15.480	0.104	37.691	0.259
	52.28	47.72					18.344	0.129	15.547	0.106	37.940	0.265
	53.60	46.40					18.085	0.136	15.318	0.114	37.349	0.284
	52.94	47.06					18.334	0.141	15.517	0.119	37.866	0.303
	52.79	47.21					18.379	0.162	15.552	0.135	37.955	0.339
	52.22	47.78					18.442	0.204	15.602	0.167	38.076	0.424
	52.41	47.59				Pyrite	18.579	0.226	15.711	0.188	38.338	0.486
	53.01	46.99					18.213	0.306	15.377	0.254	37.585	0.641
	52.86	47.14					18.427	0.354	15.614	0.295	37.933	0.712
	53.11	46.89					18.338	0.490	15.495	0.423	37.876	1.047
	52.68	47.32					18.275	0.506	15.545	0.413	37.604	1.054
	52.67	47.33					18.097	0.774	15.326	0.623	37.534	1.634
	52.06	47.94					18.147	0.794	15.310	0.650	37.468	1.501
ZF-18/467.5	33.66	28.01		38.33		Pentlandite	18.169	0.022	15.477	0.019	37.680	0.047
	47.90	42.62			9.48	Chalcopyrite	18.167	0.023	15.473	0.020	37.663	0.049
	52.56	46.55		0.89		Pyrite	18.176	0.025	15.484	0.022	37.703	0.052
	34.92	33.10			31.98	Chalcopyrite	18.185	0.025	15.484	0.022	37.683	0.053
	36.16	33.63			30.21		18.162	0.032	15.467	0.027	37.659	0.066
	53.05	46.95				Pyrite	18.189	0.032	15.478	0.028	37.697	0.068
	53.65	46.35					18.150	0.036	15.454	0.031	37.633	0.078
	38.43	35.33			26.24	Chalcopyrite	18.111	0.048	15.422	0.040	37.572	0.104
	34.98	33.14			31.87		18.107	0.088	15.411	0.075	37.559	0.189
	53.26	46.74				Pyrite	18.142	0.209	15.442	0.176	37.599	0.438
	53.05	46.95					18.070	0.212	15.239	0.177	37.291	0.451
	52.86	45.78		1.37			18.248	0.256	15.044	0.213	37.129	0.531
	53.50	46.50					18.417	0.326	15.127	0.243	37.425	0.664
	52.69	45.57	1.74			Pentlandite	18.646	0.328	15.820	0.265	38.918	0.698
	35.13	33.70			31.18	Chalcopyrite	18.489	0.332	15.715	0.271	38.255	0.684
	53.31	46.69				Pyrite	18.954	0.395	14.254	0.326	36.691	0.824
	37.19	44.62			18.19	Chalcopyrite	18.569	0.445	15.804	0.372	38.446	0.930

(continued)

Table 3 (continued)

Borehole number/Depth, m	S	Fe	Co	Ni	Cu	Mineral, formula	$^{206}Pb/^{204}Pb$	2σ	$^{207}Pb/^{204}Pb$	2σ	$^{208}Pb/^{204}Pb$	2σ
	53.41	44.48	2.11			Pyrite	18.427	0.495	15.693	0.409	38.491	1.080
	43.35	40.23			16.42	Chalcopyrite	18.037	0.680	15.339	0.555	37.463	1.433
KZ-931/643.9	36.80	63.20			31.24		18.307	0.033	15.519	0.028	37.884	0.071
	37.06	62.94			31.48		18.337	0.064	15.543	0.054	37.891	0.133
	34.29	34.47			30.81		18.292	0.070	15.507	0.058	37.806	0.148
	34.14	34.38			30.90		18.626	0.110	15.783	0.091	38.467	0.230
	34.89	34.30				Pyrrhotite	18.285	0.026	15.507	0.022	37.794	0.057
	37.24	62.76					18.326	0.032	15.558	0.027	37.963	0.066
	37.40	62.60					18.494	0.075	15.699	0.063	38.306	0.161
	38.45	61.55					18.227	0.089	15.488	0.074	37.702	0.190
	37.48	62.52					18.348	0.091	15.582	0.077	37.996	0.187
	36.70	63.30					18.428	0.093	15.634	0.078	38.152	0.193
	37.22	62.78					18.401	0.102	15.596	0.089	38.045	0.211
	34.67	34.43					18.436	0.104	15.633	0.090	38.190	0.222
	37.27	62.73					18.508	0.110	15.695	0.095	38.308	0.231
	36.46	63.54					18.314	0.110	15.595	0.093	37.989	0.229
	37.09	62.34		0.57			18.468	0.111	15.666	0.093	38.235	0.236
	37.52	62.48					18.437	0.112	15.621	0.093	38.181	0.238
	37.12	62.88					18.199	0.115	15.430	0.100	37.642	0.238
	37.26	62.74					18.358	0.120	15.572	0.100	38.027	0.250
	37.62	62.38					18.424	0.121	15.629	0.103	38.131	0.256
	37.33	62.67					18.388	0.123	15.599	0.100	38.064	0.256
	37.63	62.37					18.302	0.123	15.562	0.103	37.927	0.258
	37.46	62.54					18.472	0.124	15.682	0.103	38.223	0.261
	37.82	62.18					18.560	0.128	15.754	0.108	38.427	0.267
	37.00	63.00					18.404	0.128	15.615	0.108	38.080	0.274
	36.83	63.17					18.563	0.133	15.748	0.111	38.459	0.275

(continued)

Table 3 (continued)

Borehole number/Depth, m	S	Fe	Co	Ni	Cu	Mineral, formula	$^{206}Pb/^{204}Pb$	2σ	$^{207}Pb/^{204}Pb$	2σ	$^{208}Pb/^{204}Pb$	2σ
	37.25	62.75					18.655	0.135	15.835	0.117	38.614	0.282
	37.93	62.07					18.405	0.136	15.638	0.113	38.083	0.284
	37.55	61.53		0.91			18.663	0.138	15.832	0.114	38.660	0.283
	37.10	62.90					18.747	0.144	15.922	0.120	38.860	0.301
	37.16	62.84					18.415	0.146	15.633	0.125	38.120	0.306
	38.00	62.00					18.503	0.147	15.719	0.126	38.366	0.315
	37.27	62.73					18.347	0.148	15.590	0.125	37.961	0.309
	36.79	62.62		0.59			18.422	0.152	15.650	0.127	38.097	0.316
	37.64	62.36					18.565	0.152	15.773	0.128	38.414	0.317
	37.22	62.78					18.209	0.154	15.464	0.127	37.718	0.319
	36.95	63.05					18.478	0.154	15.673	0.131	38.231	0.324
	37.38	62.62					18.477	0.168	15.684	0.141	38.255	0.349
	37.10	62.90					18.578	0.169	15.725	0.138	38.387	0.352
	36.98	63.02					18.458	0.171	15.655	0.140	38.173	0.364
	37.49	62.51					18.777	0.173	15.930	0.143	38.872	0.361
	37.74	62.26				Pyrrhotite	18.523	0.176	15.729	0.148	38.358	0.369
	36.93	61.66					18.328	0.177	15.547	0.147	37.967	0.372
	37.22	62.78					18.633	0.178	15.840	0.151	38.569	0.374
	37.36	62.64					18.521	0.179	15.651	0.154	38.154	0.386
	36.88	63.12					18.639	0.181	15.812	0.151	38.614	0.389
	37.75	62.25					18.755	0.187	15.929	0.160	38.819	0.392
	36.80	61.66		1.55			18.563	0.196	15.734	0.166	38.364	0.406
	37.33	62.67					18.671	0.200	15.830	0.159	38.687	0.393
	37.29	62.71					18.580	0.210	15.805	0.174	38.414	0.437
	37.23	62.77					18.663	0.212	15.843	0.177	38.672	0.455
	36.89	63.11					18.806	0.225	15.917	0.190	38.920	0.466
	37.20	62.80					18.695	0.228	15.920	0.196	38.790	0.478
	36.51	63.49					18.559	0.230	15.746	0.193	38.347	0.483
	36.77	63.23					18.811	0.233	15.915	0.196	38.801	0.492
	36.73	63.27					18.676	0.243	15.866	0.207	38.673	0.510
	37.01	62.99					18.564	0.252	15.705	0.206	38.256	0.518

(continued)

Table 3 (continued)

Borehole number/Depth, m	S	Fe	Co	Ni	Cu	Mineral, formula	$^{206}Pb/^{204}Pb$	2σ	$^{207}Pb/^{204}Pb$	2σ	$^{208}Pb/^{204}Pb$	2σ
	37.74	62.26					18.734	0.344	15.919	0.296	38.775	0.727
	36.76	63.24					18.736	0.503	15.922	0.409	38.553	1.058
KZ-952/1010.4	33.28	33.19			31.21	Chalcopyrite	18.176	0.039	15.555	0.043	37.619	0.181
	35.21	33.58			30.68		18.223	0.073	15.557	0.062	37.862	0.151
	33.90	30.72			31.42		18.139	0.075	15.478	0.064	37.676	0.160
	35.02	34.31			30.45		18.188	0.080	15.529	0.068	37.771	0.174
	34.06	34.52			30.94		18.138	0.081	15.476	0.067	37.694	0.184
	34.78	34.77			31.49		18.341	0.083	15.629	0.071	38.129	0.180
	34.38	34.68			30.89		18.175	0.084	15.518	0.072	37.767	0.176
	34.69	33.83			30.04		18.341	0.090	15.642	0.074	38.114	0.189
	34.80	34.32			31.39		18.365	0.095	15.652	0.082	38.161	0.202
	32.93	33.44			30.75		18.230	0.099	15.563	0.083	37.874	0.211
	35.07	34.89			30.89		18.301	0.100	15.614	0.085	38.021	0.210
	34.00	34.61			30.80		18.100	0.103	15.451	0.088	37.630	0.217
	34.49	34.76			31.31		18.230	0.105	15.546	0.089	37.885	0.221
	34.42	34.69			31.50		18.275	0.107	15.605	0.092	37.982	0.225
	34.73	34.47			29.92		18.303	0.107	15.627	0.090	38.038	0.231
	34.74	33.96			31.17		18.229	0.109	15.548	0.093	37.845	0.233
	34.03	34.48			31.41		18.418	0.111	15.707	0.095	38.254	0.237
	34.59	35.49			31.24		18.474	0.115	15.769	0.098	38.393	0.243
	34.97	33.86			31.37		18.339	0.116	15.636	0.099	38.086	0.244
	35.00	33.58			31.23		18.369	0.117	15.675	0.099	38.183	0.245
	32.85	32.86			31.37		18.343	0.121	15.665	0.102	38.170	0.258
	35.15	33.61			33.22		18.198	0.122	15.516	0.102	37.782	0.258
	35.17	33.46			31.04		18.342	0.123	15.653	0.103	38.112	0.259
	35.34	33.43			30.85		18.434	0.127	15.742	0.112	38.314	0.271
	35.07	33.56			29.97		18.339	0.127	15.659	0.107	38.114	0.274

(continued)

Table 3 (continued)

Borehole number/Depth, m	S	Fe	Co	Ni	Cu	Mineral, formula	$^{206}Pb/^{204}Pb$	2σ	$^{207}Pb/^{204}Pb$	2σ	$^{208}Pb/^{204}Pb$	2σ
	33.99	32.79			30.98		18.270	0.127	15.588	0.108	37.941	0.268
	34.25	34.70			32.27		18.251	0.127	15.562	0.109	37.917	0.280
	34.58	34.57			30.76		18.287	0.128	15.629	0.109	37.997	0.277
	35.34	34.69			31.13		18.435	0.131	15.726	0.111	38.288	0.276
	34.60	34.42			31.54		18.464	0.132	15.738	0.111	38.347	0.281
	34.14	33.58			31.33		18.191	0.132	15.506	0.112	37.783	0.272
	35.14	34.11			31.09		18.318	0.136	15.635	0.113	38.046	0.290
	35.45	33.43			31.56		18.385	0.139	15.672	0.117	38.180	0.292
	34.86	33.60			29.97		18.418	0.140	15.713	0.120	38.289	0.293
	34.36	34.31			30.85		18.384	0.145	15.689	0.126	38.215	0.309
	34.11	34.80			31.20		18.541	0.149	15.824	0.127	38.492	0.310
	34.48	33.96			31.46		18.366	0.151	15.672	0.130	38.135	0.317
	34.73	35.29			30.44		18.310	0.154	15.621	0.132	38.028	0.324
	34.17	34.97			30.90		18.389	0.161	15.719	0.136	38.193	0.340
	34.75	34.05			31.69		18.628	0.161	15.904	0.135	38.727	0.336
	34.77	33.77			31.03		18.402	0.166	15.713	0.141	38.247	0.351
	35.21	34.35			31.42		18.383	0.172	15.699	0.148	38.210	0.366
	34.57	34.53			31.26		18.291	0.174	15.597	0.146	37.947	0.362
	34.22	34.09			29.55		18.439	0.193	15.761	0.168	38.356	0.411
	36.39	55.85			30.37		18.764	0.209	15.992	0.176	38.987	0.444
	34.70	34.27			31.21		18.596	0.214	15.883	0.177	38.616	0.458
	34.30	34.28			31.10		18.693	0.245	15.976	0.218	38.826	0.519
	34.78	33.96			31.88		18.491	0.327	15.777	0.279	38.373	0.694
	35.86	34.59			30.88		17.085	0.328	14.545	0.282	35.551	0.723
	35.05	34.59		33.53		Pentlandite	18.121	0.032	15.466	0.028	37.635	0.072
	34.95	33.84		35.38			18.183	0.058	15.515	0.049	37.784	0.126
	33.09	33.06		33.62			18.209	0.089	15.530	0.075	37.821	0.189
	33.82	35.08		34.29			18.317	0.111	15.628	0.096	38.055	0.234
	34.22	33.90		33.84			18.434	0.236	15.766	0.225	38.172	0.492
	34.91	34.21		7.75		Pyrrhotite	18.446	0.164	15.640	0.136	38.119	0.344

(continued)

Table 3 (continued)

Borehole number/Depth, m	S	Fe	Co	Ni	Cu	Mineral, formula	$^{206}Pb/^{204}Pb$	2σ	$^{207}Pb/^{204}Pb$	2σ	$^{208}Pb/^{204}Pb$	2σ
Central part												
KZ-1112/1094.8	35.58	32.26			32.72	Chalcopyrite	17.955	0.150	15.480	0.043	36.844	0.311
	35.40	33.03			32.40		18.103	0.163	15.424	0.132	37.514	0.334
	35.08	32.83			31.76		17.862	0.221	15.202	0.190	36.965	0.485
	35.12	32.50			31.94		18.501	0.224	15.727	0.188	38.390	0.471
	35.16	32.62			32.15		18.147	0.234	15.438	0.189	37.682	0.501
	33.27	30.74	2.01	33.71	0.00	Pentlandite	18.189	0.244	16.403	0.220	38.697	0.510
	35.51	32.26			33.16	Chalcopyrite	18.312	0.252	15.483	0.217	37.869	0.535
	34.69	32.48			32.31		18.262	0.259	15.517	0.211	37.876	0.536
	35.41	32.28			32.25		18.589	0.263	15.803	0.216	38.530	0.568
	35.67	32.42			31.79		18.235	0.263	15.532	0.232	37.754	0.560
	35.32	32.60			32.08		18.463	0.284	15.681	0.237	38.290	0.606
	32.61	30.65	2.17	33.87	0.00	Pentlandite	18.724	0.321	15.949	0.273	38.839	0.690
	35.45	32.74			31.95	Chalcopyrite	17.722	0.326	15.085	0.264	36.679	0.683
	34.40	30.96	1.69	26.77	6.47	Pentlandite	18.487	0.334	15.627	0.275	41.937	0.730
	35.26	32.61			32.23	Chalcopyrite	18.105	0.354	15.427	0.289	37.622	0.759
	35.55	32.18			32.40		18.424	0.355	15.702	0.312	38.201	0.764
	34.59	31.76			32.16		18.631	0.372	15.810	0.332	38.583	0.808
	35.17	32.42			32.14		18.597	0.400	15.840	0.329	38.548	0.837
	34.79	32.10			32.55		18.576	0.446	15.809	0.376	38.594	0.907
	34.74	32.14			32.48		18.198	0.463	15.455	0.385	37.740	1.013
	35.26	32.18			32.23		18.559	0.485	15.783	0.420	38.213	0.979
	35.27	32.93			32.76		18.409	0.489	15.740	0.412	38.129	1.059
	35.10	32.37			31.80		18.248	0.498	15.397	0.436	37.631	0.911
	35.29	32.17			31.93		18.963	0.500	16.166	0.431	39.373	1.060
	35.33	32.65			32.00		18.732	0.510	15.946	0.427	38.955	1.092

(continued)

Table 3 (continued)

Borehole number/Depth, m	S	Fe	Co	Ni	Cu	Mineral, formula	$^{206}Pb/^{204}Pb$	2σ	$^{207}Pb/^{204}Pb$	2σ	$^{208}Pb/^{204}Pb$	2σ
	35.75	31.97			32.65		18.049	0.543	15.119	0.415	36.735	1.064
	35.51	32.82			32.36		18.633	0.555	15.924	0.447	38.806	1.124
	35.17	32.00			32.28		18.392	0.563	15.661	0.465	38.064	1.183
	35.30	32.60			32.19		18.328	0.584	15.538	0.530	37.927	1.275
	35.61	32.23			32.50		18.634	0.623	15.836	0.492	38.490	1.336
	35.04	32.71			32.75		17.503	0.629	14.924	0.532	36.345	1.350
	35.93	32.23			32.03		18.144	0.771	15.714	0.718	37.843	1.686
	35.37	32.69			32.31		18.404	0.791	15.866	0.656	38.322	1.615
	34.99	32.14			31.82		18.741	0.855	15.999	0.710	38.908	1.796
	35.67	32.42			31.81		17.511	0.876	14.942	0.742	36.186	1.868
	35.19	32.57			33.12		17.992	0.922	15.225	0.718	37.110	1.699
	35.51	32.66			32.58		18.152	1.166	15.336	0.985	37.415	2.534
	35.49	32.26			32.31		18.679	1.665	15.814	1.399	38.269	3.490
	35.46	32.06			32.58		18.764	2.767	15.789	2.494	38.354	6.296
KZ-1112/1098.4	34.91	32.58			32.67		18.124	0.034	15.441	0.028	37.583	0.072
	35.41	32.28			32.25		18.143	0.039	15.459	0.033	37.610	0.082
	34.97	31.98			32.60		18.135	0.041	15.453	0.034	37.584	0.083
	34.82	32.48			32.66		18.152	0.041	15.467	0.035	37.633	0.086
	35.50	32.49			32.97		18.090	0.043	15.429	0.036	37.518	0.088
	35.18	32.77			32.50		18.187	0.045	15.485	0.037	37.706	0.098
	35.00	32.70			31.90		18.194	0.048	15.505	0.041	37.784	0.151
	35.60	32.20			32.29		18.175	0.052	15.493	0.042	37.689	0.107
	35.02	32.35			32.47		18.221	0.053	15.538	0.046	37.835	0.113
	35.54	32.33			32.76		18.064	0.057	15.408	0.049	37.494	0.121
	35.06	32.09			32.29		18.220	0.062	15.427	0.052	37.683	0.130
	34.96	32.41			32.17		18.215	0.063	15.525	0.053	37.754	0.131
	34.77	32.31			32.61		18.124	0.066	15.450	0.057	37.592	0.140
	35.80	32.21			32.42		18.355	0.069	15.633	0.057	38.131	0.151
	35.20	32.96			32.40		18.135	0.080	15.459	0.067	37.615	0.172
	35.39	31.93			32.86		18.068	0.082	15.385	0.073	37.452	0.183

(continued)

Table 3 (continued)

Borehole number/Depth, m	S	Fe	Co	Ni	Cu	Mineral, formula	$^{206}Pb/^{204}Pb$	2σ	$^{207}Pb/^{204}Pb$	2σ	$^{208}Pb/^{204}Pb$	2σ
	34.83	32.45			32.02		18.133	0.091	15.461	0.077	37.614	0.194
	35.11	32.46			32.19		18.222	0.095	15.533	0.082	37.760	0.200
	35.67	32.42			31.79		18.171	0.113	15.461	0.096	37.659	0.239
	34.69	32.48			32.31		18.633	0.182	15.603	0.147	38.836	0.379
KZ-1112/1100.4	34.58	31.64			34.53	Chalcopyrite	18.189	0.029	15.532	0.025	37.856	0.062
	34.35	31.88			34.09		18.173	0.030	15.521	0.025	37.827	0.063
	34.16	31.66			34.29		18.180	0.031	15.525	0.026	37.827	0.065
	34.51	31.54			34.36		18.157	0.032	15.508	0.027	37.791	0.067
	34.32	30.93			33.27		18.169	0.032	15.518	0.027	37.805	0.069
	34.08	31.82			34.29		18.185	0.033	15.530	0.028	37.860	0.070
	34.45	32.08			33.74		18.196	0.034	15.543	0.029	37.872	0.071
	34.67	31.72			33.62		18.162	0.034	15.514	0.028	37.803	0.072
	34.67	31.53			34.09		18.187	0.035	15.529	0.030	37.853	0.074
	34.15	31.57			33.96		18.137	0.036	15.490	0.031	37.757	0.078
	34.48	31.41			34.17		18.174	0.040	15.528	0.034	37.834	0.083
	35.14	31.42			34.02		18.145	0.041	15.495	0.035	37.769	0.087
	35.31	30.84			34.80		18.215	0.042	15.549	0.036	37.899	0.087
	34.44	31.47			33.77		18.159	0.042	15.508	0.036	37.801	0.090
	34.46	31.52			34.31		18.164	0.043	15.512	0.036	37.815	0.090
	34.64	31.38			33.76		18.263	0.043	15.588	0.037	38.013	0.090
	34.69	31.55			33.95		18.203	0.043	15.547	0.036	37.894	0.090
	34.70	31.46			33.84		18.267	0.043	15.597	0.037	38.019	0.090
	33.80	31.69			33.46		18.185	0.044	15.530	0.037	37.860	0.091
	34.99	31.79			33.97		18.220	0.044	15.558	0.037	37.928	0.092
	34.70	31.46			33.98		18.165	0.045	15.516	0.038	37.816	0.095
	33.99	31.74			34.12		18.263	0.045	15.592	0.038	38.010	0.097
	34.88	31.25			33.68		18.163	0.046	15.507	0.039	37.810	0.096
	34.63	31.08			33.84		18.213	0.046	15.557	0.039	37.917	0.095
	34.21	31.62			33.71		18.161	0.047	15.505	0.040	37.794	0.099
	35.10	31.16			33.44		18.257	0.050	15.591	0.043	37.994	0.106
	33.94	31.26			34.88		18.201	0.052	15.541	0.044	37.883	0.111

(continued)

Table 3 (continued)

Borehole number/Depth, m	S	Fe	Co	Ni	Cu	Mineral, formula	$^{206}Pb/^{204}Pb$	2σ	$^{207}Pb/^{204}Pb$	2σ	$^{208}Pb/^{204}Pb$	2σ
	34.08	31.05			34.50		18.214	0.053	15.551	0.045	37.898	0.113
	33.91	31.72			34.53		18.204	0.053	15.544	0.045	37.880	0.113
	34.47	31.69			34.78		18.236	0.054	15.570	0.046	37.971	0.113
	34.86	31.05			33.80		18.259	0.055	15.598	0.047	38.051	0.129
	34.55	31.33			34.50		18.365	0.057	15.681	0.048	38.263	0.124
	34.26	32.47			34.65		18.253	0.063	15.585	0.054	37.976	0.135
	34.44	31.85			33.94		18.199	0.064	15.540	0.054	37.890	0.136
	34.58	31.48			33.85		18.181	0.064	15.520	0.055	37.845	0.135
	34.64	31.55			34.51		18.093	0.067	15.457	0.055	37.654	0.142
	34.06	31.29			33.22		18.310	0.071	15.630	0.061	38.118	0.149
	34.30	31.17			33.87		18.308	0.074	15.635	0.064	38.083	0.158
	34.64	30.86			33.77		18.320	0.075	15.644	0.063	38.122	0.155
	34.91	31.45			34.57		18.325	0.076	15.648	0.064	38.176	0.161
	34.15	31.66			33.24		18.225	0.076	15.568	0.064	37.952	0.159
	34.39	31.99			34.75		18.418	0.079	15.714	0.067	38.305	0.168
	34.07	31.40			33.81		18.360	0.080	15.686	0.069	38.193	0.170
	34.34	31.08			33.84		18.242	0.081	15.569	0.068	37.951	0.169
	34.31	31.73			34.19		18.325	0.082	15.651	0.070	38.146	0.175
	33.92	31.30			34.28		18.455	0.103	15.583	0.083	38.220	0.215
	34.34	31.21			34.28		18.333	0.114	15.664	0.096	38.165	0.236
	33.98	31.52			33.64		18.221	0.116	15.566	0.097	37.918	0.243
	34.59	31.45			34.45		18.499	0.142	15.801	0.121	38.470	0.304
	34.91	31.41			34.10		18.578	0.183	15.882	0.159	38.686	0.388
	34.98	31.77			34.18		18.474	0.188	15.777	0.162	38.460	0.398
	34.20	31.51			33.62		18.816	0.201	16.150	0.301	38.746	0.368
	34.30	31.38			34.11		18.612	0.300	15.962	0.258	38.780	0.647
	34.49	32.06			33.47		18.282	0.568	15.526	0.470	37.858	1.160
	39.89	60.11				Pyrrhotite	18.683	0.199	15.932	0.155	38.943	0.430
KZ-1112/1102.4	34.61	31.14			34.25	Chalcopyrite	18.171	0.033	15.520	0.028	37.824	0.070
	34.56	31.63			33.81		18.195	0.036	15.536	0.030	37.871	0.075

(continued)

Table 3 (continued)

Borehole number/Depth, m	S	Fe	Co	Ni	Cu	Mineral, formula	$^{206}Pb/^{204}Pb$	2σ	$^{207}Pb/^{204}Pb$	2σ	$^{208}Pb/^{204}Pb$	2σ
	35.12	31.84			33.04		18.161	0.036	15.511	0.031	37.805	0.076
	35.55	30.69			33.76		18.169	0.038	15.518	0.032	37.817	0.080
	34.81	31.78			33.41		18.181	0.038	15.527	0.032	37.845	0.081
	34.13	32.20			33.67		18.159	0.041	15.511	0.035	37.794	0.086
	34.63	31.41			33.96		18.126	0.042	15.481	0.036	37.722	0.089
	34.38	31.29			34.33		18.191	0.042	15.534	0.036	37.854	0.088
	34.74	31.22			34.03		18.149	0.042	15.498	0.035	37.776	0.089
	34.21	31.66			34.13		18.195	0.043	15.537	0.036	37.866	0.091
	34.67	31.67			33.66		18.173	0.043	15.511	0.038	37.802	0.095
	34.86	31.57			33.57		18.171	0.043	15.520	0.037	37.823	0.091
	34.48	32.01			33.51		18.189	0.045	15.534	0.038	37.859	0.095
	34.89	31.53			33.59	Chalcopyrite	18.180	0.046	15.525	0.040	37.843	0.098
	34.53	31.99			33.48		18.224	0.048	15.566	0.041	37.937	0.101
	34.35	31.51			34.14		18.264	0.048	15.591	0.042	37.994	0.103
	34.52	31.62			33.87		18.132	0.048	15.492	0.040	37.736	0.101
	34.98	31.14			33.88		18.110	0.049	15.460	0.041	37.672	0.103
	35.01	31.07			33.52		18.149	0.049	15.498	0.042	37.790	0.110
	34.93	31.55			33.93		18.249	0.050	15.586	0.042	37.975	0.104
	34.75	31.50			33.75		18.229	0.055	15.567	0.047	37.949	0.115
	35.30	31.07			33.63		18.197	0.055	15.541	0.047	37.876	0.118
	34.51	32.01			33.48		18.073	0.055	15.442	0.048	37.628	0.117
	35.20	31.83			32.97		18.194	0.061	15.533	0.051	37.850	0.128
	34.31	30.97			34.72		18.205	0.065	15.536	0.056	37.892	0.137
	34.15	31.84			34.01		18.162	0.076	15.516	0.065	37.807	0.162
	34.79	31.26			33.94		18.264	0.078	15.593	0.066	38.012	0.167
	34.91	31.18			33.90		18.156	0.080	15.509	0.069	37.780	0.169
	33.72	31.45			34.83		18.448	0.091	15.762	0.078	38.417	0.193
	34.30	32.00			33.71		18.169	0.094	15.513	0.079	37.791	0.202
	33.93	31.62			34.46		18.185	0.098	15.528	0.082	37.849	0.204
	34.30	31.60			34.10		18.329	0.102	15.645	0.086	38.133	0.213

(continued)

Table 3 (continued)

Borehole number/Depth, m	S	Fe	Co	Ni	Cu	Mineral, formula	$^{206}Pb/^{204}Pb$	2σ	$^{207}Pb/^{204}Pb$	2σ	$^{208}Pb/^{204}Pb$	2σ
	34.99	31.45			33.56		18.276	0.103	15.602	0.087	38.016	0.220
	33.98	32.03			33.99		18.256	0.107	15.594	0.091	37.966	0.226
	34.62	32.19			33.18		18.241	0.113	15.580	0.095	37.968	0.237
	33.53	32.16			34.31		18.311	0.114	15.628	0.097	38.085	0.240
	34.59	32.00			33.41		18.450	0.117	15.753	0.097	38.397	0.251
	33.86	31.54			34.59		18.102	0.117	15.446	0.099	37.636	0.248
	35.00	31.56			33.44		18.311	0.119	15.625	0.099	38.091	0.254
	34.10	31.22			34.68		18.333	0.126	15.651	0.105	38.122	0.261
	34.22	31.47			34.31		18.303	0.128	15.627	0.107	38.081	0.273
	34.38	31.54			34.08		18.435	0.143	15.726	0.120	38.327	0.300
	34.35	30.97			34.68		18.563	0.155	15.852	0.135	38.737	0.326
	34.35	30.96			34.69		18.388	0.157	15.685	0.132	38.196	0.330
	34.51	32.32			33.17		18.456	0.165	15.311	0.130	37.579	0.346
	34.11	31.52			34.36		18.293	0.170	15.610	0.143	37.966	0.355
	34.10	31.27			34.63		18.460	0.177	15.759	0.150	38.401	0.371
	34.33	31.60			34.07		18.298	0.182	15.614	0.155	38.013	0.381
	34.70	31.52			33.78		18.557	0.185	15.865	0.159	38.632	0.389
	34.60	31.44			33.96		18.442	0.186	15.730	0.159	38.348	0.402
	34.73	31.44			33.82		18.571	0.197	15.857	0.166	38.616	0.419
	34.51	31.14			34.36		18.404	0.205	15.724	0.179	38.236	0.436
	34.03	31.34			34.63		18.586	0.210	15.827	0.179	38.631	0.453
	34.96	31.64			33.40		18.601	0.228	15.935	0.191	38.739	0.491
	34.76	31.25			33.99		18.413	0.238	15.718	0.205	38.249	0.497
	39.58	59.67		0.75		Pyrrhotite	18.363	0.243	15.775	0.228	38.407	0.543
	34.74	31.60			33.66	Chalcopyrite	18.580	0.255	15.862	0.224	38.696	0.553
	33.94	31.76			34.30		18.546	0.274	15.849	0.203	38.708	0.491
	34.17	31.89			33.93		18.377	0.373	15.580	0.317	38.111	0.791
	35.38	32.49			32.13		18.429	0.396	15.756	0.333	38.225	0.842
	34.47	31.85			33.68		18.335	0.417	15.698	0.366	38.094	0.898
	34.69	31.99			33.32		18.244	0.756	15.396	0.630	37.720	1.563

(continued)

Table 3 (continued)

Borehole number/Depth, m	S	Fe	Co	Ni	Cu	Mineral, formula	$^{206}Pb/^{204}Pb$	2σ	$^{207}Pb/^{204}Pb$	2σ	$^{208}Pb/^{204}Pb$	2σ
KZ-1112/1103.8	34.51	32.35			33.14		18.124	0.039	15.466	0.033	37.674	0.079
	34.37	31.63			34.00		18.324	0.082	15.644	0.068	38.074	0.170
	35.54	30.83			33.63		18.224	0.087	15.544	0.075	37.859	0.184
	33.86	31.14			35.00		18.009	0.100	15.373	0.084	37.458	0.208
	34.63	31.41			33.97		18.111	0.105	15.448	0.087	37.627	0.222
	34.04	31.45			34.51		18.253	0.105	15.562	0.089	37.947	0.216
	34.27	32.04			33.69		18.186	0.107	15.508	0.090	37.809	0.221
	34.27	31.34			34.39		18.486	0.109	15.726	0.093	38.392	0.230
	34.62	32.55			32.83		18.144	0.115	15.447	0.097	37.681	0.245
	34.85	31.67			33.48		18.089	0.117	15.396	0.100	37.554	0.251
	33.68	32.11			34.21		18.445	0.128	15.736	0.108	38.308	0.266
	33.84	32.10			34.07		18.468	0.129	15.762	0.112	38.370	0.272
	34.57	31.14			34.29		18.235	0.133	15.545	0.114	37.896	0.279
	33.83	31.87			34.3		18.368	0.134	15.689	0.114	38.183	0.283
	33.17	32.39			34.44		18.499	0.152	15.794	0.127	38.521	0.318
	34.42	31.03			34.56		18.154	0.154	15.485	0.133	37.705	0.327
	34.09	32.10			33.81		18.456	0.158	15.722	0.131	38.308	0.335
	34.69	31.10			34.21		18.694	0.163	15.958	0.134	38.835	0.342
	34.79	31.04			34.18		18.291	0.165	15.612	0.137	38.005	0.350
	34.58	30.52			34.90	Chalcopyrite	18.563	0.169	15.840	0.143	38.596	0.358
	34.83	31.12			34.05		18.266	0.172	15.563	0.135	37.976	0.341
	33.03	31.71			35.26		18.457	0.173	15.769	0.146	38.362	0.360
	33.34	31.78			34.88		18.454	0.174	15.769	0.146	38.357	0.361
	33.64	31.63			34.73		18.523	0.177	15.782	0.147	38.471	0.372
	34.37	31.94			33.69		18.311	0.180	15.609	0.148	38.036	0.371
	33.77	31.56			34.67		18.510	0.186	15.790	0.154	38.488	0.389
	34.25	30.17			35.58		18.346	0.186	15.631	0.156	38.127	0.391
	34.08	31.42			34.50		18.509	0.191	15.777	0.162	38.459	0.406
	34.28	31.12			34.60		18.605	0.194	15.874	0.167	38.686	0.418

(continued)

Table 3 (continued)

Borehole number/Depth, m	S	Fe	Co	Ni	Cu	Mineral, formula	$^{206}Pb/^{204}Pb$	2σ	$^{207}Pb/^{204}Pb$	2σ	$^{208}Pb/^{204}Pb$	2σ
	33.77	32.66			33.56		18.624	0.196	15.865	0.161	38.748	0.414
	34.08	32.22			33.70		18.558	0.209	15.827	0.175	38.549	0.444
	35.33	31.75			32.92		18.493	0.213	15.784	0.177	38.481	0.442
	33.27	32.33			34.41		18.476	0.213	15.750	0.183	38.453	0.459
	34.69	31.28			34.03		18.744	0.214	15.948	0.183	38.939	0.455
	34.40	31.70			33.90		18.654	0.232	15.887	0.191	38.786	0.478
	33.68	31.63			34.69		18.266	0.257	15.573	0.217	37.971	0.553
	34.30	31.88			33.82		18.536	0.264	15.799	0.220	38.549	0.558
	35.36	31.03			33.62		18.597	0.278	15.832	0.228	38.640	0.588
	34.44	30.95			34.60		18.526	0.282	15.786	0.240	38.476	0.588
	34.41	31.36			34.23		18.678	0.294	15.933	0.251	38.850	0.617
	34.02	32.53			33.45		18.633	0.305	15.891	0.262	38.773	0.648
	33.77	31.99			34.24		18.419	0.313	15.706	0.265	38.288	0.668
	34.70	32.03			33.27		18.658	0.325	15.938	0.277	38.759	0.685
	34.32	31.85			33.83		18.452	0.327	15.734	0.273	38.368	0.688
	33.96	31.61			34.43		18.544	0.328	15.818	0.272	38.556	0.691
	33.74	33.42			32.84		18.620	0.354	15.889	0.301	38.614	0.760
	34.45	31.72			33.83		18.711	0.380	15.995	0.323	38.902	0.796
	34.73	30.81			34.46		18.627	0.401	15.869	0.337	38.671	0.835
	34.73	31.41			33.86		18.708	0.414	15.893	0.353	38.804	0.869
	34.34	32.91			32.75		18.757	0.993	15.970	0.843	38.852	2.090
KZ-981/1.122.4–1123	34.53	33.10			32.37		18.135	0.022	15.456	0.019	37.641	0.047
	34.88	32.75			32.37		18.141	0.023	15.469	0.019	37.683	0.047
	34.37	33.11			32.52		18.154	0.023	15.480	0.020	37.682	0.083
	35.16	32.76			32.09		18.130	0.024	15.457	0.020	37.653	0.049
	54.19	45.81				Pyrite	18.136	0.024	15.462	0.021	37.671	0.054
	34.68	33.18			32.14	Chalcopyrite	18.141	0.025	15.465	0.022	37.664	0.052
	35.48	32.67			31.84		18.154	0.026	15.480	0.022	37.715	0.053
	35.17	33.20			31.64		18.160	0.026	15.485	0.022	37.725	0.054
	34.97	32.14			32.89		18.132	0.026	15.460	0.022	37.654	0.056
	35.24	32.25			32.51		18.142	0.026	15.469	0.023	37.682	0.056
	35.19	31.62			33.19		18.118	0.027	15.448	0.023	37.632	0.058
	34.87	32.88			32.25		18.147	0.028	15.468	0.024	37.680	0.058
	34.88	32.41			32.71		18.147	0.028	15.476	0.024	37.672	0.065

(continued)

Table 3 (continued)

Borehole number/Depth, m	S	Fe	Co	Ni	Cu	Mineral, formula	$^{206}Pb/^{204}Pb$	2σ	$^{207}Pb/^{204}Pb$	2σ	$^{208}Pb/^{204}Pb$	2σ
	35.30	32.37			32.34		18.139	0.029	15.462	0.024	37.672	0.059
	35.16	32.62			32.22		18.112	0.033	15.446	0.028	37.624	0.070
	35.19	32.63			32.18		18.141	0.034	15.467	0.028	37.677	0.098
	34.68	32.69			32.63		18.183	0.034	15.502	0.029	37.763	0.072
	35.04	31.77			33.18		18.103	0.034	15.434	0.029	37.587	0.073
	53.54	46.46				Pyrite	18.122	0.034	15.441	0.029	37.618	0.074
	53.49	46.51					18.105	0.035	15.432	0.029	37.586	0.074
	53.10	46.90					18.142	0.035	15.460	0.030	37.672	0.073
	34.89	32.58			32.53	Chalcopyrite	18.163	0.037	15.489	0.032	37.725	0.093
	35.35	32.83			31.82		18.167	0.037	15.486	0.032	37.720	0.079
	34.45	33.44			32.11		18.149	0.038	15.475	0.032	37.696	0.079
	34.34	32.98			32.68		18.169	0.038	15.490	0.032	37.733	0.080
	52.89	47.11				Pyrite	18.093	0.042	15.415	0.036	37.575	0.088
	34.42	32.80			32.78	Chalcopyrite	18.145	0.043	15.470	0.036	37.684	0.089
	53.35	46.65				Pyrite	18.138	0.046	15.454	0.040	37.655	0.099
	35.94	32.25			31.81	Chalcopyrite	18.095	0.048	15.416	0.041	37.548	0.100
	35.34	33.96			30.69		18.131	0.049	15.457	0.041	37.655	0.104
	54.16	45.84				Pyrite	18.188	0.052	15.496	0.044	37.776	0.111
	53.55	46.45					18.205	0.052	15.515	0.044	15.410	0.113
	53.29	46.71					18.177	0.056	15.465	0.047	37.722	0.120
	35.12	32.10			32.78	Chalcopyrite	18.105	0.058	15.433	0.049	37.619	0.121
	34.52	32.45			33.02		18.122	0.060	15.458	0.051	37.650	0.127
	53.90	46.10				Pyrite	18.130	0.063	15.450	0.054	37.646	0.136
	53.69	46.31					18.045	0.064	15.381	0.054	37.473	0.135
	53.14	46.86				Pyrite	18.098	0.065	15.422	0.056	37.582	0.135
	53.50	46.50					18.184	0.068	15.503	0.058	37.765	0.141
	53.04	46.96					18.103	0.073	15.397	0.062	37.533	0.156
	53.46	46.54					18.162	0.081	15.475	0.069	37.723	0.168
	53.37	46.63					18.335	0.125	15.606	0.105	38.041	0.264
	53.28	46.72					18.519	0.164	15.783	0.140	38.440	0.347
	53.27	46.73					18.394	0.190	15.475	0.159	38.154	0.390

(continued)

Table 3 (continued)

Borehole number/Depth, m	S	Fe	Co	Ni	Cu	Mineral, formula	$^{206}Pb/^{204}Pb$	2σ	$^{207}Pb/^{204}Pb$	2σ	$^{208}Pb/^{204}Pb$	2σ
	51.77	48.23					18.247	0.197	15.510	0.165	37.862	0.412
	53.11	46.89					18.987	0.594	15.622	0.491	38.663	1.223
	53.04	46.96					18.851	0.694	15.876	0.569	39.030	1.446
KZ-1089/1154	34.96	32.49	0.00		32.41	Chalcopyrite	18.162	0.042	15.457	0.036	37.614	0.088
	35.00	32.05	0.00		32.16		18.129	0.046	15.413	0.039	37.482	0.092
	38.31	60.46	0.00	0.44		Pyrrhotite	18.174	0.048	15.478	0.040	37.631	0.102
	38.87	60.84	0.00	0.54			18.169	0.051	15.467	0.044	37.638	0.107
	35.13	31.93	0.00		32.83	Chalcopyrite	18.354	0.055	15.617	0.044	38.016	0.117
	34.86	32.78	0.00		32.61		18.213	0.071	15.504	0.061	37.697	0.150
	34.91	32.07	1.44	26.86	4.60	Pentlandite	18.226	0.093	15.509	0.079	19.322	0.181
	34.71	32.43	0.00		32.00	Chalcopyrite	18.430	0.099	15.690	0.083	38.156	0.213
	38.61	60.76	0.00	0.63		Pyrrhotite	18.064	0.153	15.403	0.131	37.438	0.317
	38.42	60.07	0.00	0.51			17.587	0.228	14.943	0.198	36.493	0.492
	38.08	59.49	0.00	0.63			18.311	0.262	15.579	0.212	37.897	0.526
	38.34	60.44	0.48	0.37			17.952	0.303	15.368	0.255	37.264	0.636
	38.47	60.30	0.00	0.54			18.291	0.324	15.617	0.292	37.970	0.658
	38.34	60.44	0.48	0.37			18.209	0.341	15.537	0.305	37.800	0.734
	38.83	60.16	0.00	0.44			18.776	0.351	15.901	0.312	38.881	0.773
	38.73	60.64	0.00	0.38			18.859	0.359	16.051	0.308	39.055	0.745
	38.80	61.57	0.00	0.57			18.595	0.365	15.855	0.323	38.566	0.798
	38.13	60.60	0.00	0.66			18.293	0.393	15.585	0.339	37.938	0.822
	38.52	60.35	0.00	0.67			18.632	0.409	15.896	0.346	38.602	0.850
	38.71	61.15	0.00	0.45			18.695	0.466	15.896	0.389	38.683	0.967
	38.87	60.84	0.00	0.54			18.457	0.519	15.808	0.462	38.346	1.090
	38.74	60.27	0.00	0.51			18.514	0.690	15.807	0.578	38.200	1.466
	38.01	60.01	0.40	0.56			18.629	0.820	15.914	0.688	38.759	1.712
KZ-1089/1155.6	34.88	33.96			31.16	Chalcopyrite	18.224	0.084	15.528	0.071	37.828	0.178
	34.70	34.94			30.36		18.276	0.096	15.579	0.082	37.935	0.203

(continued)

Table 3 (continued)

Borehole number/Depth, m	S	Fe	Co	Ni	Cu	Mineral, formula	$^{206}Pb/^{204}Pb$	2σ	$^{207}Pb/^{204}Pb$	2σ	$^{208}Pb/^{204}Pb$	2σ
	35.07	33.93			31.00		18.264	0.107	15.567	0.092	37.934	0.226
	34.64	33.91			31.46		18.120	0.182	15.455	0.149	37.645	0.382
	34.93	34.53			30.54		18.600	0.195	15.857	0.169	38.593	0.406
	34.78	34.30			30.91		18.291	0.197	15.586	0.165	38.005	0.418
	34.18	34.65			31.17		18.693	0.247	15.924	0.205	38.828	0.509
	35.06	34.40			30.54		18.480	0.250	15.749	0.209	38.423	0.529
	34.24	34.17			31.59		18.669	0.269	15.852	0.477	38.726	1.284
	34.29	34.17			31.54		18.628	0.320	15.877	0.262	38.746	0.682
	34.32	33.82			31.85		18.465	0.325	15.733	0.284	38.416	0.689
	36.89	35.27	1.95	25.88		Pentlandite	18.119	0.079	15.443	0.066	37.616	0.165
	34.32	42.92	1.28	21.48			18.075	0.096	15.396	0.080	37.520	0.203
	33.29	34.53	1.63	30.55			18.345	0.113	15.400	0.090	37.653	0.244
	33.06	33.42	1.62	31.89			18.199	0.119	15.542	0.098	37.825	0.255
	33.65	33.52		32.83			18.665	0.136	15.910	0.114	38.719	0.296
	33.74	34.01		32.25			18.468	0.144	15.792	0.119	38.358	0.302
	32.56	33.48	1.77	32.19			18.385	0.154	15.688	0.130	38.201	0.323
	33.36	33.68	1.74	31.22			18.263	0.174	15.570	0.142	37.977	0.351
	32.65	33.26	1.66	32.42			18.711	0.264	16.011	0.224	38.861	0.559
	32.93	33.59	1.60	31.87			18.628	0.285	15.919	0.249	38.716	0.599
	32.71	33.78	1.94	31.57			18.502	0.435	15.696	0.357	38.283	0.911
	34.13	44.27	1.27	20.33			18.630	0.526	15.910	0.447	38.726	1.122
	37.59	62.41				Pyrrhotite	18.085	0.106	15.457	0.090	37.593	0.228
	36.75	63.25					18.371	0.236	15.599	0.286	39.452	1.097
	37.52	62.48					18.647	0.264	15.886	0.226	38.670	0.561
	37.60	62.40					18.240	0.301	15.775	0.341	37.993	0.640
	37.29	62.71					18.863	0.308	15.820	0.385	38.453	0.636
	37.51	62.49					18.707	0.411	16.056	0.359	38.923	0.857

(continued)

Table 3 (continued)

Borehole number/Depth, m	S	Fe	Co	Ni	Cu	Mineral, formula	$^{206}Pb/^{204}Pb$	2σ	$^{207}Pb/^{204}Pb$	2σ	$^{208}Pb/^{204}Pb$	2σ
Southern flank												
KZ-1084/1146.9	34.75	31.45			33.81	Chalcopyrite	18.258	0.031	15.507	0.027	37.801	0.067
	34.88	31.51			33.61		18.298	0.032	15.521	0.027	37.862	0.067
	34.88	31.51			33.61		18.315	0.035	15.534	0.029	37.902	0.074
	34.28	32.16			33.57		18.266	0.038	15.494	0.033	37.797	0.079
	34.01	31.75			34.23	Chalcopyrite	18.408	0.040	15.611	0.034	38.086	0.086
	34.09	31.93			33.98		18.324	0.042	15.544	0.035	37.920	0.088
	33.94	32.12			33.94		18.373	0.042	15.586	0.036	38.026	0.091
	34.73	31.62			33.65		18.394	0.045	15.606	0.038	38.079	0.095
	35.02	31.06			33.93		18.317	0.045	15.543	0.039	37.909	0.095
	38.82	61.18				Pyrrhotite	18.445	0.051	15.641	0.045	38.199	0.122
	34.16	30.85			34.99	Chalcopyrite	18.402	0.053	15.611	0.044	38.094	0.108
	34.26	32.02			33.72		18.366	0.054	15.573	0.045	38.000	0.112
	34.80	31.50			33.70		18.441	0.057	15.648	0.049	38.170	0.122
	31.49	33.82	1.82	30.18	2.68	Pentlandite	18.329	0.079	15.539	0.066	37.940	0.167
	32.48	30.67	2.56	34.29			18.438	0.091	15.629	0.078	38.151	0.191
	35.23	31.69			33.08	Chalcopyrite	18.393	0.114	15.603	0.095	38.096	0.236
	38.53	60.31		1.16		Pyrrhotite	18.500	0.164	15.728	0.138	38.328	0.346
	39.21	60.79					18.295	0.172	15.533	0.149	37.796	0.381
	32.11	31.44	2.00	34.45		Pentlandite	18.684	0.184	15.833	0.154	38.677	0.389
	38.73	60.51		0.77		Pyrrhotite	18.713	0.197	15.856	0.169	38.691	0.412
	31.88	30.29	2.24	35.60		Pentlandite	18.472	0.217	15.643	0.165	38.149	0.402
	33.86	32.22		0.89	33.03	Chalcopyrite	18.732	0.372	15.894	0.331	38.688	0.796
	38.34	61.66				Pyrrhotite	18.898	1.273	15.765	1.163	39.294	2.647
KZ-361bis/1064.4	34.72	33.93			31.35	Chalcopyrite	18.098	0.046	15.423	0.038	37.588	0.095
	34.57	33.34			32.09		18.191	0.048	15.491	0.041	37.785	0.098
	33.77	34.00			32.24		18.030	0.049	15.365	0.041	37.437	0.105
	33.99	35.55			30.46		18.155	0.050	15.473	0.042	37.701	0.105
	34.37	34.14			31.49		18.123	0.052	15.451	0.044	37.642	0.110
	34.16	34.25			31.60		18.121	0.053	15.446	0.045	37.613	0.114

(continued)

Table 3 (continued)

Borehole number/Depth, m	S	Fe	Co	Ni	Cu	Mineral, formula	$^{206}Pb/^{204}Pb$	2σ	$^{207}Pb/^{204}Pb$	2σ	$^{208}Pb/^{204}Pb$	2σ
	34.36	34.62			31.02		18.097	0.053	15.426	0.045	37.587	0.112
	33.84	34.11			32.05		18.166	0.054	15.489	0.046	37.741	0.113
	34.04	34.16			31.80		18.090	0.054	15.414	0.045	37.552	0.114
	32.27	32.68	2.29		32.76		18.089	0.057	15.418	0.049	37.565	0.118
	33.53	33.82			32.65		18.090	0.058	15.426	0.049	37.563	0.120
	33.59	34.28			32.13		18.072	0.059	15.410	0.049	37.544	0.127
	34.47	33.90			31.64		18.102	0.059	15.432	0.050	37.583	0.123
	34.34	34.55			31.11		18.106	0.060	15.435	0.051	37.616	0.130
	34.46	34.81			30.72		18.298	0.063	15.581	0.053	37.980	0.131
	33.66	35.04			31.30		18.060	0.064	15.400	0.055	37.503	0.135
	33.72	34.42			31.86		18.098	0.064	15.430	0.054	37.583	0.137
	34.12	34.64			31.24		18.021	0.069	15.359	0.058	37.395	0.148
	33.78	34.28			31.94		18.191	0.070	15.501	0.058	37.774	0.147
	34.54	34.00			31.46		18.133	0.070	15.459	0.060	37.646	0.144
	34.14	33.71			32.14		18.039	0.071	15.388	0.060	37.448	0.147
	34.52	34.52			30.96		18.027	0.073	15.365	0.062	37.428	0.153
	33.87	35.16			30.98		18.190	0.074	15.509	0.062	37.770	0.157
	34.09	34.69			31.22		18.168	0.079	15.482	0.065	37.709	0.164
	34.28	34.28			31.44		18.161	0.080	15.478	0.068	37.740	0.171
	34.55	34.01			31.44		18.091	0.081	15.412	0.071	37.563	0.172
	33.66	34.54			31.80		18.238	0.081	15.547	0.071	37.866	0.176
	33.66	34.82			31.52		18.116	0.083	15.433	0.070	37.602	0.176
	34.16	34.51			31.34		18.181	0.086	15.500	0.073	37.770	0.180
	34.64	33.78			31.58		18.115	0.086	15.433	0.073	37.608	0.185
	37.16	62.84				Pyrrhotite	18.120	0.088	15.439	0.074	37.624	0.185
	33.95	34.09			31.96	Chalcopyrite	17.933	0.095	15.284	0.078	37.222	0.199
	34.34	33.44			32.23		18.187	0.100	15.509	0.085	37.761	0.214
	34.02	33.85			32.13		17.939	0.105	15.289	0.088	37.228	0.224
	34.24	34.49			31.27		18.087	0.108	15.425	0.092	37.579	0.229
	34.91	35.04			30.06		18.194	0.112	15.512	0.095	37.780	0.234
	34.13	34.11			31.77		17.999	0.112	15.347	0.096	37.358	0.237

(continued)

Table 3 (continued)

Borehole number/Depth, m	S	Fe	Co	Ni	Cu	Mineral, formula	$^{206}Pb/^{204}Pb$	2σ	$^{207}Pb/^{204}Pb$	2σ	$^{208}Pb/^{204}Pb$	2σ
	34.28	33.88			31.83		18.013	0.113	15.363	0.093	37.392	0.239
	33.97	34.05			31.97		17.966	0.216	15.311	0.184	37.272	0.456
	34.08	34.16			31.77		18.740	0.226	15.965	0.194	38.894	0.472
Maslovsky deposit												
OM-123/1033.7	35.38	43.12			21.50		18.141	0.065	15.504	0.056	37.769	0.138
	34.34	49.72		1.14	14.81		18.059	0.090	15.436	0.076	37.590	0.191
	32.54	41.46		26.00		Pentlandite	18.324	0.163	15.662	0.138	38.192	0.342
	35.21	64.79				Pyrrhotite	18.878	0.181	16.083	0.154	39.319	0.395
	35.92	64.08					17.709	0.189	15.155	0.161	36.890	0.405
	35.99	64.01					18.689	0.203	15.991	0.171	38.999	0.424
	36.29	63.71					17.659	0.256	15.120	0.221	36.756	0.546
	35.78	64.22					17.850	0.256	15.263	0.217	37.218	0.546
	36.06	63.94					18.422	0.356	15.786	0.299	38.264	0.734
	35.87	64.13					18.988	0.398	16.275	0.334	39.579	0.832
	36.22	63.78					18.672	0.548	15.928	0.470	38.836	1.135
OM-10/1072.9–1073	32.57	28.81	2.09	36.54		Pentlandite	17.847	0.033	15.446	0.029	37.584	0.078
	35.67	1.62	0.72	61.99			17.861	0.045	15.465	0.039	37.629	0.097
	33.30	26.58	0.83	34.79	4.51		18.061	0.366	15.645	0.311	38.034	0.812
OM-10/1082.4–1083	55.45	44.55				Pyrite	17.942	0.023	15.445	0.020	37.581	0.054
	54.60	45.40					17.941	0.025	15.443	0.022	37.581	0.055
	54.34	45.66					17.945	0.025	15.446	0.022	37.575	0.054
	54.82	45.18					17.943	0.026	15.447	0.022	37.598	0.057
	54.64	45.36					17.961	0.028	15.465	0.024	37.633	0.064
	54.39	45.61					17.955	0.030	15.457	0.026	37.629	0.067
	54.31	45.69					17.929	0.030	15.433	0.026	37.572	0.065
	54.17	45.83					17.933	0.031	15.440	0.026	37.583	0.067
	55.54	44.46					17.933	0.031	15.424	0.026	37.546	0.064
	54.65	45.35					17.927	0.033	15.415	0.029	37.518	0.071
	54.33	45.67					17.920	0.034	15.427	0.029	37.557	0.072
	33.60	29.56	1.13	35.70		Pentlandite	17.928	0.035	15.429	0.030	37.543	0.077

(continued)

Table 3 (continued)

Borehole number/Depth, m	S	Fe	Co	Ni	Cu	Mineral, formula	$^{206}Pb/^{204}Pb$	2σ	$^{207}Pb/^{204}Pb$	2σ	$^{208}Pb/^{204}Pb$	2σ
	54.15	45.85				Pyrite	17.960	0.038	15.447	0.033	37.588	0.082
	54.37	45.63					17.979	0.040	15.474	0.034	37.658	0.086
	54.77	45.23					17.926	0.042	15.426	0.036	37.534	0.091
	54.43	45.57					17.926	0.042	15.426	0.036	37.534	0.091
	54.42	45.58					17.895	0.045	15.408	0.039	37.494	0.095
	54.33	45.67					17.942	0.047	15.423	0.040	37.540	0.098
	54.63	45.37					17.896	0.048	15.401	0.041	37.494	0.100
	54.87	45.13					17.931	0.052	15.410	0.045	37.495	0.111
	54.75	45.25					17.986	0.053	15.472	0.045	37.639	0.113
	54.60	45.40					17.937	0.059	15.430	0.051	37.566	0.126
	55.10	44.90					17.882	0.070	15.381	0.061	37.438	0.152
	54.96	45.04					17.942	0.072	15.425	0.062	37.566	0.156
	54.60	45.40					17.933	0.083	15.420	0.071	37.520	0.179
	55.09	44.91					18.063	0.085	15.526	0.072	37.792	0.177
	54.73	45.27					17.831	0.085	15.328	0.072	37.338	0.181
	53.94	46.06					18.025	0.086	15.517	0.075	37.749	0.184
	54.78	45.22					17.931	0.088	15.428	0.074	37.548	0.189
	54.51	45.49					18.056	0.092	15.639	0.079	37.771	0.191
	54.70	45.30					17.691	0.104	15.227	0.090	37.056	0.224
	54.71	45.29					17.909	0.121	15.405	0.103	37.505	0.258
	54.47	45.53					17.934	0.154	15.428	0.134	37.480	0.329
	54.97	45.03					18.709	0.189	15.499	0.152	37.718	0.384
	54.73	45.27					18.198	0.201	15.657	0.171	38.084	0.423
	55.03	44.97					17.653	0.265	15.199	0.228	36.954	0.587
	54.31	45.69					18.013	0.267	15.506	0.221	37.784	0.541
Vologochan area												
OV-28/813.8–814.2	37.67	29.65			32.68	Chalcopyrite	18.255	0.047	15.507	0.039	37.782	0.099
	38.21	28.84			32.95		18.222	0.059	15.490	0.049	37.732	0.124
	37.56	29.76			32.68		18.231	0.074	15.484	0.062	37.728	0.157
	37.91	29.30			32.79		18.319	0.084	15.552	0.069	37.904	0.174

(continued)

Table 3 (continued)

Borehole number/Depth, m	S	Fe	Co	Ni	Cu	Mineral, formula	$^{206}Pb/^{204}Pb$	2σ	$^{207}Pb/^{204}Pb$	2σ	$^{208}Pb/^{204}Pb$	2σ
	38.18	28.71			33.11		18.259	0.105	15.511	0.089	37.792	0.219
	38.66	29.31			32.02		18.147	0.120	15.417	0.101	37.520	0.260
	35.69	31.65		32.66		Pentlandite	18.229	0.039	15.476	0.033	37.701	0.086
	34.84	31.53		33.63			18.239	0.055	15.491	0.046	37.680	0.094
	35.05	31.74		33.21			18.344	0.078	15.580	0.065	37.968	0.159
	36.01	31.51		32.48			18.234	0.078	15.493	0.065	37.751	0.159
	35.44	30.26	1.36	32.95			18.138	0.202	15.419	0.174	37.506	0.429
	39.98	60.02				Pyrrhotite	18.153	0.099	15.389	0.084	37.536	0.208
Norilsk-1 deposit												
N1-0	34.88	41.74			23.37	Chalcopyrite	18.021	0.028	15.447	0.024	37.592	0.060
	35.11	42.07			22.82		18.027	0.029	15.448	0.024	37.580	0.060
	34.84	42.09			23.07		18.031	0.029	15.455	0.024	37.632	0.070
	35.6	42.29			22.11		18.027	0.029	15.452	0.025	37.594	0.063
	34.3	31.65			34.05		18.020	0.030	15.445	0.025	37.580	0.062
	34.15	31.78			34.07		18.039	0.032	15.458	0.028	37.633	0.074
	34.95	41.63			23.42		18.010	0.033	15.438	0.028	37.562	0.071
	34.23	31.69		4.10	29.99		18.043	0.034	15.463	0.029	37.610	0.072
	35.00	40.54			24.46	Chalcopyrite	18.015	0.034	15.443	0.029	37.584	0.073
	35.00	42.19			22.81		18.050	0.036	15.470	0.030	37.644	0.076
	35.10	42.60			22.29		17.994	0.037	15.425	0.032	37.523	0.078
	34.47	31.71			33.82		18.023	0.037	15.450	0.032	37.598	0.078
	33.92	31.62			34.46		18.033	0.038	15.450	0.032	37.594	0.080
	35.01	42.37			22.62		18.065	0.038	15.482	0.032	37.671	0.081
	34.21	43.09			22.70		18.016	0.039	15.441	0.033	37.573	0.083
	34.54	31.61			33.84		18.046	0.039	15.463	0.033	37.625	0.082
	35.23	42.02			22.75		18.063	0.040	15.486	0.034	37.683	0.084
	34.64	33.72			31.64		18.065	0.040	15.480	0.034	37.659	0.084
	34.26	32.11		1.24	32.40	Cubanite	18.058	0.041	15.477	0.035	37.666	0.086
	34.94	42.16			22.90	Chalcopyrite	18.050	0.041	15.473	0.035	37.635	0.086
	35.38	41.82			22.80		18.054	0.041	15.477	0.035	37.652	0.090

(continued)

Table 3 (continued)

Borehole number/Depth, m	S	Fe	Co	Ni	Cu	Mineral, formula	$^{206}Pb/^{204}Pb$	2σ	$^{207}Pb/^{204}Pb$	2σ	$^{208}Pb/^{204}Pb$	2σ
	35.21	41.76			23.03		18.034	0.042	15.456	0.035	37.608	0.088
	35.28	42.11			22.61		18.063	0.042	15.478	0.035	37.663	0.088
	34.88	42.31			22.81		18.012	0.043	15.436	0.036	37.564	0.088
	34.33	31.52			34.15		18.017	0.043	15.442	0.037	37.567	0.092
	34.60	41.79			23.61		18.040	0.043	15.465	0.037	37.604	0.091
	35.33	41.98			22.69		18.059	0.043	15.483	0.037	37.667	0.091
	34.01	31.35			34.64		18.062	0.045	15.479	0.038	37.671	0.095
	34.18	34.20		4.79	26.83	Cubanite	18.048	0.046	15.467	0.039	37.628	0.095
	35.02	42.65			22.33	Chalcopyrite	18.069	0.046	15.487	0.040	37.673	0.097
	34.47	32.21			33.32		18.058	0.052	15.470	0.044	37.655	0.107
	34.20	32.34			33.46		18.027	0.052	15.450	0.044	37.564	0.109
	34.76	42.43			22.80		18.014	0.054	15.442	0.047	37.556	0.116
	34.36	31.84			33.80		18.041	0.055	15.458	0.047	37.607	0.116
	33.31	32.60		30.64	3.450	Pentlandite	18.061	0.068	15.487	0.058	37.659	0.150
	34.71	41.91			23.38	Chalcopyrite	18.113	0.070	15.527	0.059	37.752	0.145
	34.99	42.14			22.87		18.076	0.073	15.491	0.061	37.686	0.155
	35.12	31.02			33.86		18.077	0.074	15.491	0.063	37.676	0.163
	33.11	33.09		33.8		Pentlandite	18.073	0.077	15.494	0.066	37.702	0.175
N1-6	34.13	30.82			35.04	Chalcopyrite	18.130	0.097	15.583	0.081	37.939	0.213
	34.43	31.29			34.28		17.719	0.123	15.225	0.105	36.992	0.258
	33.87	31.62			34.51		18.124	0.136	15.598	0.120	37.978	0.282
	33.05	32.73		14.01	20.20	Cubanite	18.362	0.139	15.780	0.120	38.364	0.297
	34.37	30.96			34.67	Chalcopyrite	18.038	0.223	15.503	0.193	37.756	0.480
	34.07	31.49			34.44		18.301	0.238	15.707	0.203	38.248	0.506
	34.47	31.84			33.69		18.249	0.251	15.712	0.218	38.206	0.551
	33.57	31.50			34.93		18.698	0.259	16.084	0.227	39.101	0.566
	35.39	30.98		0.93	32.70		18.704	0.281	16.076	0.239	39.127	0.605
	33.94	32.07			34.00		18.431	0.282	15.828	0.218	38.565	0.536
	33.57	31.77			34.65		18.640	0.285	15.962	0.247	38.898	0.605
	34.59	31.24			34.17		18.553	0.302	15.979	0.253	38.810	0.632

(continued)

Table 3 (continued)

Borehole number/Depth, m	S	Fe	Co	Ni	Cu	Mineral, formula	$^{206}Pb/^{204}Pb$	2σ	$^{207}Pb/^{204}Pb$	2σ	$^{208}Pb/^{204}Pb$	2σ
	33.75	32.48			33.76		18.519	0.354	15.901	0.303	38.654	0.742
	33.99	31.55			34.46		18.175	0.367	15.615	0.315	38.011	0.785
	32.46	35.71		31.83		Pentlandite	18.543	0.386	15.966	0.321	39.012	0.833
	34.26	32.20			33.53	Chalcopyrite	18.648	0.429	15.993	0.359	39.032	0.888
	34.48	31.81			33.71		18.573	0.446	15.958	0.379	38.790	0.961
	33.17	37.16		29.67		Pentlandite	18.194	0.449	15.588	0.377	38.148	0.935
	33.82	31.51			34.68	Chalcopyrite	18.648	0.497	16.008	0.426	38.964	1.081
	34.74	37.86		2.02	25.37		18.569	0.502	15.968	0.421	38.903	1.063
	33.83	32.67			33.50		18.448	0.509	15.867	0.418	38.547	1.068
	34.63	31.41		2.19	31.77		18.679	0.530	15.980	0.460	38.949	1.118
	33.82	32.95		2.10	31.13		18.451	0.562	15.910	0.489	38.632	1.201
	30.00	36.86		33.14		Pentlandite	18.778	0.562	16.151	0.480	39.229	1.190
	34.12	31.17			34.71	Chalcopyrite	18.668	0.595	16.156	0.516	39.218	1.298
	34.83	31.37			33.79		18.836	0.649	16.248	0.560	39.691	1.409
	34.89	31.87			33.24		18.580	0.664	15.954	0.568	38.875	1.418
	34.64	30.80			34.56		18.634	0.681	16.091	0.563	39.136	1.440
	34.62	31.33			34.05		18.681	0.683	16.074	0.582	39.233	1.450
	34.54	31.16			34.30		18.387	0.849	15.708	0.731	38.528	1.840
	33.87	31.90			34.24		18.380	0.876	15.771	0.728	38.295	1.820
	34.20	31.55			34.25		18.872	1.309	16.211	1.107	39.583	2.817
	34.14	31.74			34.12		18.051	1.337	15.776	1.188	37.765	2.704
Talnakh intrusive												
OUG-2	33.74	34.39		7.52	24.34	Talnakhite(?)	17.867	0.058	15.220	0.049	37.081	0.126
	31.76	34.34	1.31		32.59	Chalcopyrite	18.089	0.060	15.404	0.051	37.535	0.130
	34.31	34.30			31.40	Chalcopyrite	18.099	0.075	15.405	0.063	37.538	0.165
	34.32	34.27			31.41		18.031	0.092	15.359	0.078	37.428	0.196
	34.05	34.50			31.44		18.181	0.106	15.463	0.087	37.690	0.220
	34.00	33.76			32.24		18.077	0.114	15.398	0.096	37.491	0.248
	34.34	34.38			31.28		17.865	0.134	15.224	0.113	37.061	0.285
	33.75	34.10			32.15		17.902	0.153	15.255	0.130	37.191	0.325

(continued)

Table 3 (continued)

Borehole number/Depth, m	S	Fe	Co	Ni	Cu	Mineral, formula	$^{206}Pb/^{204}Pb$	2σ	$^{207}Pb/^{204}Pb$	2σ	$^{208}Pb/^{204}Pb$	2σ
	34.62	34.23			31.15		17.756	0.163	15.123	0.135	36.839	0.343
	34.18	34.41			31.41		17.801	0.166	15.160	0.140	36.967	0.346
KZ-774/1032.4–1033	39.17	60.83				Pyrrhotite	18.084	0.061	15.448	0.052	37.631	0.131
	37.53	44.26		12.59	5.62	Chalcopyrite	18.154	0.066	15.515	0.055	37.788	0.143
	33.74	35.39	1.42	29.45		Pentlandite	18.148	0.085	15.502	0.074	37.738	0.185
	39.57	60.43				Pyrrhotite	18.033	0.120	15.426	0.102	37.542	0.251
	35.59	32.13			32.29	Chalcopyrite	18.093	0.126	15.433	0.104	37.607	0.268
	35.72	31.36			32.92		17.954	0.136	15.345	0.113	37.392	0.293
	34.68	39.85		25.46		Pentlandite	18.091	0.177	15.442	0.148	37.601	0.357
KZ-774/1042.2–1049.7	52.89	47.11				Pyrite	18.153	0.019	15.480	0.016	37.710	0.042
	52.74	47.26					18.149	0.020	15.479	0.017	37.708	0.045
	52.82	47.18					18.159	0.020	15.480	0.017	37.720	0.044
	53.32	46.68					18.159	0.021	15.484	0.018	37.749	0.054
	52.79	47.21					18.140	0.021	15.479	0.018	37.707	0.046
	52.95	47.05					18.163	0.022	15.482	0.018	37.716	0.047
	53.27	46.73					18.160	0.022	15.488	0.019	37.729	0.049
	52.77	47.23					18.135	0.022	15.478	0.019	37.708	0.047
	52.18	47.82					18.154	0.023	15.481	0.019	37.711	0.050
	53.97	46.03					18.140	0.023	15.489	0.019	37.715	0.049
	52.88	47.12					18.143	0.023	15.471	0.019	37.683	0.051
	54.20	45.80					18.153	0.024	15.478	0.020	37.729	0.055
	52.87	47.13					18.150	0.025	15.476	0.021	37.705	0.055
	52.43	46.61		0.96		Pyr-(ni)	18.162	0.025	15.481	0.022	37.725	0.056
	53.36	46.64				Pyrite	18.123	0.026	15.471	0.022	37.697	0.055
	53.16	46.84					18.111	0.026	15.472	0.022	37.680	0.055
	53.24	46.76					18.156	0.026	15.480	0.022	37.711	0.054
	53.18	46.82					18.126	0.026	15.478	0.022	37.699	0.059
	53.37	44.21		2.41		Pyr-(ni)	18.165	0.028	15.490	0.023	37.739	0.057
	53.86	46.14				Pyrite	18.145	0.035	15.473	0.030	37.713	0.077
	52.36	47.64					18.166	0.036	15.487	0.030	37.724	0.073

(continued)

Table 3 (continued)

Borehole number/Depth, m	S	Fe	Co	Ni	Cu	Mineral, formula	$^{206}Pb/^{204}Pb$	2σ	$^{207}Pb/^{204}Pb$	2σ	$^{208}Pb/^{204}Pb$	2σ
	53.51	46.49					18.181	0.038	15.492	0.032	37.781	0.081
	52.52	47.48					18.184	0.044	15.506	0.037	37.775	0.093
	53.77	46.23					18.316	0.061	15.605	0.051	38.031	0.133
	53.04	45.52					18.086	0.062	15.418	0.051	37.576	0.130
	53.58	46.42					18.166	0.113	15.479	0.095	37.685	0.239
	53.50	46.50					18.061	0.115	15.399	0.098	37.510	0.244
	53.85	46.15					18.331	0.154	15.595	0.129	38.080	0.325
	52.75	47.25					18.810	0.229	16.010	0.196	39.118	0.488
	53.54	46.46					18.144	0.246	15.492	0.214	37.688	0.529
	53.63	46.37					18.057	0.347	15.373	0.288	37.531	0.721
	53.28	45.52		1.20		Pyr-(ni)	18.051	0.805	15.456	0.676	37.222	1.679
Burkan deposit, sample 657-26												
	37.63	29.21			33.16	Chalcopyrite	17.989	0.031	15.446	0.026	37.670	0.065
	37.73	29.06			33.21		17.961	0.036	15.442	0.031	37.669	0.081
	37.37	28.97			33.66		17.978	0.037	15.436	0.031	37.631	0.079
	37.35	29.34			33.31		17.985	0.039	15.445	0.034	37.676	0.090
	37.64	29.02			33.34		17.970	0.039	15.449	0.033	37.675	0.085
	37.57	29.94			32.49		17.915	0.042	15.381	0.036	37.503	0.088
	37.62	29.18			33.20		17.977	0.044	15.443	0.037	37.646	0.104
	38.21	28.77			33.02		17.901	0.047	15.360	0.040	37.458	0.103
	37.56	28.85			33.59		18.043	0.048	15.396	0.040	37.531	0.105
	38.40	28.69			32.91		17.988	0.048	15.442	0.041	37.662	0.103
	38.18	28.55			33.27		17.949	0.050	15.420	0.043	37.614	0.109
	37.32	28.63			34.05		18.011	0.062	15.462	0.052	37.707	0.135
	37.98	29.47			32.55		17.964	0.065	15.440	0.057	37.652	0.136
	37.32	30.19			32.48		17.873	0.096	15.340	0.081	37.376	0.201
	36.98	28.92			34.10		18.060	0.171	15.510	0.142	37.893	0.411
	34.8	30.47		34.72		Pentlandite	17.947	0.060	15.423	0.050	37.582	0.176
	34.47	30.36	2.07	33.11			18.041	0.065	15.482	0.055	37.727	0.145
	37.61	28.94			33.45	Pentlandite	18.226	0.201	15.650	0.170	38.165	0.438

(continued)

Table 3 (continued)

Borehole number/Depth, m	S	Fe	Co	Ni	Cu	Mineral, formula	$^{206}Pb/^{204}Pb$	2σ	$^{207}Pb/^{204}Pb$	2σ	$^{208}Pb/^{204}Pb$	2σ
	42.13	57.87				Pyrrhotite (?)	18.082	0.094	15.542	0.082	37.854	0.219
	39.07	60.93					17.717	0.135	15.213	0.112	37.099	0.292
Zub-Marksheidersky intrusive, MP-25												
	56.46	42.41			1.13	Pyrrhotite	18.242	0.021	15.504	0.017	37.813	0.058
	38.10	29.05			32.86	Chalcopyrite	18.396	0.026	15.488	0.022	37.975	0.055
	57.51	42.49				Pyrite	18.226	0.027	15.490	0.023	37.735	0.058
	57.51	42.49					18.233	0.027	15.492	0.024	37.739	0.060
	57.56	42.44					18.246	0.029	15.500	0.024	37.785	0.064
	57.94	42.06					18.270	0.029	15.501	0.025	37.813	0.064
	58.18	41.82					18.210	0.034	15.478	0.029	37.703	0.076
	57.84	41.41	1.3				18.338	0.035	15.474	0.029	37.866	0.072
	57.32	40.50		1.07			18.266	0.043	15.503	0.036	37.838	0.094
	57.03	42.97					18.325	0.043	15.561	0.036	37.946	0.092
	57.93	42.07					18.244	0.056	15.494	0.047	37.785	0.118
	57.82	42.18					18.146	0.057	15.421	0.048	37.571	0.125
	57.66	42.34					18.203	0.085	15.465	0.071	37.717	0.179

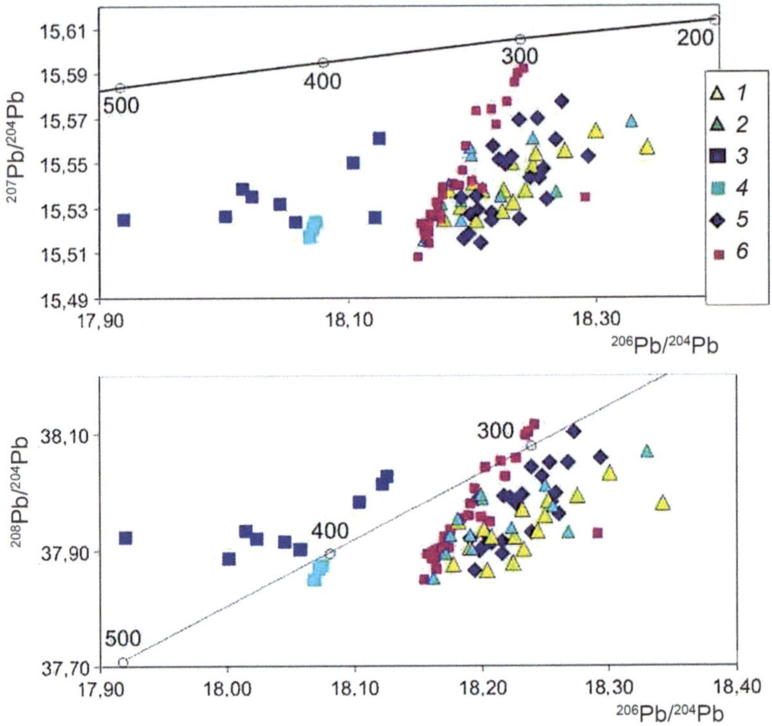

Fig. 3 Pb–Pb isotope systematics of sulphides from commercial-ore-bearing intrusives of Norilsk Province. Figures near Stacey-Kramers curves—age of the corresponding model reservoir, Ma. 1—massive and 2—disseminated sulphide ores of Kharaelakh intrusive; 3—massive and 4—disseminated sulphide ores of Talnakh intrusive; 5—sulphide ores and 6—low-sulphide horizon of Norilsk intrusive

silicates are not syngenetic, but paragenetic. This conclusion is consistent with the results of the previous studies [5], according to which in commercially mineralized massifs of the Norilsk type, nickel from olivine was not extracted by ore matter despite the highest affinity of this element to sulphides. In other words, sulphide ore is not in chemical and isotopic equilibrium with silicate rocks. Sulphide and silicate melts were not isotopically uniform, and, hence, were never a homogeneous liquid phase.

Isotopic composition of lead in sulphides of the medium-grade mineralized intrusions (Chernogorsky, Zub-Marksheidersky and Vologochan) as well as in poor ones (weakly mineralized Lower Norilsk and Lower Talnakh) has no obvious differences from the composition of sulphides from the massifs with high-grade mineralization in the Talnakh and Kharaelakh intrusions. For diagnosing the prospects of intrusions, lead isotopic composition can be used only as an additional indicator, for example, in the $^{207}Pb/^{204}Pb$–$\delta^{34}S$ system (Fig. 4). This approach was

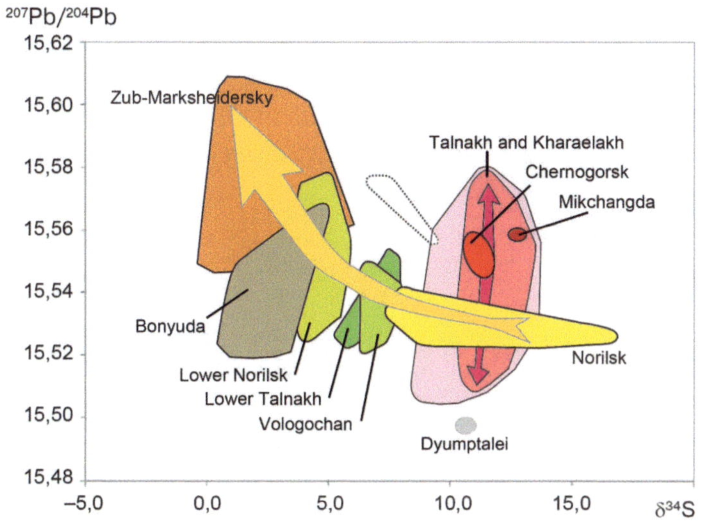

Fig. 4 $^{207}Pb/^{204}Pb$–$\delta^{34}S$ diagram for the studied intrusives of Norilsk and-Taimyr provinces

realized when determining the prospects of the inadequately studied intrusions (Binyuda, Mikchangda, Dyumptalei).

In the $^{207}Pb/^{204}Pb$–$\delta^{34}S$ diagram (Fig. 4), isotope analogues of sulphides in high-grade mineralized Talnakh and Kharaelakh intrusions are only ores of the Chernogorsky and Mikchanda intrusions. Two trends are revealed (shown with arrows): Talnakh-Kharaelakh and Norilsk-Zub-Marksheidersky, presumably, corresponding to mixtures of different sources of ore matter. Location of the Zelenaya Griva intrusion (unfilled field, dotted line) due to non-representative data remains uncertain; the affiliation of the Dyumptalei intrusion is also not clear (only one analysis of Pb and S isotopic composition). Arrows indicate the position of the main mixed components: the mantle one in the Zub-Marksheidersky intrusion and two crustal ones—one in the Kharaelakh deposit and another one in Norilsk.

Consideration of the combined isotopic S–Pb system leads to the conclusion that sulphides of the Mikchangda intrusion are analogous to the high-grade ores of the Talnakh and Kharaelakh deposits. Judging by the isotopic characteristics of sulphides, the Binyuda intrusion, similar to the Zub-Marksheidersky and the Vologochan intrusions in the previous group of intrusive bodies, should be assigned to a group of objects differing from the high-grade ones, and, thus, having lower prospects for revealing massive sulphide ores.

In the Lower Norilsk and the Lower Talnakh intrusions, $^{206}Pb/^{204}Pb$ ratio in Cu–Ni sulphides varies significantly along with a relative constancy of $^{207}Pb/^{204}Pb$ ratio in sulphides of the Lower Talnakh intrusion and considerable fluctuations of this ratio in the Lower Norilsk intrusion. This feature distinguishes them from Cu–Ni sulphides of the Talnakh and Kharaelakh deposits and makes them somewhat similar to the Norilsk deposit, which also shows elevated variations of $^{206}Pb/^{204}Pb$

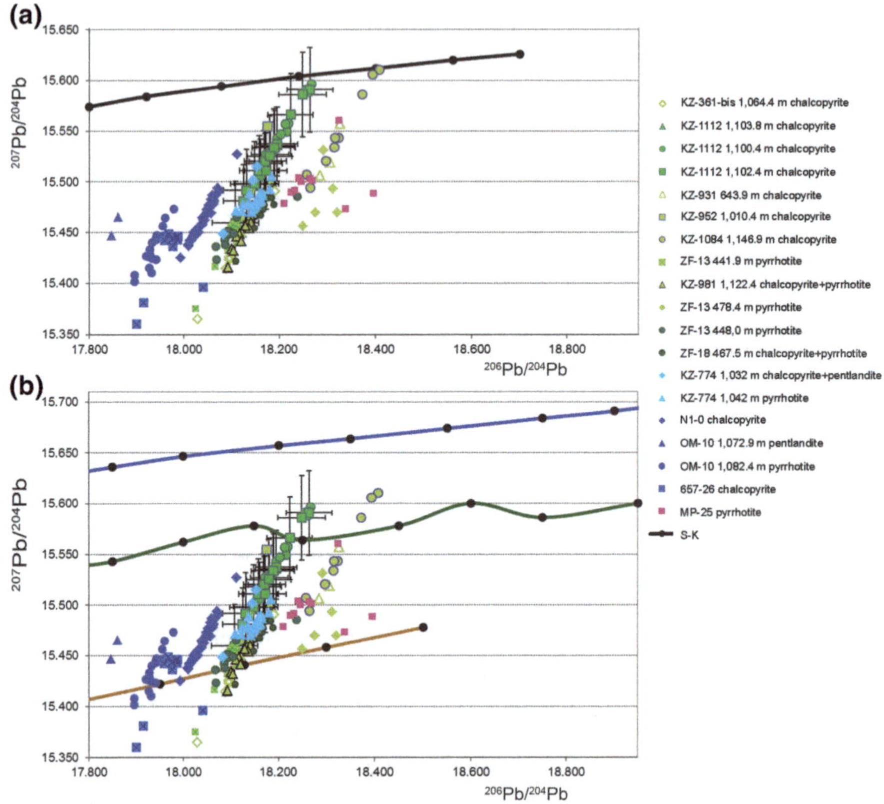

Fig. 5 Isotopic composition of lead in sulphides of Norilsk District (28 samples from boreholes), evolution line of "ordinary" lead according to Stacey-Kramers model (**a**) and evolution line of "ordinary" lead in geochemical reservoirs according to Doe and Zartman plumbotectonics model (**b**). For one of the samples, typical values of error of a single determination are shown (1σ)

ratio. However, such a similarity in variations does not determine their identity, since lead isotopic composition of sulphides in these intrusions mainly lies outside the sulphide field at the Norilsk deposit; and if this difference is primary, it points to different sources of ore matter, more precisely, lead.

Difference between lead isotopic composition in sulphides and plagioclases (Tables 1 and 2) clearly indicates that in the commercially mineralized intrusions the ore matter and silicates are not syngenetic, but paragenetic. This conclusion is consistent with the results of the previous studies, according to which in the commercially mineralized massifs of the Norilsk type nickel was not extracted from olivine by ore matter, despite the highest affinity of this element to sulphides. Thus, sulphide ore is neither in chemical, nor isotopic equilibrium with silicate rocks. If, according to the magmatic model of mineralization, ore in the form of sulphide liquid really settled in the silicate melt, then this liquid came from a different silicate melt, and the melts with which it arrived to the magma chambers of the Norilsk intrusions, acted only as a delivery vehicle for it.

Abundance of sulphide ore matter in the commercial deposits excludes its appearance as a result of liquation of the homogeneous mantle melt within the intrusive body.

Basic process of crust-mantle interaction resulting in matter formation had, apparently, occurred before the melt arrived to the magma chamber. The melt could be contaminated with crustal sulphur, lead and other elements also in situ. Apart from these two stages of rock and ore formation, there was, at least, the third process represented by Ni–Co–arsenide mineralization in arsenide–carbonate veins with sulphides. Under the influence of this late metasomatic process, the geochemical closure of isotope systems could be disturbed.

3 Local Studies

ICPMS method (inductively coupled plasma mass spectrometry) is widely used in the study of different isotopic systems due to a high rate and simplicity of sample introduction. In combination with laser sampling (LA—laser ablation), when analysing sulphides, the method has an additional advantage, since it does not require chemical separation of lead from the sample. This method also allows to determine the isotopic composition of lead in the selected grain of a particular sulphide type and compare the results for each type of sulphide from the same sample. It is also possible to compare lead isotopic composition in each type of sulphides from all the samples from a certain ore district, for example, Norilsk.

Local analysis of lead isotopic composition in sulphides comprises a series of successive procedures including mineralogical separation of the electromagnetic and non-magnetic components of sulphide fraction, making a microslide from epoxy resin (disk), into which individual sulphide grains are placed. After polishing the disk, its review surveying is conducted using an optical microscope and determination of concentrations of the main chemical elements comprised into the largest sulphide grains for each sample, which allows identifying the type of sulphide for each specific grain (electron microprobe).

A set of analytical equipment for the studies applying LA-ICPMS method at CIR VSEGEI comprises a system of laser sampling DUV-193 equipped with an excimer laser COMPEX-102 and a multicollector inductively coupled plasma mass spectrometer Neptune.

Configuration of collectors when measuring Pb isotopic composition allowed a simultaneous recording of ionic currents of isotopes ^{202}Hg–^{203}Tl–$^{204}(Hg + Pb)$–^{205}Tl–^{206}Pb–^{207}Pb–^{208}Pb. A relatively high Pb content in certain types of sulphides in the Norilsk district makes the LA-ICPMS method of lead isotope analysis in this case an express and precise enough procedure. Internal measurement error of 0.1–0.2% can be achieved in the integration time of about 30–40 s. NIST-611 glass was used as a standard.

For the correction of mass discrimination in this standard, normalization of ratios measured after the known $^{203}Tl/^{205}Tl$ ratio was used. The resulting adjusted values within the measurement accuracy coincide with the published data for NIST-611. The average value of the mass discrimination factor for $^{207}Pb/^{206}Pb$ ratio based on the multiple results was 1.00456 ± 0.00030. The correction of lead isotopic composition in respect of mass discrimination was carried out using the correction coefficient values calculated when measuring the standard (international standard glass NIST-611) for an individual measuring session. Correction for isobar overlap of ^{204}Hg isotope contained in NIST-611 was performed by subtracting the signal from ^{204}Hg (^{202}Hg isotope free from overlapping was measured). This allowed obtaining a correct value of all lead isotope ratios to ^{204}Pb in a wide range of signal intensities (at different values of laser beam diameter and laser pulse frequency).

A typical value of laser beam diameter in the analysis of sulphides is ~ 125 μm; the frequency of laser pulses is 7 Hz. In the analysis of the laser beam diameter standard ~ 90 μm pulse frequency is 7 Hz. Laser pulses energy is ~ 200 mJ; power of ICP generator ~ 1050 W.

In case of a possible uncertainty of the mass discrimination factor, the final measurement error of isotope ratios did not exceed ~ 0.1–0.2% (1σ) at a sufficient concentration of lead in grains.

Pb isotopic composition was investigated in individual sulphide grains of 28 samples from drill-holes in different regions of commercially mineralized Talnakh, Kharaelakh and Norilsk intrusions (Table 3).

A more contrasting picture of variations of sulphide isotopic composition compared with the earlier known data is revealed by the analysis of individual sulphide grains using LA-ICPMS.

Sulphides in the samples investigated by LA-IPMS method are represented by chalcopyrite, pyrrhotite, pentlandite and cubanite (Table 3). Comparison of the data of element and local isotopic analyses shows that most of lead in many sulphide samples is contained as an admixture in chalcopyrite. However, in samples OM-10-1082.4-1083.0; ZF-13-441.9-442.5; ZF-13-478,4-479; ZF-13-448.0; KZ-774-1042-21049.7-EM, KZ-981-1122-4-1123; MP-25_23, where the sulphide component consists almost exclusively of pyrite, relatively high lead concentrations are also recorded (Table 4).

The vast majority of data points of lead isotopic composition in the studied sulphide samples in the diagram are between the lines of lead evolution in the mantle and upper crustal model reservoirs (Fig. 5). The most "ancient" values in terms of model age were obtained for sulphide samples OM-10 (horizon 1082.4—pyrite; horizon 1072—pentlandite), 657-26 (chalcopyrite), N1-0 (chalcopyrite). Thus, lead isotopic composition in sulphide samples (OM-10 (horizon 1082.4—pyrite), 657-26 (chalcopyrite), N1-0 (chalcopyrite) coincides with or is close to the evolution line of isotopic composition of the model mantle reservoir, which may indicate a significant contribution of the mantle component in the formation of these sulphides. The youngest sulphides (with the greatest amount of radiogenic lead) correspond to the samples ZF-13 (horizon 478.4—pyrite), KZ-1084 (horizon 1146.9—chalcopyrite), MR-25-23 (pyrite), KZ-931 (horizon 643.9 chalcopyrite).

Table 4 Isotopic composition of lead in sulphides of Norilsk Region

Sample number	Depth (m)	Mineral	Number of grains	$^{206}Pb/^{204}Pb$	Err	$^{207}Pb/^{204}Pb$	Err	$^{208}Pb/^{204}Pb$	Err
Kharaelakh intrusive, Oktyabrskoe deposit									
Western flank									
ZF-13	441.9–442.5	total	34	18.101	0.013	15.449	0.011	37.621	0.029
		Ccp	1	18.087	0.066	15.429	0.056	37.589	0.142
		Py	33	18.101	0.013	15.449	0.011	37.621	0.029
ZF-13	448	Py	15	18.091	0.013	15.449	0.011	37.620	0.029
ZF-13	478.4–479	Py	23	18.292	0.028	15.480	0.024	37.744	0.061
ZF-18	467.5	total	19	18.170	0.019	15.471	0.022	37.671	0.040
		Ccp	8	18.166	0.028	15.470	0.024	37.660	0.058
		Pn	2	18.170	0.410	15.480	0.310	37.700	1.100
		Py	9	18.176	0.033	15.468	0.057	7.678	0.071
KZ-931	643.9	total	58	18.374	0.032	15.588	0.028	38.021	0.070
		Ccp	4	18.330	0.140	15.540	0.120	37.910	0.260
		Po	54	18.386	0.036	15.639	0.031	38.148	0.076
KZ-952	1010.4	total	55	18.247	0.035	15.576	0.030	37.924	0.077
		Ccp	49	18.278	0.034	15.604	0.027	38.004	0.082
		Pn	5	18.156	0.083	15.493	0.044	37.710	0.110
		Po	1	18.446	0.164	15.640	0.136	38.119	0.344
Central part									
KZ-1112	1094.8	total	39	18.250	0.110	15.531	0.063	37.790	0.230
		Ccp	36	18.230	0.110	15.504	0.065	37.710	0.230
		Pn	3	18.410	0.330	16.100	1.000	39.500	4.100
KZ_1112	1098.4	Ccp	20	18.158	0.031	15.470	0.020	37.630	0.052

(continued)

Table 4 (continued)

Sample number	Depth (m)	Mineral	Number of grains	$^{206}Pb/^{204}Pb$	Err	$^{207}Pb/^{204}Pb$	Err	$^{208}Pb/^{204}Pb$	Err
KZ-1112	1100.4	total	55	18.205	0.018	15.545	0.014	37.890	0.035
		Ccp	54	18.204	0.017	15.545	0.014	37.889	0.035
		Po	1	18.683	0.199	15.932	0.155	38.943	0.430
KZ-1112	1102.4	total	62	18.194	0.019	15.536	0.016	37.864	0.039
		Ccp	61	18.194	0.019	15.536	0.016	37.864	0.039
		Po	1	18.363	0.243	15.775	0.228	38.407	0.543
KZ-1112	1103.8	Ccp	50	18.299	0.053	15.608	0.045	38.020	0.110
KZ-981	1122.4–1123	total	47	18.141	0.010	15.465	0.009	37.669	0.022
		Ccp	26	18.141	0.012	15.467	0.010	37.673	0.025
		Py	21	18.138	0.022	15.455	0.019	37.655	0.049
KZ_1089	1154	total	23	18.206	0.047	15.497	0.040	37.680	0.110
		Ccp	5	18.210	0.140	15.500	0.120	37.690	0.300
		Pn	1	18.226	0.093	15.509	0.079	19.322	0.181
		Po	17	18.179	0.075	15.481	0.063	37.660	0.130
KZ-1089	1155.6	total	29	18.288	0.073	15.572	0.064	37.940	0.150
		Ccp	11	18.310	0.089	15.598	0.076	38.000	0.190
		Pn	12	18.280	0.130	15.550	0.120	37.900	0.270
		Po	6	18.260	0.300	15.570	0.210	37.930	0.590
Southern flank									
KZ-1084	1146.9	total	23	18.342	0.030	15.561	0.024	37.959	0.063
		Ccp	14	18.331	0.034	15.553	0.027	37.939	0.070
		Pn	4	18.410	0.110	15.605	0.090	38.100	0.220
		Po	5	18.454	0.090	15.653	0.078	38.210	0.210

(continued)

Table 4 (continued)

Sample number	Depth (m)	Mineral	Number of grains	$^{206}Pb/^{204}Pb$	Err	$^{207}Pb/^{204}Pb$	Err	$^{208}Pb/^{204}Pb$	Err
KZ-361bis	1064.4	total	40	18.115	0.024	15.439	0.019	37.616	0.050
		Ccp	39	18.115	0.024	15.439	0.020	37.616	0.051
		Po	1	18.120	0.088	15.439	0.074	37.624	0.185
Maslovsky deposit									
OM-123	1033.7	total	11	18.170	0.200	15.540	0.170	37.850	0.420
		Ccp	2	18.110	0.100	15.480	0.089	37.710	0.220
		Pn	1	18.324	0.163	15.662	0.138	38.192	0.342
		Po	8	18.300	0.470	15.650	0.390	38.130	0.990
OM10	1072.9–1073	Pn	3	17.853	0.052	15.454	0.046	37.600	0.120
OM-10	1082.4–1083	Py	37	17.940	0.014	15.439	0.012	37.571	0.029
Vologochan area									
OV-28	813.8–814.2	total	12	18.242	0.028	15.493	0.025	37.737	0.064
		Ccp	6	18.245	0.043	15.500	0.023	37.764	0.090
		Pn	5	18.245	0.055	15.494	0.048	37.730	0.130
		Po	1	18.153	0.099	15.389	0.084	37.536	0.208
Norilsk-1									
N1-0		total	39	18.037	0.012	15.460	0.011	37.615	0.027
		Ccp	35	18.036	0.013	15.458	0.011	37.612	0.027
		Cb	2	18.053	0.060	15.473	0.051	37.650	0.120
		Pn	2	18.070	0.100	15.490	0.085	37.680	0.220
N1-6		total	33	18.230	0.100	15.670	0.090	38.130	0.230
		Ccp	29	18.200	0.120	15.650	0.100	38.080	0.250
		Cb	1	18.362	0.139	15.780	0.120	38.364	0.297
		Pn	3	18.480	0.510	15.880	0.430	38.800	1.100

(continued)

Table 4 (continued)

Sample number	Depth (m)	Mineral	Number of grains	$^{206}Pb/^{204}Pb$	Err	$^{207}Pb/^{204}Pb$	Err	$^{208}Pb/^{204}Pb$	Err
Talnakh intrusive									
OUG-2		total	10	18.000	0.091	15.329	0.074	37.350	0.180
		Ccp (*)	1	17.867	0.058	15.220	0.049	37.081	0.126
		Ccp	9	18.044	0.089	15.364	0.072	37.430	0.140
KZ-774	1029.0–1029.6	total	3	18.180	0.100	15.462	0.089	37.670	0.230
		Ccp	2	18.190	0.110	15.467	0.093	37.690	0.250
		Pn	1	18.039	0.181	15.410	0.156	37.476	0.369
KZ-774	1032.4–1033	total	7	18.102	0.067	15.465	0.057	37.660	0.140
		Ccp	3	18.110	0.100	15.473	0.088	37.690	0.230
		Pn	2	18.140	0.150	15.490	0.130	37.710	0.320
		Po	2	18.070	0.110	15.444	0.090	37.610	0.230
KZ-774	1042.2–1049.7	total	32	18.151	0.010	15.481	0.008	37.717	0.022
		Py-(Ni)	3	18.163	0.037	15.485	0.031	37.731	0.079
		Py	29	18.150	0.010	15.481	0.009	37.715	0.022
Burkan deposit									
657-26		total	20	17.972	0.022	15.431	0.019	37.626	0.048
		Ccp	15	17.970	0.023	15.428	0.020	37.623	0.050
		Pn	3	18.000	0.200	15.460	0.150	37.670	0.900
		Po	2	18.000	2.200	15.400	2.000	37.600	4.600
Zub-Marksheidersky intrusive									
MP-25		total	13	18.266	0.037	15.495	0.015	37.810	0.058
		Po	1	18.242	0.021	15.504	0.017	37.813	0.058
		Ccp	1	18.396	0.026	15.488	0.022	37.975	0.055
		Py	11	18.251	0.031	15.491	0.010	37.787	0.049

Most of the investigated samples have isotopic composition of lead, which can be correlated with the interval of model ages 200–300 Ma. This does not contradict the generally accepted results of dating rock and ore samples in the studied massifs on the basis of different isotopic systems including U-Pb dating of zircons and model Re–Os age for massive sulphide ores 227–322 Ma.

New data are a methodologically independent confirmation of the results obtained by SIMS method, which showed significant differences in Pb isotopic composition in different intrusions, and, as a consequence, different sources of ore matter. The most "ancient" values in terms of model age were obtained for sulphide samples from Norilsk-1 deposits, the most radiogenic ones correspond to the sample of disseminated ore KZ-1084 from the southern Oktyabrsky deposit.

References

1. Stacey JS, Kramers JD (1975) Approximation of terrestrial lead isotope evolution by a two-stage model. Earth Planet Sci Lett 26(2)207–221
2. Zartman, RE Stacey JS (1979) General theory of plumbotectonics
3. Arndt NT, Czamanske GK, Walker RJ et al (2003) Geochemistry and origin of the intrusion hosts of the Noril'sk-Talnakh Cu–Ni–PGE sulphide deposit. Econ Geol 98:495–515
4. Zartman RE, Doe BR (1981) Plumbotectonics – the model/Zartman RE, Doe BR U.S. Geol Surv, Denver, CO 80225 (U.S.A.). Tectonophysics 75:135–162
5. Doe BR, Zartman RE, Stacey JS (1979) General theory of plumbotectonics. Nauka

Lutetium and Hafnium Isotopes in Zircons

Igor Kapitonov, Kirill Lokhov, Dmitriy Sergeev, Elena Adamskaya, Nikolay Goltsin and Sergey Sergeev

Abstract Lu–Hf isotope is an informative geochemical tool. Comprehensive in situ study of U–Pb and Lu–Hf isotope systems in zirconium minerals (zircon, baddeleyite) by LA-ICP-MS allows obtaining characteristics of the initial substance and evolution of the earth's crust. The data bear the crucial important information about the parent rocks and ore source in the Norilsk-Taymyr district. This chapter presents data obtained in VSEGEI isotope laboratory and in Australia. In Norilsk ores, significant variations in Hf isotopic composition have been revealed. Zircons of commercially ore-bearing intrusions have an increased value of εHf (to +10, weighted average εHf(T) = 8.2 ± 1.8). Compared with them, the Lower Talnakh low ore-bearing intrusion has practically zero εHf values. Thus, according to the results of the studied zircon sample, we can say that high εHf values are a necessary isotope criterion of ore content.

In the course of studying the characteristics of Lu–Hf isotopic system in zircons extracted from samples from different formations in the Norilsk ore district, new data on the genesis of rocks in mineralized massifs of Norilsk type were obtained. Characteristics of the Lu–Hf system for commercially mineralized and poorer intrusions were compared.

^{176}Lu isotope decays to produce hafnium isotope ^{176}Hf (decay constant $\sim 1.865 \times 10^{-11}$ years^{-1}). Hf and Zr are the elements of one group, that is why Hf content in zirconium minerals (zircon, baddeleyite) is sufficiently high, which makes it possible to determine hafnium isotopic composition using LA-ICPMS method.

As a rule, Lu–Hf ratio in zircons is very low; zircon actually fixes the isotopic composition of Hf captured at the time of formation. The necessary correction (usually minor) for the decay of ^{176}Lu which directly occurs in zircon can be easily introduced.

I. Kapitonov (✉) · K. Lokhov · D. Sergeev · E. Adamskaya · N. Goltsin · S. Sergeev
Russian Geological Research Institute (VSEGEI), 74 Sredny Prospect,
199106 St. Petersburg, Russia
e-mail: Igor_Kapitonov@vsegei.ru

O. Petrov (ed.), *Isotope Geology of the Norilsk Deposits*, Springer Geology,
https://doi.org/10.1007/978-3-030-05216-4_6

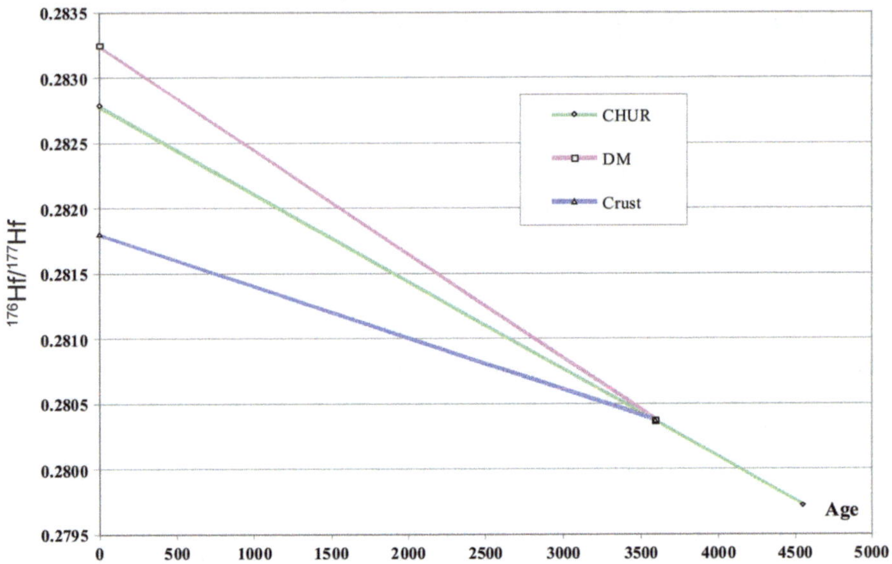

Fig. 1 Evolution of isotopic composition of hafnium in model reservoirs [1]

Figure 1 shows the evolution line of hafnium isotopic composition for three model reservoirs resulting from differentiation of the initial matter [1].

Red line corresponds to the depleted mantle with an elevated Lu/Hf ratio; blue line, to the crustal reservoir with a low Lu/Hf ratio, and, as a consequence, accumulation of less radiogenic ^{176}Hf, which leads to lower ^{176}Hf/^{177}Hf isotope ratios. Green line corresponds to the evolution of matter with chondrite composition. Thus, hafnium isotopic composition in combination with the data on the age of zircons indicates the type of the initial parent matter within the given model.

1 Methodology, Samples

Analytical approach to determining hafnium isotopic composition. Determination of parameters of Lu–Hf isotopic system in zircons was performed by laser sampling (laser ablation) in combination with mass spectrometric measurements with sample ionization in inductively coupled plasma (LA-ICP-MS).

Zircon grains had previously been implanted into epoxy resin (a disk 2.5 cm across). Review surveying was performed using an optical microscope (Fig. 2).

Grains of control standards were also placed into the same disk. Using the attachments to the electron microscope (Camscan), cathode luminescent images of grains were obtained and an image in the flow of return electrons for distinguishing the least disturbed grain zones.

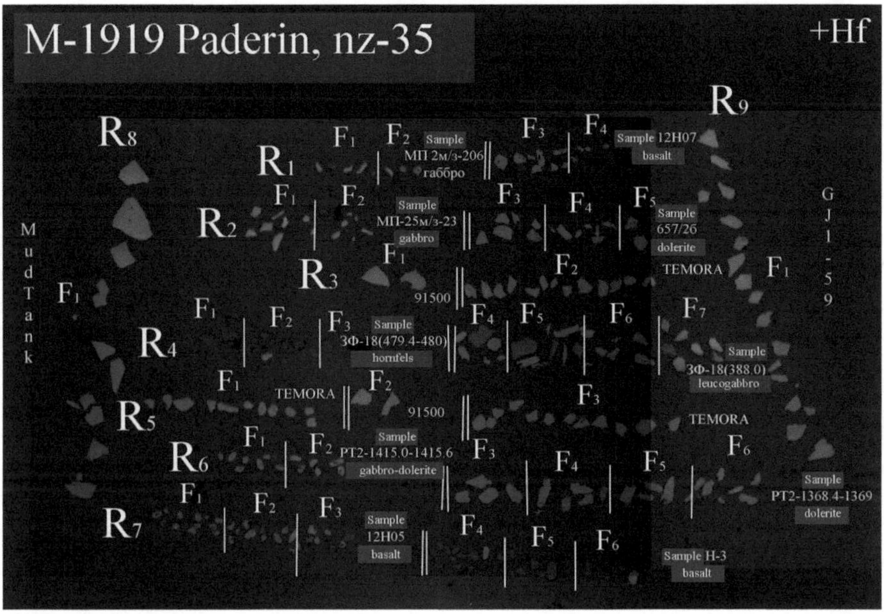

Fig. 2 Overview of disc surface

After this, U–Pb dating of zircons was performed on the secondary-ion mass spectrometer SHRIMP II (CIR VSEGEI).

Study of hafnium isotope systematics at the I stage (2005–2008) was accomplished in Australia (“GEMOS”, Sydney). The applied method is described in detail [2, 3].

In 2012–2014, all the analytical studies of zircons from the samples taken in different intrusions in the Norilsk district were carried out at CIR VSEGEI.

An analytical complex for studying hafnium isotopic composition at CIR VSEGEI comprises a laser ablation system DUV-193 (New Wave Research) based on the 193 nm ArF excimer laser COMPex-102 (Lambda Physik) and a multicollector mass spectrometer ThermoFinnigan Neptune. To study the isotopic systematics of Lu–Hf on the basis of “GEMOS” methodology [2], its modification was developed adapted to the set of analytical equipment available at CIR. Configuration of the collectors of mass spectrometer allowed a simultaneous recording of ^{172}Yb, ^{174}Yb, ^{175}Lu, ^{176}Hf, ^{177}Hf, ^{178}Hf, ^{179}Hf isotopes. $^{178}Hf/^{177}Hf$ normalizing ratio was used to correct mass discrimination. A correct value of ^{176}Hf signal was obtained by deducting from total intensity of the ion current at 176 fraction weight corresponding to ^{176}Yb and ^{176}Lu (for this purpose, ^{172}Yb and ^{175}Lu free from overlaps were measured).

A typical value of laser beam diameter in the analysis of zircons is ~50–70 mcm; laser pulse frequency is 7 Hz; energy of laser pulses ~200 mJ; power of ICP-generator ~1150 W.

When analyzing zircons extracted from the samples, a set of several types of zircons with the known $^{176}Hf/^{177}Hf$ isotopic ratios was additionally measured (measurement of "standards"). The meaning of such a measurement is to control the correctness of settings and operation of the analytical complex and the accuracy of measuring hafnium isotopic composition in the broadest possible range of $^{176}Hf/^{177}Hf$, Yb/Hf and Lu/Hf ratios.

For a reliable control of the correctness of analytical data, we constantly use three types of the internationally recognized zircon standard: GJ-1, MudTank, Temora. They are widely used to control the correctness of measurements in the analytical practice of different laboratories in the world. Hf isotope ratio values obtained at the CIR in these standards with an accuracy of more than 0.01% coincide with published data.

Standards also cover a broad range of Lu/Hf and Yb/Hf element ratios. In a typical multi-day measuring session, they cover the following range of ratios: $^{176}Yb/^{177}Hf = 0.00058–0.049$, $^{176}Lu/^{177}Hf = 0.000015–0.0018$.

For a correct determination of Lu, Yb and Hf concentrations, in addition to the above standards, we have used the international zircon standard 91500 certified for Hf and Lu content.

Limitations of the method. It should be noted that the change of $^{176}Hf/^{177}Hf$ isotopic composition in the Earth's evolution history gives a value of about 1%. Consequently, to obtain a meaningful result, the precision of not more than a few hundredths of a percent is required. The main contribution to an isobaric overlapping of ^{176}Hf is made by ^{176}Yb and ^{176}Lu. Undoubtedly, any correction of such overlaps can give a correct result only within certain limits. In any case, a reliable control of the measurement correctness is only possible within the range of $^{176}Yb/^{177}Hf$ and $^{176}Lu/^{177}Hf$ ratios in the standards that are used for control. In addition to the problem of correcting ^{176}Hf for the main isobaric overlaps, there is also a number of limitations, which are, as a rule, not taken into account. In particular, these are isobaric overlaps on 176 and 177 mass numbers, oxides of barium isotopes, which are often contained in crystals with a disturbed lattice and a high content of uranium and rare earths.

2 Results and Discussion

Results of the first stage of analytical studies of Lu–Hf isotopic system of zircons in the Norilsk district were obtained in Australia and have already been presented in a number of publications [4, 5].

At the I stage, data on concentrations of rare earth elements in zircons of commercially mineralized intrusions were also obtained. Variations of total REE (ΣREE) concentrations in zircon are from 300 to 38,500 ppm for Norilsk-1; from 400 to 20,900 ppm for the Talnakh; from 1200 to 15,500 ppm for the Kharaelakh intrusions.

It should be noted that the upper limits of concentration values are extremely high. It was shown that the investigated zircons also have elevated (to more than +10) εHf(T) values. The value of εHf(T) is calculated as multiplied by 10,000 value of deviation of the initial $^{176}Hf/^{177}Hf$ ratio (corresponding to the formation time of this zircon) from $^{176}Hf/^{177}Hf$ for the chondrite reservoir at the same time.

At the II stage, we analysed 27 grains from 8 zircon samples from wells in different areas of the Norilsk district. As the grains of many zircons used for dating were very small in size, it was impossible to determine hafnium isotopic composition (Table 1).

The newly obtained results lie in the same range of $^{176}Hf/^{177}Hf$ ratios and complement the values obtained at the I stage.

One should note high concentrations of heavy rare earth elements and, consequently, high Yb/Hf and Lu/Hf ratio in most of the studied zircon grains (after LA-ICPMS data) and high U and Th concentrations (data obtained in the course of SHRIMP dating). In some cases, there is a more than four-fold excess of Yb/Hf

Table 1 Isotopic composition of hafnium in zircons of Norilsk Region (all data by years)

Sample number	Age	Error	$^{176}Hf/^{177}Hf$	Error	$^{176}Lu/^{177}Hf$	$^{176}/^{177}Hf$ (T)	εHf (T)
Burkan deposit (2014)							
657-2b_N12-2	126	4	0.281952	0.000023	0.001265	0.281949	−26.4
657-2b_N1	132	3	0.281951	0.000029	0.002079	0.281946	−26.3
657-2b_N6-1	125	5	0.282115	0.000031	0.003264	0.282107	−20.8
657-2b_N3	126	4	0.282080	0.000057	0.003234	0.282073	−22.0
657-2b_N11-2	128	3	0.282181	0.000049	0.003311	0.282173	−18.4
Vologochan area (2008)							
424_29-17_2.1	226.5	1.0	0.282829	0.000055	0.003117	0.282816	6.5
M423_29-9_1.1	260.0	14.0	0.282900	0.000064	0.003189	0.282885	9.7
424_29-9_7.1	259.8	1.5	0.282817	0.000083	0.003446	0.282800	6.7
424_29-17_1.2	245.6	1.9	0.282949	0.000058	0.003681	0.282932	11.1
M423_29-9_19.1	265.0	15.0	0.282834	0.000240	0.003703	0.282816	7.4
424_29-16_4.1	228.6	1.0	0.282888	0.000062	0.003805	0.282872	8.6
424_29-16_2.1	246.0	1.3	0.282835	0.000042	0.003898	0.282817	7.0
424_29-17_1.1	222.1	1.6	0.282896	0.000072	0.003973	0.282880	8.7
M423_29-9_16.1	252.0	14.0	0.282934	0.000540	0.004804	0.282911	10.5
Zelenaya Griva intrusive (2008)							
F233-7_1.1	268.1	4.9	0.282631	0.000028	0.001504	0.282623	0.6
F233-2_2.1	264.9	4.5	0.282611	0.000033	0.002083	0.282601	−0.2
F233-4_17.1	266.8	2.0	0.282524	0.000049	0.002093	0.282514	−3.3
F233-2_1.1	279	4.2	0.282654	0.000030	0.002293	0.282642	1.5
Zub-Marksheidersky intrusive (2014)							
MP-25_mz23_N1	237	6	0.283017	0.00004	0.003128	0.283003	13.4
MP-25_mz23_N2	237	7	0.282799	0.00005	0.003233	0.282785	5.7

(continued)

Table 1 (continued)

Sample number	Age	Error	$^{176}Hf/^{177}Hf$	Error	$^{176}Lu/^{177}Hf$	$^{176/177}Hf$ (T)	εHf (T)
Listvyanka deposit (2014)							
12N07-N3	150	5	0.282664	0.00006	0.000690	0.282662	−0.6
12N07-N1	150	5	0.282774	0.00003	0.001143	0.282771	3.2
12N07-N2	230	6	0.282867	0.00003	0.002063	0.282858	8.1
12N05_N9	240	6	0.282621	0.00006	0.001388	0.282615	−0.3
MP-2_f/g-206-N6	244	6	0.283214	0.00020	0.010334	0.283166	19.3
Mikchangda intrusive (2008)							
536_48-30_1.1_ (first)	263.3	2.6	0.283012	0.000023	0.004948	0.282988	13.4
536_48-30_1.1_(all)	263.3	2.6	0.282970	0.000030	0.005227	0.282944	11.9
Lower Talnakh intrusive (2008)							
31-7_1.1_2	293.0	8.1	0.282588	0.000028	0.000212	0.282587	−0.1
31-7_1.1_1	293.0	8.1	0.282491	0.000037	0.000231	0.282490	−3.6
31-7_9.2_2	237.3	5.7	0.282687	0.000014	0.000344	0.282685	2.2
31-10_4.1_all	221.8	5.4	0.282553	0.000049	0.000400	0.282551	−2.9
31-7_2.1_2	248.0	7.0	0.282698	0.000045	0.000449	0.282696	2.8
31-13_28.1_1	269.0	7.6	0.282599	0.000028	0.000482	0.282597	−0.3
31-7_9.2_1	237.3	5.7	0.282643	0.000019	0.000487	0.282641	0.6
31-13_42.1_2	242.0	6.8	0.282704	0.000030	0.000489	0.282702	2.8
31-13_9.2	288.9	9.0	0.282563	0.000310	0.000507	0.282560	−1.2
31-13_9.1_2	254.2	8.1	0.282623	0.000025	0.000528	0.282620	0.2
31-16_26.2	270.0	15.0	0.282672	0.000027	0.000539	0.282669	2.3
31-13_28.2_1	245.0	7.0	0.282520	0.000020	0.000561	0.282517	−3.6
31-7_6.1_2	233.7	5.7	0.282607	0.000020	0.000573	0.282604	−0.8
31-13_8.1_1	257.8	7.8	0.282612	0.000026	0.000576	0.282609	−0.1
31-13_8.1_2	257.8	7.8	0.282570	0.000038	0.000586	0.282567	−1.6
31-13_9.1_1	254.2	8.1	0.282634	0.000024	0.000587	0.282631	0.6
31-16_20.1	230.0	13.0	0.282564	0.000027	0.000594	0.282561	−2.4
31-9_2.1_2	217.6	5.3	0.282642	0.000020	0.000608	0.282640	0.1
31-13_42.1_3	242.0	6.8	0.282738	0.000015	0.000627	0.282735	4.0
31-7_2.1_1	248.0	7.0	0.282630	0.000041	0.000639	0.282627	0.3
31-10_1.1	238.4	5.8	0.282636	0.000025	0.000647	0.282633	0.3
31-11_1.1_2	217.6	5.3	0.282669	0.000029	0.000678	0.282666	1.0
31-13_54.3_2	266.3	8.1	0.282562	0.000030	0.000692	0.282559	−1.7
31-13_54.3_1	266.3	8.1	0.282672	0.000017	0.000698	0.282669	2.2
31-13_50	258.6	7.9	0.282635	0.000019	0.000750	0.282631	0.7
31-7_6.1_1	233.7	5.7	0.282523	0.000035	0.000751	0.282520	−3.8
31-13_42.1_1	242.0	6.8	0.282743	0.000031	0.000772	0.282740	4.2
31-13_28.2_2	245.0	7.0	0.282542	0.000035	0.000817	0.282538	−2.9
31-16_13.1	265.9	6.1	0.282645	0.000017	0.000872	0.282641	1.2
31-13_20.1	260.8	8.0	0.282671	0.000023	0.000890	0.282667	2.0
31-13_19.1	256.7	7.8	0.282604	0.000018	0.000893	0.282600	−0.5

(continued)

Table 1 (continued)

Sample number	Age	Error	$^{176}Hf/^{177}Hf$	Error	$^{176}Lu/^{177}Hf$	$^{176}/^{177}Hf$ (T)	εHf (T)
31-13_28.1_2	269.0	7.6	0.282560	0.000040	0.000959	0.282555	−1.8
31-11_1.1_1	217.6	5.3	0.282620	0.000038	0.000972	0.282616	−0.7
31-13_8.2_2	304.7	9.5	0.282441	0.000037	0.001024	0.282435	−5.2
31-13_8.2_1	304.7	9.5	0.282443	0.000033	0.001043	0.282437	−5.2
31-13_23.1	258.8	7.9	0.282653	0.000017	0.001064	0.282648	1.3
31-16_67.1	233.0	13.0	0.282656	0.000028	0.001098	0.282651	0.8
31-13_6.1	263.2	8.1	0.282578	0.000016	0.001118	0.282572	−1.3
31-7_5.1	230.5	5.6	0.282646	0.000024	0.001175	0.282641	0.4
31-13_11.1_2	234.0	6.6	0.282560	0.000047	0.001214	0.282555	−2.6
31-13_14.2	250.1	7.6	0.282642	0.000023	0.001250	0.282636	0.7
31-16_78.1	255.0	14.0	0.282667	0.000039	0.001261	0.282661	1.7
31-13_48.1_1	275.9	8.4	0.282555	0.000022	0.001303	0.282548	−1.9
31-13_48.1_2	275.9	8.4	0.282562	0.000029	0.001325	0.282555	−1.6
31-16_26.1	231.0	13.0	0.282759	0.000026	0.001364	0.282753	4.4
31-9_3.1_2	223.9	5.5	0.282647	0.000021	0.001392	0.282641	0.3
31-13_4.2	273.1	8.4	0.282620	0.000019	0.001417	0.282613	0.4
31-9_3.1_1	223.9	5.5	0.282644	0.000034	0.001420	0.282638	0.2
31-16_19.2_2	240.8	5.8	0.282702	0.000035	0.001466	0.282695	2.6
31-16_25.2	295.0	16.0	0.282620	0.000026	0.001474	0.282612	0.8
31-16_19.2_1	240.8	5.8	0.282693	0.000035	0.001485	0.282686	2.3
31-16_22.2	267.0	15.0	0.282791	0.000039	0.001509	0.282783	6.3
31-16_37.1	271.0	15.0	0.282645	0.000031	0.001658	0.282637	1.2
31-13_11.1_1	234.0	6.6	0.282500	0.000037	0.001712	0.282493	−4.8
31-16_47.1	247.0	14.0	0.282629	0.000030	0.001749	0.282621	0.1
31-3_2.1_2	226.5	5.5	0.282677	0.000018	0.001985	0.282669	1.3
31-16_44.1_2	262.5	6.1	0.282688	0.000019	0.002008	0.282678	2.4
31-3_2.1_1	226.5	5.5	0.282667	0.000023	0.002017	0.282658	1.0
31-16_73.1	254.7	6.2	0.282682	0.000041	0.002121	0.282672	2.1
31-13_5.1	240.0	7.0	0.282736	0.000039	0.002167	0.282726	3.7
31-16_49.1	253.0	14.0	0.282718	0.000035	0.002375	0.282707	3.3
31-16_39.2	213.2	5.6	0.282654	0.000023	0.002459	0.282644	0.2
31-13_21.1	220.0	6.2	0.282649	0.000062	0.002475	0.282639	0.1
31-16_31.1	262.0	15.0	0.282585	0.000051	0.002559	0.282572	−1.3
31-16_39.1	254.2	6.2	0.282805	0.000059	0.002675	0.282792	6.3
31-16_6.1	263.2	8.1	0.282567	0.000039	0.002791	0.282553	−2.0
31-16_54.1	256.0	14.0	0.282753	0.000035	0.002795	0.282740	4.5
31-16_44.1_1	262.5	6.1	0.282652	0.000025	0.002874	0.282638	1.0
31-16_10.1	260.0	14.0	0.282661	0.000023	0.002875	0.282647	1.3
31-13_12.1	225.0	6.4	0.282739	0.000093	0.002908	0.282727	3.3
31-16_38.1	261.0	14.0	0.282803	0.000046	0.003563	0.282786	6.2
Norilsk-1 deposit (2008)							
N1-8.17.1	271.8		0.282864	0.000045	0.000792	0.282860	9.1
N1-8.(17).3.1	239.3		0.282826	0.000045	0.001004	0.282822	7.0

(continued)

Table 1 (continued)

Sample number	Age	Error	$^{176}Hf/^{177}Hf$	Error	$^{176}Lu/^{177}Hf$	$^{176}/^{177}Hf$ (T)	εHf (T)
N1-8.10.1	256.4		0.282857	0.000045	0.001274	0.282851	8.4
N1-9.17.1	231.1		0.282890	0.000045	0.001542	0.282883	9.0
N1-7.19.1	248.4		0.282856	0.000045	0.001624	0.282848	8.2
N1-7.9.8.1	245.8		0.282896	0.000045	0.001628	0.282889	9.5
N1-9.25.1	229.1		0.282853	0.000045	0.001737	0.282846	7.6
N1-7.17.3.1	265.9		0.282774	0.000045	0.001767	0.282765	5.6
N1-7.(23).1.1	252.1		0.282870	0.000045	0.001789	0.282862	8.7
N1-8.(3).10.1	250.0		0.282905	0.000045	0.002018	0.282896	9.9
N1-9.17.2	246.7		0.282910	0.000045	0.002085	0.282900	10.0
N1-10.2.6.1	266.4		0.282804	0.000045	0.002325	0.282792	6.6
N1-9.5.1	218.3		0.282880	0.000045	0.002508	0.282870	8.3
N1-1.12.1.1	232.6		0.282670	0.000045	0.002601	0.282659	1.1
N1-8.(6).8.1	257.2		0.282941	0.000045	0.002764	0.282828	11.2
N1-7.(18).5.1	263.4		0.283035	0.000001	0.002876	0.283021	14.6
N1-7.18.1	239.2		0.282975	0.000004	0.002988	0.282962	12.0
N1-5.13.3.1	256.3		0.283133	0.000008	0.003103	0.283118	17.9
N1-5.17.1	260.0		0.282977	0.000012	0.003239	0.282961	12.4
N1-3.5.2	246.9		0.282914	0.000036	0.003996	0.282896	9.8
N1-6.11.1	249.0		0.282908	0.000036	0.004004	0.282889	9.6
N1-1.5.7.1	239.5		0.282814	0.000041	0.004162	0.282795	6.1
N1-6.11.2	229.9		0.283106	0.000051	0.004474	0.283087	16.2
N1-5.4.8.1	252.2		0.282960	0.000064	0.004893	0.282937	11.4
N1-6.(27).6.1	243.7		0.283045	0.000074	0.005213	0.283021	14.2
N1-6.(18).1.4	231.4		0.283142	0.000087	0.005595	0.283118	17.3
N1-3.5.1	237.9		0.282946	0.000091	0.005725	0.282921	10.5
N1-4.12.1	237.4		0.283066	0.000092	0.005781	0.283040	14.7
N1-5.(1).7.1	238.2		0.283041	0.000096	0.005888	0.283015	13.8
N1-3.2.1	238.7		0.283084	0.000104	0.006153	0.283057	15.3
N1-6.26.1	231.6		0.283201	0.000112	0.006395	0.283173	19.3
N1-5.(22).1.1	277.7		0.283133	0.000126	0.006858	0.283097	17.6
N1-4.17.1	232.1		0.283179	0.000134	0.007091	0.283148	18.4
N1-9.4.9.1	250.8		0.283045	0.000134	0.007096	0.283012	14.0
N1-2.19.1	262.4		0.283262	0.000143	0.007384	0.283226	21.8
N1-5.(17).2.1	245.7		0.283057	0.000168	0.008171	0.283019	14.2
N1-3.3.1.1	243.6		0.283115	0.000168	0.008191	0.283078	16.2
N1-4.6.6.1	253.6		0.283131	0.000241	0.010489	0.283081	16.5
Norilsk-1 deposit (2014)							
N-3_N2-2	210	10	0.282937	0.00007	0.000540	0.282935	10.4
Talnakh intrusive (2008)							
T-17_1.2	251.2		0.282867	0.000027	0.000376	0.282865	8.8
T-17_1.1	238.0		0.282807	0.000025	0.000520	0.282805	6.4
T-13_5.2	256.3		0.283093	0.000029	0.001487	0.283086	16.7
T-12_6.1	265.0		0.282928	0.000150	0.001786	0.282919	11.0

(continued)

Table 1 (continued)

Sample number	Age	Error	$^{176}Hf/^{177}Hf$	Error	$^{176}Lu/^{177}Hf$	$^{176}/^{177}Hf$ (T)	εHf (T)
T-1_26.1-2	245.3		0.282870	0.000025	0.001873	0.282861	8.6
T-1_26.1-1	245.3		0.282857	0.000026	0.001943	0.282848	8.1
T-15_1.1	261.0		0.282875	0.000033	0.002091	0.282865	9.0
T-13_5.1	272.3		0.282858	0.000098	0.002112	0.282847	8.6
T-1_24.1	212.0		0.282834	0.000035	0.002228	0.282825	6.5
T-13_7.1	239.5		0.282971	0.000025	0.002237	0.282961	12.0
T-2_8.1	240.4		0.282871	0.000036	0.002256	0.282861	8.4
T-18_22.2-2	237.3		0.282808	0.000047	0.002513	0.282797	6.1
T-3_10.1	268.9		0.283008	0.000036	0.002608	0.282995	13.8
T-13_5.1	272.3		0.282877	0.000053	0.002662	0.282863	9.2
T-18_22.1	224.6		0.282931	0.000053	0.002889	0.282919	10.1
T-16_1.1	258.0		0.282874	0.000032	0.002937	0.282860	8.8
T-8_16.1	240.0		0.282951	0.000020	0.003152	0.282937	11.1
T-18_22.2-1	237.3		0.282956	0.000063	0.003209	0.282942	11.2
T-3_27.1	285.4		0.282937	0.000022	0.003672	0.282917	11.4
T-10_18.1	259.3		0.282989	0.000027	0.003738	0.282971	12.7
T-18_11.2-1	235.9		0.282778	0.000065	0.003767	0.282761	4.8
T-6_6.1-2	263.0		0.282962	0.000019	0.003780	0.282943	11.8
T-13_12.1	262.0		0.282926	0.000041	0.003799	0.282907	10.6
T-1_26.2	232.6		0.282938	0.000029	0.003940	0.282921	10.4
T-10_26.1	260.5		0.282777	0.000038	0.003974	0.282758	5.2
T-10_6-1	252.1		0.282978	0.000021	0.004021	0.282959	12.2
T-18_11.2-2	235.9		0.282905	0.000032	0.004027	0.282887	9.3
T-6_2.1	269.0		0.283034	0.000020	0.004122	0.283013	14.5
T-12_8(2).1	256.0		0.282890	0.000060	0.004141	0.282870	9.1
T-2_32.1	237.0		0.282987	0.000026	0.004171	0.282969	12.2
T-6_14.1	255.0		0.282828	0.000050	0.004237	0.282808	6.9
T-3_28.1	282.7		0.282975	0.000022	0.004315	0.282952	12.6
T-6_5.2	239.0		0.282910	0.000049	0.004359	0.282891	9.4
T-12_2.2	257.0		0.282840	0.000036	0.004433	0.282819	7.3
T-6_6.1-1	263.0		0.282905	0.000032	0.004495	0.282883	9.7
T-13_16.1	261.0		0.282857	0.000034	0.004613	0.282834	7.9
T-12_6.2	251.0		0.283047	0.000050	0.004991	0.283024	14.4
T-13_2	268.0		0.282879	0.000033	0.005063	0.282854	8.8
T-15_4.1	262.0		0.282824	0.000048	0.005158	0.282799	6.7
T-10_6-2	252.1		0.282911	0.000033	0.005179	0.282887	9.6
T-2_2.1	240.4		0.282857	0.000034	0.005607	0.282832	7.4
T-12_6.2	251.0		0.282930	0.000053	0.005695	0.282903	10.2
T-10_2.1	253.0		0.282808	0.000048	0.005825	0.282780	5.9
T-13_11.1	261.0		0.282850	0.000039	0.006042	0.282821	7.5

(continued)

Table 1 (continued)

Sample number	Age	Error	$^{176}Hf/^{177}Hf$	Error	$^{176}Lu/^{177}Hf$	$^{176}/^{177}Hf$ (T)	εHf (T)
Kharaelakh intrusive (2008)							
844_6_35.1	248.2		0.282856	0.000045	0.001537	0.282849	8.2
844_1_24.1	239.6		0.282917	0.000045	0.001702	0.282909	10.1
844_1_14.1	248.5		0.282913	0.000045	0.001899	0.282904	10.1
844-6_28.1	257.4		0.282767	0.000045	0.001908	0.282758	5.1
844_1_20.1	248.9		0.282962	0.000045	0.002050	0.282952	11.9
844_1_10.1	338.0		0.282820	0.000045	0.002142	0.282806	8.7
844_1_3_1-2	244.5		0.282832	0.000045	0.002207	0.282822	7.1
844_1_19.1	267.2		0.282939	0.000045	0.002228	0.282928	11.4
844_1_22.1	234.7		0.283007	0.000045	0.002232	0.282997	13.1
844_6_33.1	242.8		0.283066	0.000045	0.002251	0.283056	15.4
844_1_10.2	264.9		0.282765	0.000045	0.002332	0.282753	5.2
844_7_37.1	355.0		0.282796	0.000045	0.002336	0.282780	8.1
844_7_37.2	270.0		0.282692	0.000045	0.002383	0.282680	2.7
844-10,11_41.2	249.0		0.283078	0.000045	0.002461	0.283067	15.9
844-6_28.2	233.2		0.282777	0.000045	0.002509	0.282766	4.9
844-10,11_40.1	249.0		0.282825	0.000045	0.002710	0.282812	6.9
844_1_26.1	237.0		0.282983	0.000045	0.002817	0.282971	12.2
844_1_7.1	229.4		0.282907	0.000004	0.002987	0.282894	9.4
844-1_13.1	255.3		0.282887	0.000006	0.003035	0.282873	9.2
844_1_16.1	255.8		0.282825	0.000008	0.003106	0.282810	7.0
844-6_30	246.3		0.282803	0.000008	0.003113	0.282789	6.0
844-1_15.1	253.4		0.282992	0.000012	0.003224	0.282977	12.8
844_1_2.1	249.5		0.283073	0.000012	0.003234	0.283058	15.6
844-6_36(9).1	254.6		0.283101	0.000017	0.003374	0.283085	16.7
844_1_12.1	245.6		0.282869	0.000024	0.003614	0.282852	8.2
844_1_6.1	241.8		0.282956	0.000025	0.003641	0.282940	11.2
Kharaelakh intrusive, depth 1368.4–1369 (2014)							
PT-2-N1	246	7	0.282780	0.0000374	0.00093	0.282776	5.5
PT-2-N6	249	6	0.282775	0.0000518	0.00142	0.282769	5.4
PT-2-N5	252	8	0.282722	0.0000511	0.00240	0.282711	3.4
PT-2-N8	248	8	0.282968	0.0000423	0.00306	0.282954	11.9
PT-2-N3	252	7	0.282969	0.0000696	0.00301	0.282955	12.0
PT-2-N14	253	7	0.283277	0.0002394	0.00601	0.283249	22.4
PT-2-N2	231	6	0.282602	0.0001360	0.00516	0.282580	−1.7
PT-2-N10	241	6	0.283027	0.0002112	0.00611	0.283000	13.4
PT-2-N7	253	6	0.283118	0.0001488	0.00647	0.283088	16.7
Kharaelakh intrusive, Oktyabrskoe deposit (2014)							
ZF-18_388_N7	234	7	0.282936	0.0000364	0.00177	0.282928	10.7
ZF-18_388_N4-2	241	6	0.283148	0.0000940	0.00474	0.283127	17.9
ZF-18_388_N2	239	7	0.282594	0.0001329	0.00542	0.282570	−1.9
ZF-18_388_N1	239	5	0.283397	0.0001336	0.00902	0.283357	25.9

(continued)

Table 1 (continued)

Sample number	Age	Error	$^{176}Hf/^{177}Hf$	Error	$^{176}Lu/^{177}Hf$	$^{176}/^{177}Hf$ (T)	εHf (T)
ZF-18_388_N8-1	234	5	0.283246	0.0001665	0.00695	0.283215	20.8
Chernogorsk intrusive (2008)							
424_CH-11_4.2_(I)	248.3	1.5	0.282871	0.0000390	0.00156	0.282864	8.7
424_CH-11_3.2_old_(I)	292.5	5.4	0.282989	0.0000410	0.00159	0.282980	13.8
424_CH-11_4.2_all	248.3	1.5	0.282784	0.0000350	0.00174	0.282776	5.6
424_CH-11_4.1	245.0	2.3	0.282877	0.0000610	0.00177	0.282869	8.8
424_CH-11_2.1_(I)	242.4	2.0	0.282826	0.0000300	0.00179	0.282818	7.0
424_CH-11_2.1	242.4	2.0	0.282780	0.0000570	0.00189	0.282771	5.3
424_CH-11_7.1	251.8	1.7	0.282752	0.0000290	0.00230	0.282741	4.4
424_CH-11_6.1	242.6	2.4	0.282810	0.0000210	0.00232	0.282799	6.3
424_CH-11_5.2 (2-1)	248.7	0.9	0.282880	0.0000430	0.00237	0.282869	8.9
424_CH-11_3.1	246.2	1.7	0.282797	0.0000350	0.00241	0.282786	5.9
424_CH-11_3.1	246.2	1.7	0.282836	0.0000350	0.00272	0.282823	7.2
424_CH-11_8.1_(all)	249.6	1.1	0.282898	0.0000440	0.00543	0.282873	9.0

ratio in a number of the investigated zircon grains as compared to the maximum Yb/Hf ratio in zircons of the reference standard Temora. Such high Yb/Hf and Lu/Hf ratios require a very careful approach to data interpretation. Results of $^{176}Hf/^{177}Hf$ measurements were sorted after Yb/Hf ratio in zircons and divided into groups. They are shown in Figs. 3 and 4 in the coordinates age—$^{176}Hf/^{177}Hf$ hafnium isotope ratio.

For the highest correctness, in further discussions of Hf isotope system, only the results from the first group of zircons were used (Fig. 4). Actual data are presented in Table 3.

Zircons with high Lu/Hf and Yb/Hf ratios were rejected for guaranteeing the reliability of determining hafnium isotopic composition, carry the most important information about the parent rocks, ore sources in the Norilsk-Taimyr district. There is no doubt that the source involved in the formation of these zircons also had a high Lu/Hf ratio, which is typical of commercially mineralized intrusions.

Halogen-containing aqueous fluid, which is required to transport Zr and Hf, could be borrowed from horizons of the evaporite rocks enclosing halogenides [6]. A possibility of this mechanism is illustrated by contamination of the initial matter of intrusions by sedimentary rocks, which was previously recorded from isotopic compositions of sulphur in sulphides and argon in rocks of the intrusions [7–9].

Time and place of crustal matter assimilation by mantle magmas should affect the variations of Hf isotopic composition in zircons.

One can imagine several possible contamination stages: during the crystallization process, prior to it or in the course of postmagmatic hydrothermal alterations.

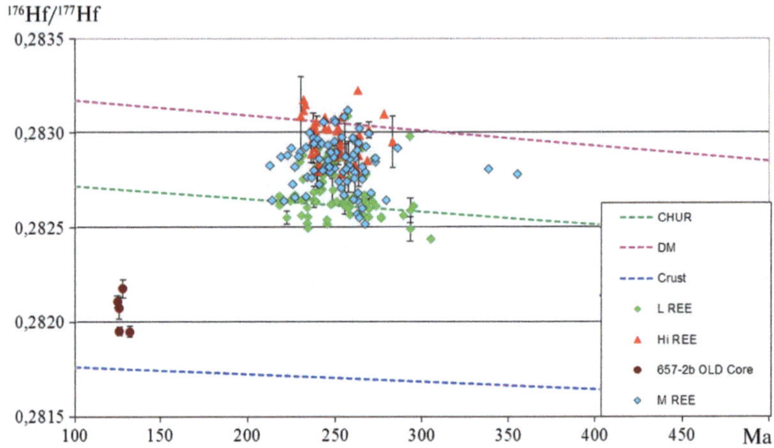

Fig. 3 Isotopic composition of Hf in zircons of Norilsk District (all data). *Green symbols* Yb/Hf and Lu/Hf ratios in zircons do not exceed the values in reference standards most rich in REE; *blue colour* shows grains with not more than a twofold excess in comparison with those measured in standards; *red dots* grains with high REE contents. Yb/Hf and Lu/Hf ratios exceed more than twofold the corresponding values in the reference standards. These grains are also characterized by higher measurement errors

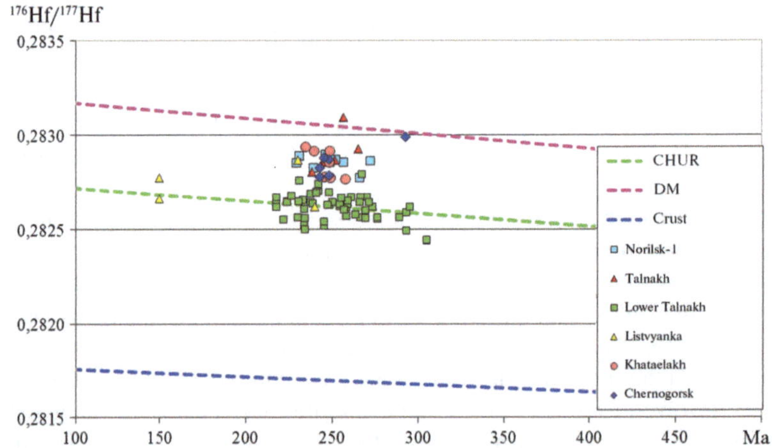

Fig. 4 Isotopic composition of Hf in zircons of Norilsk District (data for grains with Yb/Hf and Lu/Hf ratios not exceeding the maximum values in reference standards are shown)

In case of contamination at the stage of magmatic melt one can expect a higher degree of homogeneity of the characteristics including hafnium isotopic composition. When crustal material is assimilated at the postmagmatic stage in the alteration process under the influence of the fluid, significant variations are possible in the characteristics of altered rocks including hafnium isotopic composition depending on local activity of the process.

Results of the study of Hf isotopic composition (only values of the first group of zircons are used with the lowest Yb/Hf, Lu/Hf ratios in zircons of the Norilsk district) show the following (Table 2):

Table 2 Isotopic composition of hafnium in zircons of Norilsk Region: data for grains with Yb/Hf and Lu/Hf ratios not exceeding maximum values in reference standards (I group)

Sample number	Age	Error	$^{176}Hf/^{177}Hf$	Error	$^{176}Lu/^{177}Hf$	$^{176}/^{177}Hf$ (T)	εHf (T)
Listvyanka (2014)							
12N07-N3	149.7	7.0	0.28266	0.00006	0.00069	0.28266	−0.6
12N07-N1	149.7	7.0	0.28277	0.00003	0.00114	0.28277	3.2
12N07-N2	229.8	8.6	0.28287	0.00003	0.00206	0.28286	8.1
12N05_N9	239.8	8.8	0.28262	0.00006	0.00139	0.28261	−0.3
Lower Talnakh intrusive (2008)							
31-7_1.1_2	293.0	11.0	0.282588	0.000028	0.000212	0.282587	−0.1
31-7_1.1_1	293.0	11.0	0.282491	0.000037	0.000231	0.282490	−3.6
31-7_9.2_2	237.3	8.1	0.282687	0.000014	0.000344	0.282685	2.2
31-10_4.1_all	221.8	7.6	0.282553	0.000049	0.000400	0.282551	−2.9
31-7_2.1_2	248.0	9.5	0.282698	0.000045	0.000449	0.282696	2.8
31-13_28.1_1	269.0	10.3	0.282599	0.000028	0.000482	0.282597	−0.3
31-7_9.2_1	237.3	8.1	0.282643	0.000019	0.000487	0.282641	0.6
31-13_42.1_2	242.0	9.2	0.282704	0.000030	0.000489	0.282702	2.8
31-13_9.2	288.9	11.9	0.282563	0.000310	0.000507	0.282560	−1.2
31-13_9.1_2	254.2	10.6	0.282623	0.000025	0.000528	0.282620	0.2
31-16_26.2	270.0	17.7	0.282672	0.000027	0.000539	0.282669	2.3
31-13_28.2_1	245.0	9.5	0.282520	0.000020	0.000561	0.282517	−3.6
31-7_6.1_2	233.7	8.0	0.282607	0.000020	0.000573	0.282604	−0.8
31-13_8.1_1	257.8	10.4	0.282612	0.000026	0.000576	0.282609	−0.1
31-13_8.1_2	257.8	10.4	0.282570	0.000038	0.000586	0.282567	−1.6
31-13_9.1_1	254.2	10.6	0.282634	0.000024	0.000587	0.282631	0.6
31-16_20.1	230.0	15.3	0.282564	0.000027	0.000594	0.282561	−2.4
31-9_2.1_2	217.6	7.5	0.282642	0.000020	0.000608	0.282640	0.1
31-13_42.1_3	242.0	9.2	0.282738	0.000015	0.000627	0.282735	4.0
31-7_2.1_1	248.0	9.5	0.282630	0.000041	0.000639	0.282627	0.3
31-10_1.1	238.4	8.2	0.282636	0.000025	0.000647	0.282633	0.3
31-11_1.1_2	217.6	7.5	0.282669	0.000029	0.000678	0.282666	1.0
31-13_54.3_2	266.3	10.8	0.282562	0.000030	0.000692	0.282559	−1.7
31-13_54.3_1	266.3	10.8	0.282672	0.000017	0.000698	0.282669	2.2
31-13_50	258.6	10.5	0.282635	0.000019	0.000750	0.282631	0.7
31-7_6.1_1	233.7	8.0	0.282523	0.000035	0.000751	0.282520	−3.8
31-13_42.1_1	242.0	9.2	0.282743	0.000031	0.000772	0.282740	4.2

(continued)

Table 2 (continued)

Sample number	Age	Error	$^{176}Hf/^{177}Hf$	Error	$^{176}Lu/^{177}Hf$	$^{176}/^{177}Hf$ (T)	εHf (T)
31-13_28.2_2	245.0	9.5	0.282542	0.000035	0.000817	0.282538	−2.9
31-16_13.1	265.9	8.8	0.282645	0.000017	0.000872	0.282641	1.2
31-13_20.1	260.8	10.6	0.282671	0.000023	0.000890	0.282667	2.0
31-13_19.1	256.7	10.4	0.282604	0.000018	0.000893	0.282600	−0.5
31-13_28.1_2	269.0	10.3	0.282560	0.000040	0.000959	0.282555	−1.8
31-11_1.1_1	217.6	7.5	0.282620	0.000038	0.000972	0.282616	−0.7
31-13_8.2_2	304.7	12.5	0.282441	0.000037	0.001024	0.282435	−5.2
31-13_8.2_1	304.7	12.5	0.282443	0.000033	0.001043	0.282437	−5.2
31-13_23.1	258.8	10.5	0.282653	0.000017	0.001064	0.282648	1.3
31-16_67.1	233.0	15.3	0.282656	0.000028	0.001098	0.282651	0.8
31-13_6.1	263.2	10.7	0.282578	0.000016	0.001118	0.282572	−1.3
31-7_5.1	230.5	7.9	0.282646	0.000024	0.001175	0.282641	0.4
31-13_11.1_2	234.0	8.9	0.282560	0.000047	0.001214	0.282555	−2.6
31-13_14.2	250.1	10.1	0.282642	0.000023	0.001250	0.282636	0.7
31-16_78.1	255.0	16.6	0.282667	0.000039	0.001261	0.282661	1.7
31-13_48.1_1	275.9	11.2	0.282555	0.000022	0.001303	0.282548	−1.9
31-13_48.1_2	275.9	11.2	0.282562	0.000029	0.001325	0.282555	−1.6
31-16_26.1	231.0	15.3	0.282759	0.000026	0.001364	0.282753	4.4
31-9_3.1_2	223.9	7.7	0.282647	0.000021	0.001392	0.282641	0.3
31-13_4.2	273.1	11.1	0.282620	0.000019	0.001417	0.282613	0.4
31-9_3.1_1	223.9	7.7	0.282644	0.000034	0.001420	0.282638	0.2
31-16_19.2_2	240.8	8.2	0.282702	0.000035	0.001466	0.282695	2.6
31-16_25.2	295.0	19.0	0.282620	0.000026	0.001474	0.282612	0.8
31-16_19.2_1	240.8	8.2	0.282693	0.000035	0.001485	0.282686	2.3
31-16_22.2	267.0	17.7	0.282791	0.000039	0.001509	0.282783	6.3
31-16_37.1	271.0	17.7	0.282645	0.000031	0.001658	0.282637	1.2
31-13_11.1_1	234.0	8.9	0.282500	0.000037	0.001712	0.282493	−4.8
31-16_47.1	247.0	16.5	0.282629	0.000030	0.001749	0.282621	0.1
31-3_2.1_2	226.5	7.8	0.282677	0.000018	0.001985	0.282669	1.3
Norilsk-1 deposit (2008)							
N1-8.17.1	271.8	9.2	0.282864	0.000045	0.000792	0.282860	9.1
N1-8.(17).3.1	239.3	10.193	0.282826	0.000045	0.001004	0.282822	7.0
N1-8.10.1	256.4	8.064	0.282857	0.000045	0.001274	0.282851	8.4
N1-9.17.1	231.1	5.811	0.282890	0.000045	0.001542	0.282883	9.0
N1-7.19.1	248.4	8.684	0.282856	0.000045	0.001624	0.282848	8.2
N1-7.9.8.1	245.8	7.858	0.282896	0.000045	0.001628	0.282889	9.5
N1-9.25.1	229.1	10.391	0.282853	0.000045	0.001737	0.282846	7.6
N1-7.17.3.1	265.9	6.559	0.282774	0.000045	0.001767	0.282765	5.6
N1-7.(23).1.1	252.1	8.721	0.282870	0.000045	0.001789	0.282862	8.7
N-3_N2-2	209.8	12.198	0.28294	0.00007	0.00054	0.28294	10.4

(continued)

Table 2 (continued)

Sample number	Age	Error	$^{176}Hf/^{177}Hf$	Error	$^{176}Lu/^{177}Hf$	$^{176}/^{177}Hf$ (T)	εHf (T)
Talnakh intrusive (2008)							
T-17_1.2	251.2	9.012	0.282867	0.000027	0.000376	0.282865	8.8
T-17_1.1	238.0	8.88	0.282807	0.000025	0.000520	0.282805	6.4
T-13_5.2	256.3	9.063	0.283093	0.000029	0.001487	0.283086	16.7
T-12_6.1	265.0	9.15	0.282928	0.000150	0.001786	0.282919	11.0
T-1_26.1-2	245.3	8.953	0.282870	0.000025	0.001873	0.282861	8.6
T-1_26.1-1	245.3	8.953	0.282857	0.000026	0.001943	0.282848	8.1
Kharaelakh intrusive (2008)							
844_6_35.1	248.2	9.0	0.282856	0.000045	0.001537	0.282849	8.2
844_1_24.1	239.6	8.9	0.282917	0.000045	0.001702	0.282909	10.1
844_1_14.1	248.5	9.0	0.282913	0.000045	0.001899	0.282904	10.1
844-6_28.1	257.4	9.1	0.282767	0.000045	0.001908	0.282758	5.1
Kharaelakh intrusive, depth 1368.4–1369 (2014)							
PT-2-N1	245.6	9.9	0.28278	0.00004	0.00093	0.28278	5.5
PT-2-N6	249.1	9.0	0.28278	0.00005	0.00142	0.28277	5.4
Kharaelakh intrusive, Oktyabrskoe deposit (2014)							
ZF-18_388_N7	234.3	9.8	0.28294	0.00004	0.00177	0.28293	10.7
Chernogorsk intrusive (2008)							
424_CH-11_4.2_(I)	248.3	4.0	0.282871	0.000039	0.001555	0.282864	8.7
424_CH-11_3.2_(I)	292.5	8.3	0.282989	0.000041	0.001589	0.282980	13.8
424_CH-11_4.2_all	248.3	4.0	0.282784	0.000035	0.001738	0.282776	5.6
424_CH-11_4.1	245.0	4.8	0.282877	0.000061	0.001765	0.282869	8.8
424_CH-11_2.1_(I)	242.4	4.4	0.282826	0.000030	0.001792	0.282818	7.0
424_CH-11_2.1	242.4	4.4	0.282780	0.000057	0.001891	0.282771	5.3

- significant variations in Hf isotopic composition. Average epsilon values of hafnium εHf(T) are given in Table 3. For commercially mineralized intrusions (Norilsk-1, Kharaelakh, Talnakh) Hf epsilon has elevated values; weighted average for all εHf(T) = 8.2 ± 1.8. The Chernogorsky intrusion also adjoins them by this parameter. The Lower Talnakh weakly mineralized intrusion with virtually zero (within the margin of error) Hf epsilon value contrasts them. A high εHf(T) based on the results of this sample looks like a necessary isotopic criterion of ore content, which, however, is not sufficient;
- the lowest variability within this sample is characteristic of zircons from the Norilsk intrusion, which can testify in favour of magma contamination by crustal material at an earlier stage;
- significant εHf(T) variations across the district are indicative of a multistage formation process of different intrusions;
- a standalone group of grains in sample 657-2b (Fig. 5), probably, represents grains "rejuvenated" after the uranium-lead system, which preserved an

Table 3 Mean εHf(T) values for intrusives (zircons of the I group)

Intrusive	N	$\varepsilon^{176}Hf/^{177}Hf$	Standard deviation $\sqrt[2]{\frac{\sum_1^n (x_i - x_{cp})^2}{n-1}}$
Talnakh	6	9.93	3.65
Kharaelakh	7	8.39	2.35
Norilsk	10	8.13	1.22
Chernogorsk	6	8.19	3.12
Lower Talnakh	56	0.03	2.40

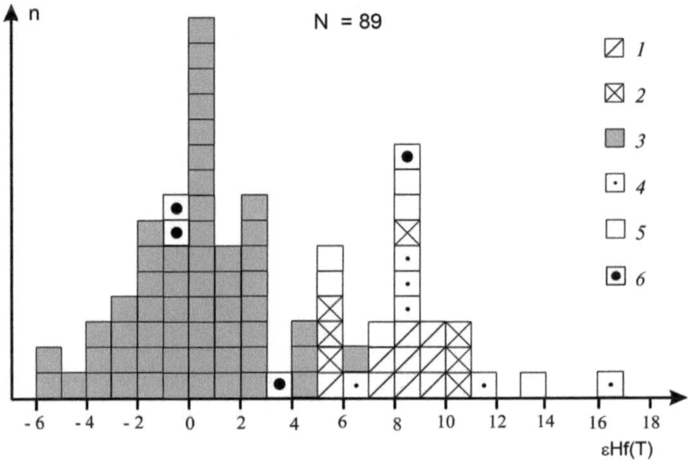

Fig. 5 Histogram of distribution Hf epsilon values in zircons of rocks of intrusives in Norilsk District. Intrusives: 1—Norilsk-1; 2—Kharaelakh; 3—Lower Talnakh; 4—Talnakh; 5—Chernogorsk; 6—Listvyanka

unchanged composition of hafnium and those initially formed ~1.9 Ga, or ancient cores with the newly formed shells. This sample is not considered, since it captured a much more ancient material.

References

1. Faure G, Mensing TM (2004) Isotopes: principles and applications 3rd Edition. USA: Wiley. 18 Oct 2004
2. Griffin WL, Pearson NJ, Belousova E et al (2000) Hf isotope composition of cratonic mantle: LAM-MC-ICPMS analysis of zircon megacrysts in kimberlites. Geochem. et Cosmochem Acta 64:133–147
3. Griffin WL, Nikolic N, O'Reilly SY, Pearson NJ (2012) Coupling, decoupling and metasomatism: Evolution of crust–mantle relationships beneath NW Spitsbergen. Lithos 149:115–135

4. Malitch KN, Belousova EA, Griffin WL et al (2008) Contrasting magma sources in ultramafic-mafic intrusions of the Norilsk area (Russia): Hf isotope evidence from zircon. Geochim. et Cosmochim. Acta 72(12S):A589
5. Malitch KN, Belousova EA, Griffin WL et al (2010) Magmatic evolution of the ultramafic-mafic Kharaelakh intrusion (Siberian Craton, Russia): insights from trace-element, U-Pb and Hf-isotope data on zircon. Contrib Min Petrol 159(6):753–768
6. Petrov OV, Lokhov KI, Kapitonov IN et al (2011) Isotopic composition of Sr, Nd, Hf and Pb as an indicator of the formation conditions of intrusions in the Norilsk ore district. Platinum Russia, Krasnoyarsk 7:458–466
7. Grinenko LN (1966) Isotopic composition of sulphur in sulphides of the Talnakh copper-nickel deposit in connection with questions of its genesis. Geol Ore Deposits 4:15–31
8. Neruchev SS, Prasolov EM (1995) Fluid-geochemical model of platinoid deposits associated with trap magmatism. Platinum of Russia. M.: Geoinformmark, pp 94–101
9. Petrov OV, Sergeev SA, Prasolov EM et al (2010) Geochronological and isotope-geochemical characteristics of mafic intrusions in the Norilsk district. Dokl RAS 434(3):388–390

Isotope Correlations in Rocks and Ores of Major Intrusions in the Norilsk District

Oleg Petrov, Edward Prasolov, Sergey Sergeev and Yury Pushkarev

Abstract He, Ar, S, and Cu isotope characteristics of the Norilsk-1, Talnakh, and Kharaelakh ore-rich intrusions have been found to diverge. The intrusions' isotope characteristics form regression lines with the Kharaelakh intrusion having the highest atmospheric contribution of Ar and the Norilsk-1—the least one: this correspond to a degree of interaction with the host-rock, believed to be a source of atmospheric Ar, though may relate to the intrusions occurrence depth. $\delta^{34}S$ has been found to decrease at growth of m ($^3He/^4He$) with the highest proportion of mantle S and He revealed in the Norilsk-1 intrusion. $\delta^{34}S$ versus $\delta^{65}Cu$ demonstrates negative correlation suggesting probable Cu income from different sources rather than its isotope fractionation: in this instance the Kharaelakh ores are dominated by the crustal Cu, Ni isotopic compositions vary narrowly, displaying no pronounced correlation with He, Ar, S, and Cu isotopes, implying: (1) a single mantle source of Ni and its minute fractionation; (2) contamination of ore system and its fluids by S, Cu and noble gases from the host-rocks; (3) the revealed isotopic variations may be accounted for two or three sources of matter.

It was found that isotopic characteristics of three richest intrusions (Norilsk-1, Talnakh, Kharaelakh) are not the same. There are correlations of He, Ar, S, Cu isotope ratios in rocks and ores of these three intrusions containing the largest Cu–Ni–PGE deposits in the Norilsk district. Here, we consider the causes of the revealed differences, find possible sources of rock and ore matter. The question of isotope correlations in rich intrusions is specially considered in [1, 2].

A close correlation is established between helium and argon isotopic composition in three intrusions richest in ore. As seen in Fig. 1a, the intrusions constitute a certain sequence. The greatest share of mantle helium **m** = 3.7% corresponds to the

O. Petrov (✉) · E. Prasolov · S. Sergeev · Y. Pushkarev
Russian Geological Research Institute (VSEGEI), St. Petersburg, Russia
e-mail: vsegei@vsegei.ru

O. Petrov (ed.), *Isotope Geology of the Norilsk Deposits*, Springer Geology,
https://doi.org/10.1007/978-3-030-05216-4_7

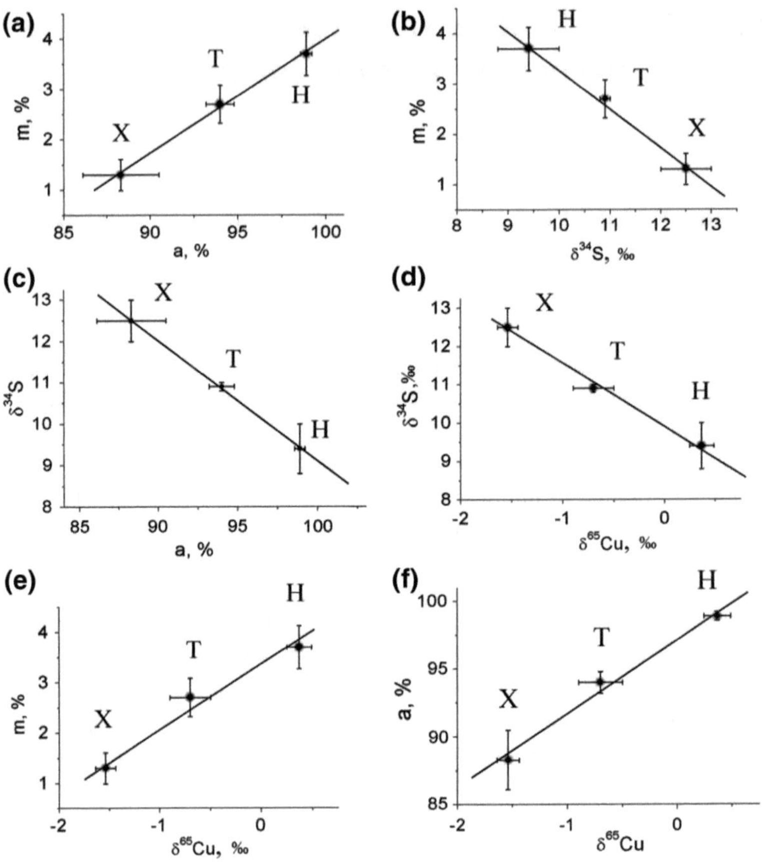

Fig. 1 Relationship of isotopic characteristics of He, Ar, S, Cu in the main ore-bearing intrusions of the Norilsk District (**a–f**) Average values and mean-square deviation of ($\pm 1\sigma$) mean are given. Intrusives: Kh, Kharaelakh; T, Talnakh; N, Norilsk; m, share (%) of mantle helium with $^{3}He/^{4}He = 1.2 \times 10^{-5}$; a, share (%) of atmospheric argon; $\delta^{34}S$, displacement (‰) of $^{34}S/^{32}S$ of the sample relative to CDT standard; $\delta^{65}Cu$, displacement of (‰) $^{65}Cu/^{63}Cu$ relative to NIST 976 standard

maximum contribution of air argon = 99% (Norilsk-1). Another extreme term of the sequence (Kharaelakh intrusion) is characterized by the values **m** = 1.3 and **a** = 88%.

Such a situation (simultaneous growth of the mantle and air components) is often recorded in the areas of modern volcanism [2], where deep fluids with a high share of mantle helium (sometimes to 100%) interact virtually on the surface with infiltration or sea water containing air argon.

In the study area, a direct source of air argon was water of the pore system of sedimentary rocks, into which magma intruded. In general, the share of air argon (**a**) in the sedimentary sequence (in gas accumulations) decreases with

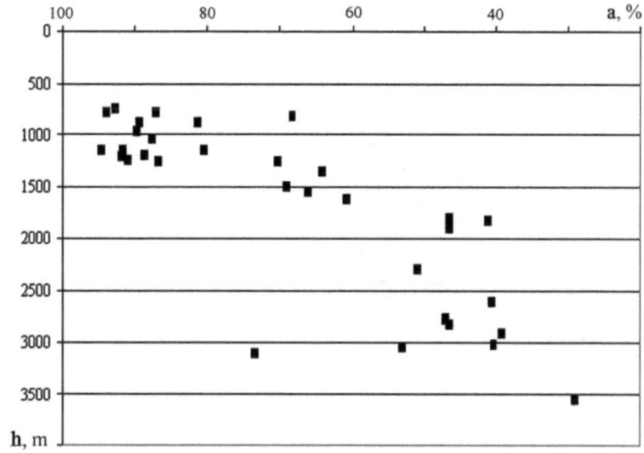

Fig. 2 Change in the share of air argon (a) in gas fields in northern West Siberia depending on their depth (h), according to data [3]

increasing occurrence depth (**h**) and accounts for, on average, about 50% [2], i.e. the role of the atmosphere is much less important than in the studied paleofluids of intrusions in the Norilsk district.

The degree of reduction idea is given by **a** versus **h** diagram for gas pools in the northern West Siberia (Fig. 2) plotted on the data of [2]. It can be seen that the depths of 1.2 – 0.7 km correspond to the range of values **a** = 85 – 100% typical of the studied intrusions. The Kharaelakh intrusion with the lowest atmospheric argon contribution should correspond to the maximum depths; and Norilsk-1, to the minimum ones. In the section (Fig. 3), the intrusions occur in this succession. The difference in the occurrence depth of the two mentioned above intrusions is about 0.6 km. It is interesting to note that in some intrusions, an larger contribution of air argon is recorded in ore intervals; the maximum one, in massive sulphide ores. Argon isotopic composition in fluids of the three intrusions is not the same; the differences are determined by local characteristics of the host rocks, especially their composition and closeness to the day surface.

In the same characteristic features, one can find the cause of helium isotope ratio growth, which determines the mantle components share increase to 3.7%. At the top of the section (Fig. 3), the Permian lavas and tuffs of the Ivakino Formation occur; in their pore fluids, volcanic gases with an admixture of mantle helium could be preserved. Other parts of the section are represented only by terrigenous rocks and evaporites and cannot deliver the mantle helium to the pore system. Ivakino Formation is in closest proximity to Norilsk 1 intrusion and farthest from the Kharaelakh intrusion, which, apparently results in a 2.4% decrease in the share of mantle helium there. We used consistently changing data bearing certain genetic information on helium and argon isotopic composition as reference characteristics.

Rather distinct associations of sulphur isotopic composition ($\delta^{34}S$) and the share of mantle helium (**m**), sulphur and the share of air argon (**a**) are revealed (Fig. 1b, c). The type of the association corresponds to the general notion of isotopically heavy crustal sulphur: $\delta^{34}S$ decreases with increasing **m** ($^{3}He/^{4}He$ isotope ratio). The largest contribution of mantle sulphur (and helium) is recorded in Norilsk-1 intrusion.

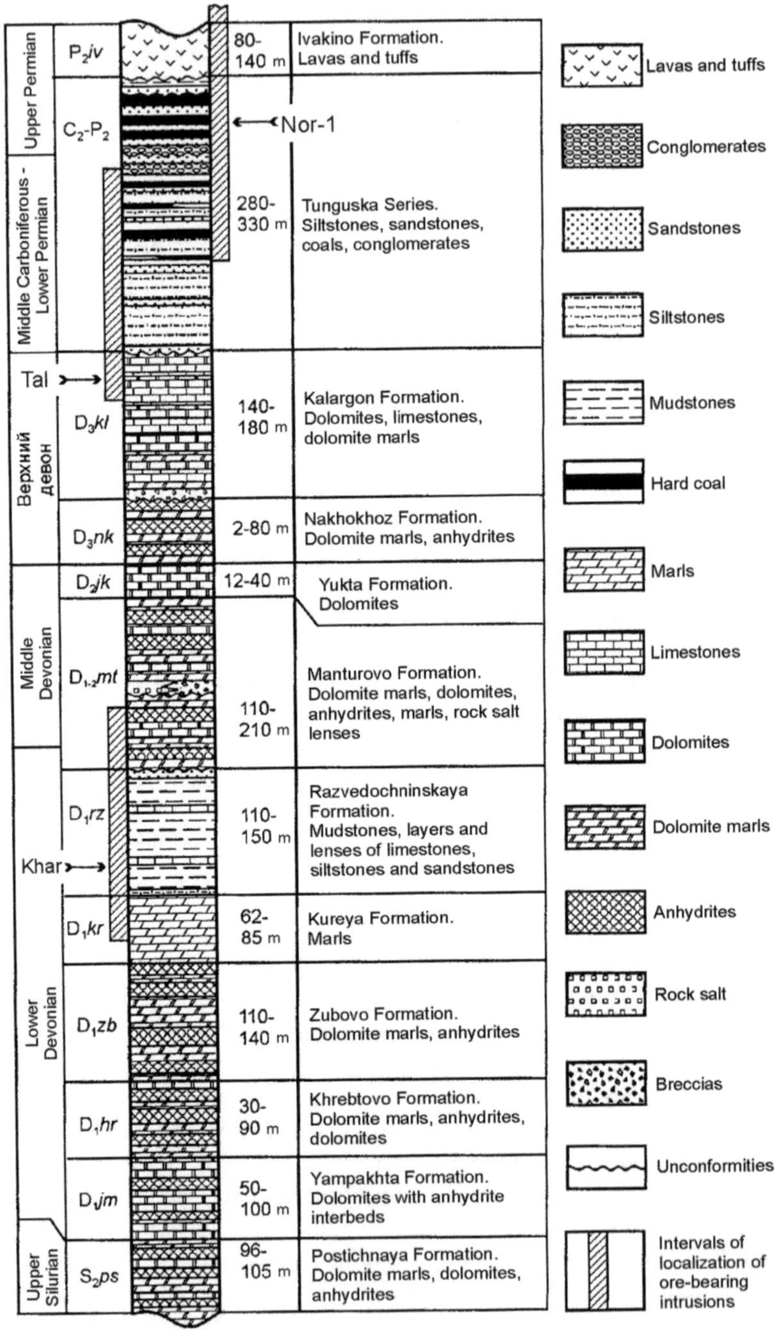

Fig. 3 Upper part of Paleozoic section in Norilsk and Talnakh ore cluster [4]

Changes in sulphur and copper isotopic composition are also interrelated (Fig. 1d). The association is negative; it shows up in copper isotopic weight decrease with increasing sulphur weight; therefore, not isotope fractionation, but the presence/mixing of copper from different sources as the cause of variations appears to be more likely. Then, copper of the Kharaelakh intrusion should be more "crustal". This conclusion is important, because, so far, clear notions about the genetic isotopic labels of copper have not yet been formulated. A clear association is also traced in the diagrams showing the relationship of isotopic composition of copper and noble gases (Fig. 1e, f). The heaviest copper δ^{65}Cu 0.37‰ corresponds to the most "mantle" Norilsk-1 intrusion.

Nickel isotopic composition (Fig. 4 in Chapter "Strontium and Neodymium Isotopes") varies within a narrow range pointing to its single source. The source is different from that of copper, since there is no interrelation of nickel isotope ratios with isotopic composition of copper and sulphur. A slight fractionation of isotopes ensuring variations of δ^{62}Ni to 1.3‰ and δ^{61}Ni to 0.65‰ in certain samples, possibly, occurred during nickel migration from the silicate matter, since nickel isotopes from olivine are also fractionated, and δ^{62}Ni value in it is slightly higher than in ore minerals.

Identified isotope correlations between such different systematics can be caused either by mixing the matter from two sources corresponding to different section intervals, or a smooth change of isotopic characteristics of three sources. In any case, these sources of matter comprising copper, sulphur, and, possibly, other ore components, should occur outside the intrusive bodies in the enclosing shallow rocks.

Thus, in the Norilsk district, the most mineralized intrusions (Kharaelakh, Talnakh and Norilsk-1) differ from the other ones not only on He, Ar, S, Cu isotopic ratios, but also on the presence of correlations between them. The absence of correlations between these elements and Ni points to a different source of this metal. These are, perhaps, silicates of mafic rocks of mantle origin indicative of an active crust-mantle interaction during emplacement of Cu–Ni–PGE deposits of a unique extent in the Norilsk district.

Existence of interrelated isotopic variations of He, Ar, S, Cu leads to the following conclusions:

- fluid components and some ore components (S, Cu) are noted for a genetic affinity;
- direct sources of fluids, sulphur and copper, apparently, occur in enclosing sedimentary rocks;
- variations in isotopic composition are caused by the presence/mixing of the matter from two or three direct sources;
- isotopic characteristics of the matter from direct sources differ due to a different contribution of the near-surface crustal and deep components.

Geochronology of the main mineralized intrusions in the Norilsk district is mainly based on the data of U–Pb method on accessory zircons. The values of zircon isotopic age were obtained by a local method (SIMS SHRIMP II). Its most important advantage is a preliminary identification of the analysed volume (~25 × 20 × 2 μm) within a zircon grain, which allows choosing for dating an undisturbed area with the known genetic affinity. In comparisons, the age of zircons and baddeleyite in leucogabbro of Norilsk-1 intrusion (251.2 ± 0.3 Ma) obtained by isotope dilution ID-TIMS was used as a "reference" [5].

In all the intrusions, magmatic zircons are present with crystallization time corresponding to the age of the silicate matrix of the rocks of intrusions. Zircons of this type are represented by large subidiomorphic crystals with a distinct oscillatory zoning. Generally, zircons of magmatic origin in mafic intrusive rocks are rare. However, in gabbro-dolerites of the considered intrusions their contents, probably, exceeds 100 ppm. Besides, they contain a lot of gas-liquid and mineral inclusions; they have unusual geochemical characteristics: a very high uranium concentration (from 1000 to 6000 ppm) with a "normal" Th/U ratio (from 1.5 to 4.5). The most high-uranium zircons were found in the Talnakh intrusion.

Population of accessory zircons also contains other types of grains—inherited, captured from different host rocks assimilated by magma during intrusion, and the secondary ones appearing in the course of metasomatic recrystallization of rocks. Sometimes, the inherited zircons can also occur as cores in magmatic crystals, and the secondary zircons can form overgrowing shells. The number and diversity of pre- and post-magmatic zircons directly depends on the intensity of secondary reworking and the extent of assimilation of the enclosing rocks. Thus, dating of zircon varieties allows to determine the time of crystallization and recrystallization of the intrusive rock and assessing the source of contaminants. In contrast to the previously presented data [1], only concordant and subconcordant (>85% concordance) values were used in the calculations, which enabled to discard some of the analyses.

Injection time of three main mineralized intrusions in the Norilsk district is 249–260 Ma. Talnakh is, apparently, slightly older than Norilsk-1. The age of Kharaelakh can be defined more precisely after additional research. Possibly, this scatter reflects the time interval of intrusion cooling. Reference points confining the time limits of magma intrusion are the age of felsic volcanics (rhyodacites) – 270 ± 3 Ma (U–Pb method, zircon, SIMS SHRIMP-RG) [3] and the age of conjugated flows of basic trap lavas –249 ± 2 Ma (Ar–Ar, [6]). Sulphide matrix is subsynchronous to the silicate one; its age is 245 – 250 Ma (Re–Os method, ore, ID-TIMS [1]).

There is a clear trend of increasing number and diversity of the inherited zircons in the succession Talnakh—Norilsk-1—Kharaelakh. In the Permo-Carboniferous zircons of this type, the uranium content is much lower. Only in Norilsk-1 intrusion, the Proterozoic zircon is found.

Age of the secondary (metasomatic) processes showing up in all the intrusions is similar (220–230 Ma) and corresponds to the intrusion time of plagiogranite massifs into rocks of the Norilsk district during the period of tectonomagmatic activation, such as, for instance, the Bolgotokh massif (229.0 ± 0.4 Ma, U–Pb method, zircon, ID-TIMS [7].

References

1. Petrov OV, Sergeev SA, Prasolov EM, Khalenev VO, Lokhov KI (2010) Geochronological and isotope-geochemical characteristics of mafic intrusions in the Norilsk district. Dokl RAS 434(3):388–390
2. Prasolov EM (1990) Isotope geochemistry and origin of natural gases. Nedra, L., 284 p
3. Czamanske GK, Wooden JL, Walker RJ et al (2000) Geochemical, isotopic, and SHRIMP age data for Precambrian basement rocks, Permian volcanic rocks, and sedimentary host rocks to the mineralized intrusions, Norilsk-Talnakh district, Siberian Russia. Int Geol Rev 42:895–927
4. Naldrett AJ (2003) Magmatic deposits of copper–nickel and platinum-metal ores. SPGU, SPb, 487 p
5. Kamo SL, Czamanske GK, Krogh TE (1996) A minimum U–Pb age for Siberian flood-basalt volcanism. Geochim et Cosmochim Acta 60:3505–3511
6. Dalrymple GB, Czamanske GK, Fedorenko A et al (1995) A reconnaissance $^{40}Ar/^{39}Ar$ geochronological study of mineralized and related rocks, Siberian Russia. Geochim et Cosmochim Acta 59:2071–2083
7. Kamo SL, Czamanske GK, Amelin Y et al (2003) Rapid eruption of Siberian flood-volcanic rocks and evidence for coincidence with the Permian-Triassic boundary. Earth Planet Sci Lett 214:75–91

Isotope Chronology of Geological Processes

O. Petrov, S. Sergeev, R. Krymsky, S. Presnyakov, N. Rodionov, A. Larionov, E. Lepekhina and D. Sergeev

Abstract The chapter presents new geochronologic results of various isotope techniques (U–Pb SIMS and Re–Os TIMS) along with their comparison with already published data. The main intrusions of the Norilsk district are demonstrated to be emplaced almost simultaneously with two possible magma intrusion pulses at 254 ± 4 and 244 ± 4 Ma (U–Pb SIMS), assuming c. 10 Ma duration of igneous activity. This is corroborated by sulfides Re–Os dating (245–250 Ma), suggestion synchroneity of intrusion and the ore formation. Some Permian and Carboniferous zircon xenocrysts have been found along with Precambrian grains (c. 1.9 and 2.7 Ga), while no Devonian xenocrysts has been revealed. A group of 145–150 Ma old mafic rocks has also been discovered: those nature and relation to the ore-bearing Norilsk intrusions yet to be studied. The geochronologic study suggests, that: (1) ore-bearing massifs belong to the early emplacement phase (250–255 Ma); (2) ore-bearing massifs contain xenogenic Palaeozoic zircons, pointing to important role of the host-rocks; (3) Late Triassic igneous activity (225–230 Ma) has not affected ore systems.

Currently, there is a great interest in precision dating of trap volcanism as an indicator of major geodynamic events, time of ore mode formation and metamorphic episodes. Acomparative study and transcontinental correlation of trap volcanism on the basis of qualitatively new isotopic data are of fundamental scientific importance and practical value.

New level of geological knowledge, especially in the field of metallogeny and geodynamics, cannot be achieved without the use of modern material research methods applying analytical capabilities of the latest generation of instrumentation, such as local methods (SIMS, LA-ICP-MS). This leads to a substantial revision of

O. Petrov (✉) · S. Sergeev · R. Krymsky · S. Presnyakov · N. Rodionov ·
A. Larionov · E. Lepekhina · D. Sergeev
Russian Geological Research Institute (VSEGEI), 74 Sredny Prospect,
St. Petersburg 199106, Russia
e-mail: vsegei@vsegei.ru

O. Petrov (ed.), *Isotope Geology of the Norilsk Deposits*, Springer Geology,
https://doi.org/10.1007/978-3-030-05216-4_8

such fundamentals as stratigraphic charts, serial legends, metallogenic models, correlation diagrams, geodynamic reconstructions, criteria for ore objects prospecting.

In recent years, CIR VSEGEI has reliably dated about 3000 major intrusive, volcanic, and metamorphic objects in Russia, which led to a serious reconsideration of regional geological structure. Geochronological work at such scale has not been performed over the whole history of domestic geology. The most important product of the Institute activities is also new isotope-geochemical methods of geological objects investigation with non-trivial isotopic systems (copper, nickel, lithium, hafnium, noble gases, etc.). In conjunction with the geochronological data, this enables to obtain previously unavailable geological information and find solutions to challenging problems.

Good example of such an integrated approach is unprecedented in scale isotopic studies of rocks and ores from the Norilsk district carried out in 2005–2008 and 2012–2014. Using the analysis of 11 isotopic systems we have examined 22 mafic intrusions, different in ore content degrees (Table 1).

We have determined isotope uranium–lead and rhenium–osmium ages of sulphide ores and accessory zircons; achieved several thousands of isotopic analyses: helium and argon from fluid microinclusions in rocks and ores, sulphur, lead, copper and nickel from sulphides. We have studied the isotope systematics of lead, sulphur, strontium, neodymium, and hafnium in different substances: in silicate material, ore (in sulphides), in paleofluids.

Table 1 Isotopic systematics studied at CIR VSEGEI in 2005–2008 and 2012–2014

Isotopic system	Diagnostic features	Number of analyses/samples	
		2008	2014
$^{3}He/^{4}He$	Crust-mantle, contribution in fluids, source rocks	85	108
$^{40}Ar/^{36}Ar$	Participation of surface fluids, contribution of atmosphere	85	108
$^{34}S/^{32}S$	Crust-mantle, source rocks	276	67
$^{65}Cu/^{63}Cu$	Sources of copper	120	44
$^{62}Ni/^{60}Ni$	Sources of nickel	79	0
$^{87}Sr/^{86}Sr$ and $^{143}Nd/^{144}Nd$	Crust-mantle, source rocks	484/121 481/120	115 115
$^{206\text{-}8}Pb/^{204}Pb$	Crust-mantle, model age	397	762/28
Hf–Nd	Source rocks	150	27/8
U–Pb	Age, source of ore matter	934/101	1020/102
Re–Os	Age of ore	662/165	–

1 Uranium–Lead Isotope Systematics

One of the leading methods in isotope-geochronological studies is the uranium–thorium–lead, using accumulation of radiogenic lead isotopes ^{206}Pb, ^{207}Pb, and ^{208}Pb in natural radioactive decay of ^{238}U, ^{235}U, and ^{232}Th. In this case, age values can be calculated from ratios of ^{206}Pb/^{238}U, ^{207}Pb/^{235}U, and ^{208}Pb/^{232}Th. Due to significant differences in ^{238}U and ^{235}U decay rate and because of the constancy in the isotope ratio of modern natural uranium in the vast majority of geological objects (^{238}U/^{235}U = 137.88), it is possible to calculate the age value from ^{207}Pb/^{206}Pb correlation from four isotope ratios (three of them are independent). It serves as a great advantage of the uranium–thorium–lead method in comparison with other isotopic methods as it enables to assess the reliability degree of age values. The similarity or identity (concordance) of values calculated from different isotope ratios indicates obtaining a valid age. Since the isotope ratios are determined by various radioactive decay series, a variety of intermediate products and their concentrations, correspondence between them is a strong evidence of a closed isotopic system of the sample and reliability of calculated age values.

In practice, used for study proper uranium minerals are often altered by imposed processes and metamictized that leads to imbalance in isotope geochronometric systems. Uranium-bearing minerals should be used in geochronological investigations of rocks and ores only after a detailed study. At the same time, some analytical and methodological aspects do not allow the use of U–Pb method for rock samples as a whole.

For the geochronology purposes in the whole range of geological time, such accessory uranium-bearing minerals as uraninite, monazite, zircon are most often used. Less suitable are sphene, orthite, tantalum-niobates, apatite. Recently, when dating mafic and ultramafic rocks, baddeleyite, pyrochlore and perovskitegroup minerals have been dealt with.

Our main method of producing isotope-geochronological data is local mass-spectrometric investigation of zircons on a unique five-collector sensitive *high-resolution* ion *microprobe* SHRIMPII, as well as Neptune ICP-mass spectrometer with laser sampling. Interpretation of the obtained age values is significantly different from the traditional methods (monofractions or single crystal analysis by isotope dilution) due to the possible localization of sampling points in the areas of growth and recrystallization of individual crystals.

SHRIMP device (Sensitive High Resolution Ion Micro Probe) has been specially designed for high-resolution isotope-geochronological studies of matter at the nano- and micro-level by isotopic systems in single crystals of zircon, monazite and other minerals. In the experiments, we have produced “local”, from 10 μm, isotopic age determinations by genetically different microzones, thanks to ultra-precise focusing of the primary beam ionizing the sample substance. Analysis of positively charged ions (using oxygen beam) and negatively charged ions (caesium beam) enables to measure the mass concentration of natural isotopes of U, Th, Pb, O, and REE.

Nine-collector high resolution i*nductively coupled plasma mass spectrometer* ICP-MS Neptune, equipped with an exclusive system of excimer laser ablation (LA) DUV having a wavelength of 193 nm, enables to analyse the majority of isotopic systems, including hardly ionized, in natural targets with laser sampling in a point from 30 μm in diameter to study migration of ore components, fluids, investigate natural heterogeneities of matter and behaviour of elements—geological processes indicators (including zircons in Pb and Hf). Among the laser varieties currently used, excimer fluorine-argon laser has the best performance. It prevents catastrophic ablation, in which large and incompletely destroyed analyte parts get in plasma, and virtually eliminates chemical fractionation effect. The most stable signal and maximum possible accuracy of determining are achieved thereby.

Method of U–Pb dating, samples. Main mineral in U–Pb dating is zircon. Zircon is found almost in all igneous, sedimentary, and metamorphic rocks. Its structure is durable and resistant to transformations; moreover, in superimposed processes zircon is often regenerated, acquiring additional peripheral zones. Exploring zircon structure, we almost always observe zonal crystals rather than homogeneous ones. It is necessary to use the local U–Pb method, which enables dating of individual growth generations in a single crystal. Amajor problem, in particular, is possibility of finding in zircons, especially from the Phanerozoic rocks, axenogenic component (e.g., in the form of older cores and relict grains) bearing older radiogenic lead, as well as finding microinclusions of minerals with higher U and Th contents than zircon incorporating them. All this determines the need for a versatile precise zircon individual study used for obtaining geochronological information. Concordant values can be considered the most reliable dating results.

Large amount of data on the age (geochronology) and geochemical features of accessory zircons from a wide variety of polygenic and polychronous rocks has been received when studying the Norilsk district.

U–Pb method using zircons is the principal one for dating intrusive and volcanic rocks. Advantages of local methods (ion microprobe, laser ablation), enabling the dating of certain areas in individual grains, make them the preferrable ones when dating felsic igneous rocks, as part of which zircon is a characteristic accessory mineral. Avery important point in selecting zircons representative for dating igneous (intrusive) events is a detailed study of all the populations present in the rock, which may include both relic protolith zircons, and xenogenic zircons trapped by magma from the enclosing rock strata.

When dating volcanic rocks, extrusive, igneous dyke bodies of any composition by the local U–Pb method by zircon, two factors should be considered:

- in thin intrusive bodies and volcanic rocks, zircons are usually very small, comparable to the size of "analytical spots" during laser or ion ablation (15–20 μm);
- in each particular case it is necessary to prove the authigenic nature of dated zircon by mineralogical and isotopic-geochemical methods.

It should be kept in mind that the age of zircon cores corresponds to the substrate age (and for para-rocks a spectrum of ages is observed), and metamorphic events may respond to zones of shells fouling the cores. Fouling rims are formed only at a high level of metamorphism not lower than the epidote-amphibolite facies and in the presence of a fluid phase or at partial melting.

If petrographic data indicate the presence of accessory minerals (zircon, sphene, monazite, etc.) in rock of different generations, it is permissible to use them for determining the age of substrate formation and its metamorphic alteration. However, the genesis of these minerals should be proved by independent methods, petrographic and petrochemical, for example, based on confinement of different mineral generations to specific structural elements (zones of intensive reworking, leucosome migmatite veins or, conversely, weakly altered substrate, paleosome).

For dating alteration processes, usually high-temperature ones, one can use the local U–Pb dating of zircon core fouling areas, as well as Pb–Pb dating of any uranium-bearing minerals (zircon, sphene, uraninite, titanates, and so on). If in such conditions Pb-containing sulphides are formed, model age can be determined by lead isotopes. However, it should be noted that in such cases the model age corresponds to the age of lead source formation (usually more ancient rocks enriched in U).

During U–Pb dating of zircons the Th/U ratio in them has been also determined. The value of this geochemical parameter markedly differs in rocks of various composition and origin, and is used as a genetic criterion. In some cases, to clarify the zircon source (to prove its authigenic or xenogenic nature), spectrum of REE distribution can be determined in the same local points that were subjected to U–Pb dating. Comparison of REE distribution in zircon and in gross rock sample provides an answer to the question of zircon crystallization in this particular rock.

U–Pb zircon dating has been carried out on secondary ion microprobe SIMS SHRIMPII at CIR VSEGEI.

Representative zircons selected manually under a microscope were implanted in epoxy resin (disk 2.5 cm in diameter) with zircon grains of international standards TEMORA and 91500, and then grinded along the elongation of half of their thickness and polished. The slide was coated with conductive lead-free gold plating in a cathode vacuum sputtering system for 1 min at a current of 20 mA. Zircon grains were documented in transmitted and reflected light, and using a scanning electron microscope CamScan MX2500 with CLI/QUA2 system for cathodoluminescent (CL) and BSE images of the internal structure and zoning of zircons.

A detailed study of zircon crystals allows to select a sufficient number of sites (points) for analysis, to the maximum extent reflecting magmatic crystallization or recrystallization. Such sites should be homogeneous, free from inclusions, imposed alterations, or mechanical damage.

Measurements of U–Pb isotope ratios and element concentrations were performed according to the procedure accepted at CIR [16, 21]. Molecular oxygen primary beam intensity of 4 nA, crater dimension of ion sampling of 30 μm at a depth to 2 μm. Data were processed using SQUID software [13]. U–Pb ratio was normalized to the value of 0.0668 assigned to the standard zircon TEMORA, which corresponds to the age of this zircon of 416.75 ± 0.24 Ma [4]. Zircon standard 91500 with uranium content of 81.2 ppm and $^{206}Pb/^{238}U$ age of 1062 Ma [20] was used as a concentration standard. Raster one-minute cleaning of a rectangular (50 × 65 μm) mineral site before dating helps to minimize surface contamination.

Errors in individual analyses (ratios and age) are all at the 1σ level; error of calculated ages at 2σ. Arens-Weatherill [19] plots with concordia were constructed using ISOPLOT/EX software [14]. Correction for non-radiogenic lead was made on measured ^{204}Pb and modern isotopic composition of lead in the Stacey-Kramers model [17].

Given the complexity of dating procedure, for correct selection and careful treatment of mineral individuals it is necessary to attract highly qualified mineralogists to the sample preparationstage. Violations of technology and errors in the choice of local sampling sites lead to serious distortions of the final results.

Results and discussion. Overall summary of geochronological data by our predecessors is presented in Table 2. Reference points limiting the injection time of Norilsk layered intrusions are the age of felsic volcanics (rhyodacites) of 270 ± 3 Ma (U–Pb method, zircon, SIMS SHRIMP-RG, Stanford) [8]) and the age of conjugate mafic trap lava flows of Ergalakh, 245 ± 1 Ma (Ar–Ar [22]).

Table 2 Early geochronological data on intrusives and accompanying rocks in Norilsk Region

Event, geological object	Age (Ma), method, material	Source
Magmatic crystallization, intrusions		
Norilsk-1, gabbro-dolerite	248.0 ± 3.7, U–Pb SIMS SHRIMPI, Zr 251.2 ± 0.3 (min) and 256.5 ± 2.5 (max), U–Pb, IDTIMS, Zr and Bd 245.9 ± 0.7, 247.2 ± 0.6, $^{40}Ar/^{39}Ar$, Bt	[6] [11, 12] [9, 10] [18, 22]
Norilsk-1 and Talnakh, picrite, ol. gabbro-dolerite	~246, Re–Os, IDTIMS, WR	
Talnakh, ol. gabbro-dolerite	249.3 ± 1.0, $^{40}Ar/^{39}Ar$, Bt	
Kharaelakh, picrite	247.6 ± 0.7, $^{40}Ar/^{39}Ar$, Bt	
L.Talnakh, picrite	248.9 ± 1.4, $^{40}Ar/^{39}Ar$, Bt	
Ergalakh, trachydolerite	245.4 ± 1.2, $^{40}Ar/^{39}Ar$, WR	
Ruinnaya Mountain, picrite	~248, $^{40}Ar/^{39}Ar$, Pl	
Dolerite dyke, Avam type	244.7 ± 1.2, $^{40}Ar/^{39}Ar$, Pl	
Zub Mountain, dolerite	234.9 ± 1.1, $^{40}Ar/^{39}Ar$, Pl	
Dolerite dyke cutting Norilsk-1	226.7 ± 1.3, $^{40}Ar/^{39}Ar$, Pl	
Bolgotokh, granodiorites (conjugation of Norilsk and Vologochan troughs)	229.0 ± 0.4, U–Pb, IDTIMS, Zr; 223.1 ± 1.1, $^{40}Ar/^{39}Ar$, Bt; 220.9 ± 1.0, 224.1 ± 1.9, $^{40}Ar/^{39}Ar$, Hb	
Bolgotokh Cu-Mo porphyry deposit	~223, $^{40}Ar/^{39}Ar$, ore	
Formation of sulphide ores		
Ore vein (Zapolyarny open pit) Norilsk-1 intrusive	248.7 ± 1.2, 249.2 ± 1.2, $^{40}Ar/^{39}Ar$, Bt; 250.1 ± 1.5, $^{40}Ar/^{39}Ar$, Bt, 249.3 ± 1.6, $^{40}Ar/^{39}Ar$, Hb	[9, 15, 18]

(continued)

Table 2 (continued)

Event, geological object	Age (Ma), method, material	Source
Sulphide ores of Norilsk-1 and Talnakh intrusives	245,7 ± 0,6, Re–Os, IDTIMS, ore	
Kharaelakh	247,0 ± 3,8, Re–Os, IDTIMS, ore	
Recrystallization of minerals and ores		
Zub-Marksheidersky intrusive	235 ± 0,8, $^{40}Ar/^{39}Ar$, Bt	[9, 10, 18]
Gabbrodolerite, Norilsk-1 intrusive	234,0 ± 1,7, $^{40}Ar/^{39}Ar$, Pl	
L.Talnakh, ol. gabbro-dolerite	214,4 ± 1,1, ($^{40}Ar/^{39}Ar$, Pl)	
Sulphide ores of Norilsk-1 intrusives	233,5 ± 0,8, Re–Os, IDTIMS, ore	
Contamination of intrusives	Inherited zircons, 288, 293, 303, 358, 362, 364, 374, 409, 493 (U–Pb, SIMS SHRIMPRG)	[8]
Contaminants (pre-Triassic rocks of Norilsk Region)		[7–9, 22]
Sedimentary volcanogenic strata (4–18 km)		
Rhyodacite, xenoliths	270 ± 3,U–Pb, SIMS SHRIMPRG, Zr	
Middle Carboniferous—Late Permian Tunguska Series with coals	320–260	
Vendian-Early Carboniferous sedimentary rocks with anhydrides	360–550	
Riphean volcanogenic sedimentary rocks, Cu-molasses	670–1600	
Leucogranite, xenoliths	~910, magmatic, PR and AR, inherited, U–Pb, SIMS SHRIMP RG, Zr	
Complex sialic basement (12–14 km)		[1, 2]
PR basement	1600–2200; 2000–2300, Pb–Pb, IDTIMS, WR	
AR basement	>2500	

Note Zr—zircon, Bd—baddeleyite, Bt—biotite, Hb—hornblende, Pl—plagioclase, Ol—olivine, WR—whole rock, PR1 and 2—Proterozoic, Early and Late, AR—Archaean, IDTIMS—isotopic dilution method, thermionization mass spectrometry, SIMS SHRIMP—secondary ion mass spectrometry, ion microprobe

Norilsk district is characterized by the presence above basalt covers of a complex rock assemblage, Paleozoic sediments to 5 km thick and more ancient (900–2300 Ma [22]) Late Proterozoic volcanogenic sedimentary series to 4 km thick and Proterozoic granite-gneiss basements. This suggests the presence in mafic intrusions of a large number of xenogenic refractory zircon trapped during penetration and partial assimilation of fusible felsic host rocks and present both in the form of individual grains and as seed cores during the growth of proper magmatic zircon.

A good benchmark for dating is the age of zircons and baddeleyite in leucogabbro from Norilsk-1 intrusion obtained in 1996 by the classical method of isotope dilution ID-TIMS [11]. This determination corresponds to 251.2 ± 0.3 Ma and enables culling out multiple "ancientized" datings, associated with a wide presence of inherited zircon phases, obtained by local SIMS.

It is obvious that most of U–Pb ratio variations reflect the heterogeneity in factors such as the nonradiogenic lead content, degree of secondary alteration, admixture of inherited component, fluctuation in measuring parameters. This follows from the identity of such variations in dating of zircons, leucogabbros and other rocks in all intrusions from the Norilsk district. In fact, only two age value-groups are real, 255–240 and 230–220 Ma. They can be traced in different intrusions and identify two principalstages in the evolution of magmatic and ore systems of Norilsk type intrusions, magmatic and postmagmatic.

Modern geochronology of the Norilsk district intrusions is mainly based on the U–Pb SIMS SHRIMP method by accessory zircons. The very presence of igneous zircons in mafic intrusive rocks is an uncharacteristic phenomenon. Nevertheless, in gabbro dolerites from Norilsk-1, Talnakh, and Kharaelakh intrusions, zircon content amounts to 50–70 ppm, corresponding to >100 ppm of zircon in the rock. The major advantage of local SIMS SHRIMP is a preliminary identification of the analysed volume ($\sim 30 \times 20 \times 2$ μm) within the zircon grain that enables to choose for dating an undisturbed area with a known genetic identity.

The most complete analytical information that affords to reconstruct all thestages of non-ferrous and precious metal deposits formation with accessory zircon has been obtained by us for the main mineralized intrusions in the Norilsk district. Consideration of the analytical results in their entirety, taking into account geochemical characteristics of the studied zircons, their internal structure and location of the source geological samples suggests the following data interpretation for the studied intrusions. The main difficulty at thisstage was to separate the real age of zircon crystallization from the measured values.

All intrusions contain "magmatic" zircons (250–240 Ma) presented by large (to 200 μm) subidiomorphic crystals with a distinct oscillatory zoning, containing a small number of gas-liquid and mineral inclusions and unusual geochemical characteristics—Th/U ratio in them is from 1.5 to 4.5 at very high uranium concentrations, generally from 1000 to 6000 ppm (Fig. 1a). Most high-uranium zircons are in the Talnakh intrusion. Age of magmatic zircon crystallization corresponds to the age of silicate matrix of intrusion rocks. This indicates a high probability of the presence of uranium mineralization in the genetic and spatial association with mafic intrusions.

Along with magmatic crystals, zircon population from intrusions rocks contains inherited grains (trapped from various assimilated host rocks during injection). Inherited (allochthonous) zircons with ages between 0.3 and 2.7 Ga may be present

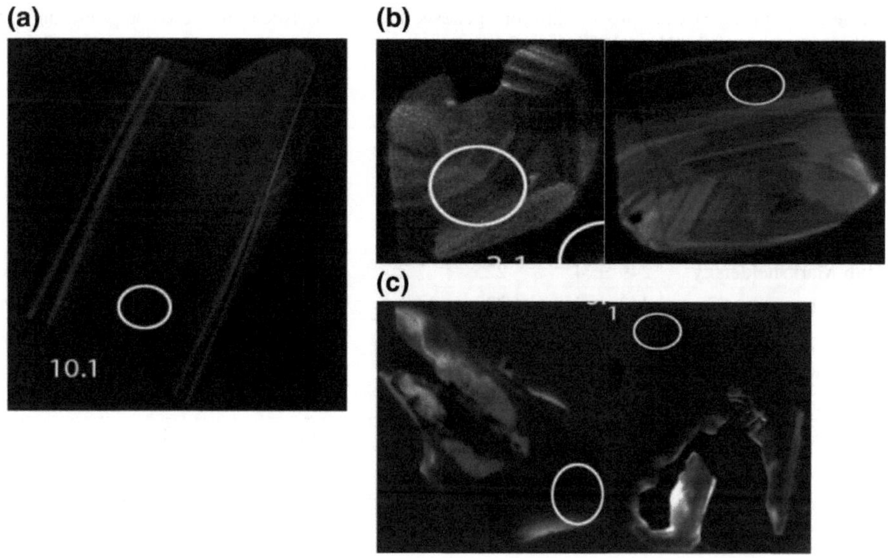

Fig. 1 Characteristic varieties of zircons from intrusives of Norilsk Region. **a** magmatic; **b** inherited; **c** secondary. Diameter of circumferences (sampling points) 30 μm

in the form of cores in magmatic crystals (Fig. 1b); recrystallized and corroded secondary zircons (220–230 Ma) arising from metasomatic recrystallization of rocks can form fouling shells (Fig. 1c). The number and variety of inherited and post-magmatic zircons are directly dependent on the intensity of secondary alteration and extent of rock strata assimilation. Thus, dating of all the discovered zircon varieties enables to characterize the time of crystallization and recrystallization of intrusive rocks and to assess the contaminant source. It is noteworthy that varieties of inherited zircons are different for the three main intrusions in the Norilsk district (Table 3).

These differences can be explained by the fact that the Norilsk district is spatially confined to the development area of near N-S Riphean (Late Proterozoic) trough, which accumulated thick sedimentary beds with volcanic formations. Igneous rocks from the Cambrian to the Upper Paleozoic have not been revealed.

As mentioned above, Paleozoic sediments, Late Proterozoic volcanic-sedimentary strata and granite-gneiss basement are present below the basalt covers. This explains the presence in mafic intrusions of a large number of xenogenic refractory zircon trapped during penetration and partial assimilation of fusible felsic host rocks. Such zircon is present both in the form of individual grains and as seed cores during the growth of proper magmatic zircon occurring at magmatic silicate melt crystallization.

A convincing illustration of the data above is a large bulk of our evidence for Norilsk-1 intrusion (Fig. 2), which allows to see a statistically reasonable distinction between younger, proper magmatic zircons, and older, metasomatic ones, the

Table 3 U–Pb age of zircons of different genesis and Re–Os isochron age of sulphide ores in mafic intrusives of the Norilsk District (Ma)

Intrusive	Magmatic zircons	Postmagmatic zircons	Inherited zircons	Massive and disseminated ores
Norilsk-1	251 ± 2	228 ± 2	1900	–
Kharaelakh	252 ± 8	230–235	290, 347 ± 16	250 ± 2
Talnakh	256 ± 1	221–231	None	251 ± 13
Zub-Marksheidersky	249 ± 4	225	None	248 ± 14
Imangda	243 ± 4	223 ± 3	320–330	–
Chernogorsk	244 ± 3	227	290	248 ± 14
Pyasina-Vologochan	242 ± 10	225	330, 2730	248 ± 14
Mikchangda	256 ± 2	230	293	–
South Pyasina	242 ± 3	210	None	248 ± 14
Kruglogorsky	241 ± 4	225	303–306	–
Zelenaya Griva	241 ± 6	None	None	250 ± 14
Lower Talnakh	254 ± 4	220–230	270, 300	247 ± 45
Lower Norilsk	247 ± 6	230	1900	251 ± 20
Binyuda (Taimyr)	249 ± 11	None	None	251 ± 13

Note For calculations, only concordant and subconcordant (>85% concordance) values of UPb age were used. Initial UPb isotope data for zircons are given in Table 6

formation of which occurred simultaneously with the injection in the Norilsk district of later granodiorite (e.g., Bolgotokh). Older datings are mostly confined to the marginal parts of grains.

The total age data aggregate also allows us to identify statistically (and, probably, geologically) a significant difference in the age of magmatic crystallization of the Norilsk district intrusions. There are "early" (255–250 Ma) and "late" (245–240 Ma) intrusions at virtually indistinguishable geochemical composition of rocks building them up and a close spatial conjugation of intrusions. The time difference between these two magmatism pulses is negligible in geological sense, no more than 5 million years, yet it would be wrong to postulate a continuous rather than a discrete magmatic cycle to 20 million years long.

Belonging to an older group may be of metallogenic significance, especially in the recognition of a large-scale assimilation of ore components from enclosing rocks. In the second scenario, a "younger" group of intrusions injects into the association of enclosing sedimentary and metamorphic rocks already depleted in ore and fluid components.

To understand the problem of sulphide ore origin from the Norilsk district, of great interest is the experience of separation and dating of zircons directly from the ore material (Fig. 3a–c). These zircons must have a direct genetic relationship either to mineralization, or to the ore component source. The first results of dating zircons separated from rich sulphide ores of the Oktyabrsky deposit (sample CG-1319, 598.4 m) have revealed the presence of as many as eight agegroups of detrital

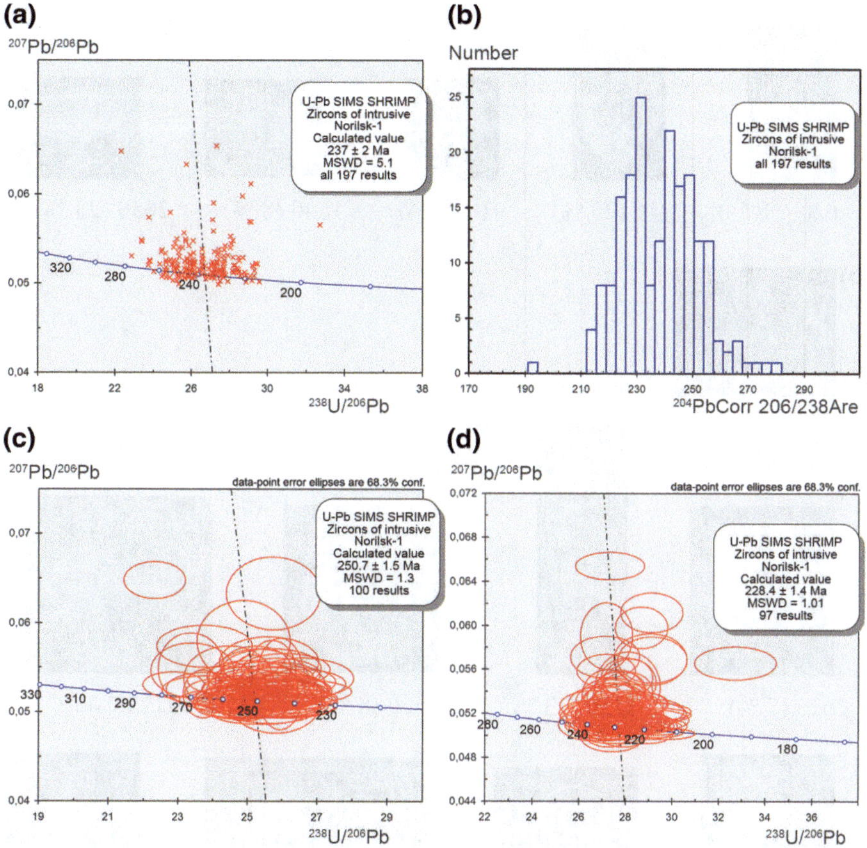

Fig. 2 Complex U–Pb isotope systematics of zircons obtained by local method of SIMS SHRIMP. As an example of presorting of two generations of zircons in rocks, a large data array is given (197 datings of SIMS SHRIMP on Norilsk-1 intrusion; rocks of Medvezhij Ruchej open pit); **a** diagram with Tera-Wasserburg concordia with all the results (without errors). Average value of 237 Ma has no geological meaning if there are several generations of zircon; **b** the same data as a histogram. Age is adjusted for common Pb^{204}. Two clusters are distinguished; **c** and **d** two data clusters on Tera-Wasserburg diagrams (errors ellipses) with ages 251 Ma (interpreted as crystallization time of Norilsk-1 intrusive) and 228 Ma (postmagmatic recrystallization and formation process of metasomatic zircon)

zircons, corresponding to different carrier rock sources. No zircons synchronous to the crystallization time of silicate matrix of layered intrusions (approx. 250 Ma) have been identified. The youngest detrital zircons (minimum age of sedimentation) are Vendian, 620 Ma, prevailing, 1940 and 2750 Ma (20–25%), the rest are intermediate Proterozoic and Late Archean (1550, 1660, 1860, 2050, and 2650 Ma).

It is noteworthy that detrital zircons separated from Igarka copper sandstone (sample I-1) and quartz geode with native copper (sample Kum-1) have very similar

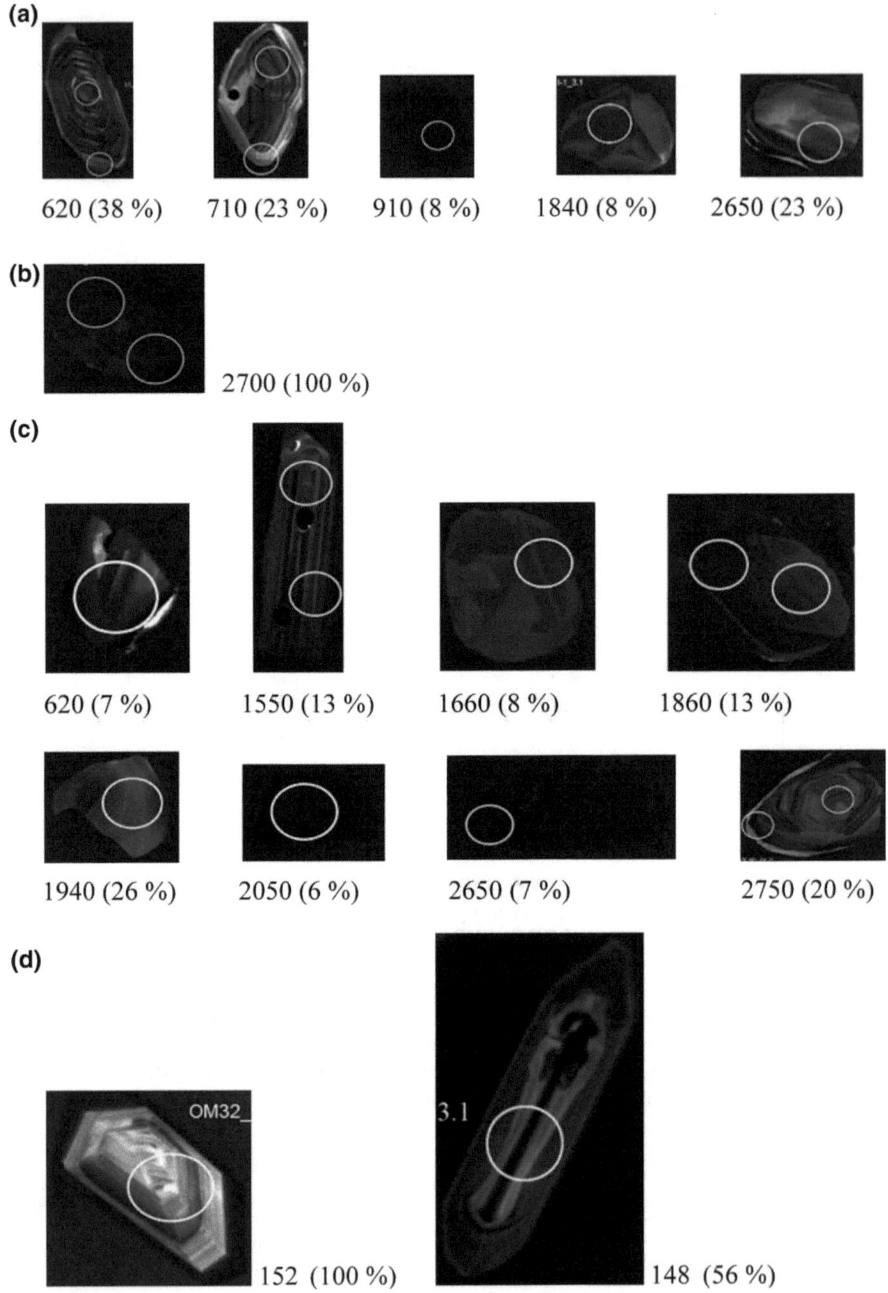
(a)
620 (38 %)
710 (23 %)
910 (8 %)
1840 (8 %)
2650 (23 %)
(b)
2700 (100 %)
(c)
620 (7 %)
1550 (13 %)
1660 (8 %)
1860 (13 %)
1940 (26 %)
2050 (6 %)
2650 (7 %)
2750 (20 %)
(d)
OM32_
152 (100 %)
3.1
148 (56 %)

◂**Fig. 3** Typical zircons from ores (**a–c**) and magmatic rocks of Cretaceous age (**d**) in Norilsk Region. Figures—concordant U–Pb age of zircons (Ma) and share of population (%). Diameter of circumferences (sampling points) 30 μm. **a** five age clusters of detritic zircons in cupriferous Igarka sandstones (sample I-1); **b** one age cluster of zircon from a quartz zeode with native Cu, Kumga R. (sample Kum-1); **c** eight age clusters of inherited zircons in rich cupriferous-noble metal ore of Oktyabrskoe deposit (sample KZ-1319, 599 m, gabbro-doleritic pyrrhotite-chalcopyrite breccia); **d** magmatic zircons from gabbro-dolerites spatially conjugated with Maslovsky deposit (sample OM-32, 1,085 m, 152 Ma), and Krasnye Kamni basalts (sample N-2, 148 Ma)

ages (620, 710, 910, 1840, 2650, 2700 Ma), but with a predominance of the Vendian-Riphean material. This is one of the compelling and visual arguments in favour of a purely crustal origin, at least, for a substantial portion of ore components.

Of particular note is that the expansion of geochronological studies in the Norilsk district in 2013–2014 enabled us to first discover signs of the presence of mafic intrusive rocks younger than the prevailing Triassic ones. Their main distribution area is around theMaslovskoye intrusion (15 km south of Norilsk), where the density of tectonic dislocations increases greatly. Zircons from gabbro-dolerite and basalt are very few, but have all the signs of crystallization from the melt and exhibit undisturbed isotope U–Pb system (Fig. 3d), which enables to assume reasonably manifestation of a magmatic event in the Norilsk ore district at the Jurassic and Cretaceous boundary. It is noteworthy that young basalts contain inherited zircons (about 40%) both of Paleozoic (280–350 Ma), and Proterozoic (1900–2000 Ma) age reflecting the composition of intruded strata. Sampling locations for zircon separation and analytical data are given in Tables 4, 5 and 6.

Table 4 Sites of taking 64 samples for distinguishing zircons in the Norilsk District (2012–2014)

Sample	Rock
Oktyabrskoe deposit, western flank, boreholes ZF 13/18/19/21/30/31/43, KZ 931/1319	
ZF 13	Gabbro-dolerite, sulph. (15%)
ZF 13 (448)	Coarse-grained olivinic gabbro, sulph. (15%)
ZF 13 (474.9–475.5)	Coarse-grained breccia, metasom., albite, gabbro, hornfels
ZF 18 (402.5)	Medium-grained olivinic dolerite, sulph. (3%)
ZF 18 (409.2)	Coarse-grained leucogabbro, sulph. (3%)
ZF 18 (419.1)	Medium-grained leucogabbro, metasom., sulph. (10%)
ZF 18 (437.2)	Medium-grained leucogabbro, sulph. (1%)
ZF 18 (450)	Coarse-grained olivinic gabbro, sulph. (1%)
ZF 18 (460.9)	Coarse-grained olivinic dolerite, sulph. (1%)
ZF 18 (471)	Pyrox. hornfels, sulph. (5%)
ZF 19 (429.7)	Olivinic gabbro, sulph. (3%)
ZF 19 (445)	Coarse-grained metagabbro, sulph. (5%)
ZF 19 (449)	The same

(continued)

Table 4 (continued)

Sample	Rock
ZF 19 (479.1)	Pl.-porphyr. coarse-grained olivinic dolerite, sulph. (1%)
ZF 19 (488.6)	Pl.-porphyr. coarse-grained olivinic dolerite, sulph. (1%)
ZF 21	Medium-grained gabbro-dolerite, carb.
ZF 21 (446.0–446.4)	Coarse-grained leucogabbro, sulph. (5%)
ZF 21 (457.6–457.9)	Coarse-grained leucogabbro, sulph. (1%)
ZF 21 (465.0–465.6)	Medium-grained dolerite, sulph. (1%)
ZF 21 (472.8–473.1)	Pyrox. hornfels, sulph. (1%)
ZF 30 (378.7)	Olivinic pyrox. leucogabbro, sulph. (5%), carb.
ZF 31 (539.8)	Hornfels, coarse-grained leucogabbro, sulph. (1%)
ZF 43 (625.5)	Fine-grained dolerite, sulph. (1%)
KZ 931 (616–622)	Pl. wehrlite, pyrox. olivinic sulph.
KZ 1319 (598.4)	Rich sulph. Ores
Oktyabrskoe deposit, central part, boreholes PT 2, KZ 981/1112	
PT 2	Gabbro-dolerite, sulph.
PT 2 (1368.4)	Medium-grained dolerite, sulph. (3%)
PT 2 (1371.2–1371.8)	The same
PT 2 (1415)	Olivinic pyrox. gabbro-dolerite,sulph. (2%)
PT 2 (1419.1)	Coarse-grained pl. wehrlite, sulph. (5%)
PT 2 (1423)	Coarse-grained olivinic leucogabbro, sulph. (1%)
PT 2 (1425)	Medium-grained Pl. wehrlite, sulph. (10%)
PT-2 (1438.7–1439.3)	Medium-grained olivinic pyrox. pl. wehrlite, sulph. (15%)
KZ 981	Medium-grained gabbrodolerite
KZ 981(1126.3–1126.9)	Medium-grained olivinic pyrox. gabbro, sulph. (1%)
KZ 1112 (1098.4)	Sulph. ore in feldspar rocks, hornfels
Oktyabrskoe deposit, southern flank, boreholes KZ 361bis/1084	
KZ 361bis	Medium-grained pyrox. gabbrodolerite
KZ 1084 (1150.1)	Gabbrodolerite, gabbro, hornfels. sulph.
Talnakh deposit	
KZ 774	Olivinic gabbro
KZ 774 (1023)	Medium-grained olivinic dolerite, sulph. (1%)
KZ 774 (1029)	Medium-grained olivinic dolerite, sulph. (10%)
KZ 774 (1032.4–1033.0)	Medium-grained wehrlite, sulph. (5%), anhydride
Maslovsky deposit, boreholes OM 10/32/123	
OM 10 (1061)	Medium-grained olivinic pyrox. pl. wehrlite, sulph. (5%)
OM 10 (1068)	The same
OM 32 (1084.8)	Medium-grained olivinic gabbro, sulph. (5%)
OM 123 (1005.6)	Medium-grained olivinic gabbro, sulph. (1%)
OM 123 (1033.7)	Medium-grained olivinic pyrox. pl. wehrlite, sulph. (5%)
Zub-Marksheidersky deposit	
MP 25 KZ (37.8)	Coarse-grained olivinic metagabbro, sulph. (15%)

(continued)

Table 4 (continued)

Sample	Rock
Chernogorsk deposit	
12 N 18	Coarse-grained metagabbro, sulph. (10%) (P)
12 N 19A	Medium-grained metagabbro, sulph. (5%) (T1)
12 N 19B	Coarse-grained metaleucogabbro, albite (T1)
12 N 21	Hornfels after qu. feldspath. sandstone (Yu) (sulph. (1%)
MP 2 206 fg	Medium-grained olivinic gabbro, sulph. (1%) (T2)
Listvyanka R. area	
12 N 05	Basalt, Syverma (T1) area
12 N 07	Leucogabbro (Yu)
12 N 08	Basalt, Tuklonskaya (P) area
Krasnye Kamni area	
N 2/N 2	Basalt (Yu)
N 3	Basalt
Vologochan area, Bh. OV 28	
OV 28 (703)	Coarse-grained gabbro, sulph. (3%)
OV 28 (711)	The same
OV 28 (827)	Coarse-grained olivinic gabbro, sulph. (5%)
Burkan M. area	
657 2b	Coarse-grained olivinic dolerite, sulph. (M)
Host rocks	
I 1	Cupriferous sandstone, Igarka
KYM 1	Qu. nodule with copper

Table 5 Sites of taking 72 samples for distinguishing zircons in Norilsk and Taimyr Districts (2005–2007)

Sample	Rock name
Norilsk intrusive, Medvezhij Ruchej, Bh. MN2	
N/N 1 3	Leucogabbro
N 1 6	Olivinic gabbro
N 1 7/8	Pl. wehrlite
N 1 1	Gabbrodiorite
N 1 2	Leucogabbro
Talnakh intrusive, Bh. OUG2	
T 2	Diorite-pegmatite
T 3	Gabbrodiorite
T 5	Leucogabbro

(continued)

Table 5 (continued)

Sample	Rock name
T 6	Olivinic gabbro
T 8	
T 10	
T 12	
T 13	Pl. wehrlite, sulph.
T 14	
T 15	
T 16	Troctolite, sulph.
T 17	Pl. pyroxenite, sulph.
T 18	Olivinic gabbro, sulph.
T 22	Pl. wehrlite, sulph.
Kharaelakh intrusive, Bh. KZ 844	
844 1	Olivinic gabbro
844 6	Troctolite, sulph.
844 7	Pl. wehrlite
844 10	Troctolite
Lower Talnakh intrusive, Bh. TG 31	
31 1	Gabbro
31 3	Troctolite
31 7	Pl. wehrlite
31 9	Pl. olivinic gabbro
31 10	Pl. wehrlite
31 11	Troctolite
31 13	Pl. wehrlite
31 16	Troctolite
Zub-Marksheidersky intrusive, Bh. 27	
27 1	Metasomatite
27 3	
27 4	
27 5	Diorite
27 6	Olivinic gabbro
27 7	
27 13	Troctolite sulph (10%)
27 14	Olivinic gabbro, sulph.
Imangda intrusive, Bh. 4	
4 3	Troctolite, sulph. (2%)
4 6	Pl. wehrlite sulph. (4%)
4 8	Troctolite, sulph. (2%)
4 9	Olivinic leucogabbro
4 10	Olivinic gabbro, sulph.

(continued)

Table 5 (continued)

Sample	Rock name
Chernogorsk intrusive, Bh. MP 2bis	
CH 9	Olivinic gabbro
CH 10	Troctolite, sulph. (2%)
CH 11	Troctolite, sulph. (20%)
South Pyasina intrusive, Bh.OV 25	
25 2	Metasomatite
25 4	Olivinic gabbro, sulph. (10%)
25 9	Olivinic gabbro, sulph. (5%)
25 31/35	Troctolite, sulph.
Vologochan intrusive, Bh.OV 29	
29 5	Olivinic gabbro
29 9	Troctolite, sulph.
29 16	Troctolite, sulph. (15%)
29 17	Troctolite, sulph.
Mikchangda intrusive, Bh. MD-48	
48 18	Olivinic gabbro
48 25(9)	Troctolite
48 30	
Binyuda intrusive, Bh. C1	
C1 4 3	Troctolite
Dyumptalej intrusive, Bh. TP 43	
43 27	Gabbro
Lower Norilsk intrusive, Bh. NP 37	
37 12	Olivinic gabbro
37 23	Pl. wehrlite
Zelenaya Griva intrusive, Bh. F 233	
233 2	Olivinic gabbro
233 4	Metasomatite
233 6	Pl. wehrlite
233 7	Troctolite
Kruglogorsky intrusive, Bh. MP 2bis	
K9	Olivinic gabbro
Morongo intrusive, Ruinnaya Mountain	
Ru2	Troctolite
Daldykan intrusive, Bh. NP 37	
37 34	Olivinic dolerite
Oganer intrusive, Bh. MD 48	
48 7	Olivinic dolerite
Ergalakh intrusive, Bh. NP 37	
37 52	Trachydolerite

Table 6 Summary of results of U–Pb SIMS SHRIMP isotope analyzes of zircons from 72 samples of 2005–2007 from rocks and ores of Norilsk and Taimyr Districts and 64 samples of 2012–2014

Crater	$^{206}Pb_c$, %	U, ppm	Th, ppm	$^{232}Th/^{238}U$	$^{206}Pb^*$, ppm	$^{206}Pb^*/^{238}U$ Age		$^{207}Pb^*/^{206}Pb^*$ Age		$^{207}Pb^*/^{235}U$	±%	$^{206}Pb^*/^{238}U$	±%	KK
ZF 13														
ZF-13_3.1	3.34	554	192	0.36	12.3	158.9	±2.9	55	±350	0.162	15.0	0.0250	1.8	0.122
ZF-13_4.1	0.18	1366	3194	2.42	47.4	254.9	±3.8	257	±46	0.286	2.5	0.0403	1.5	0.598
ZF-13_2.1	0.16	937	3734	4.12	32.8	257.0	±3.8	246	±70	0.287	3.4	0.0407	1.5	0.440
ZF-13_1.1	0.23	1134	3626	3.30	39.8	257.4	±3.8	218	±55	0.284	2.8	0.0407	1.5	0.527
ZF-13_5.1	0.12	2743	10,622	4.00	96.6	258.7	±3.7	263	±35	0.291	2.1	0.0409	1.5	0.699
ZF-13_6.1	0.22	1739	1509	0.90	61.5	259.5	±3.8	227	±50	0.287	2.7	0.0411	1.5	0.570
ZF 13 (448.0)														
ZF-13-448.0_1.1	0.38	538	1066	2.05	19.0	258.8	±2.7	–	–	0.278	4.7	0.0410	1.1	0.229
ZF-13-448.0_2.1	1.40	1286	3319	2.67	47.4	267.1	±2.6	–	–	0.299	5.8	0.0423	1.0	0.176
ZF 13 (474.9–475.5)														
ZF-13-474.9-475.5_1.1	0.21	205	137	0.69	63.3	1974.0	±18.0	1978	±24	6	1.7	0.3583	1.0	0.620
ZF 18 (402.5)														
ZF-18(402.5)_6.1	0.01	583	635	1.12	12.5	158.9	±3.0	175	±66	0.171	3.4	0.0250	1.9	0.556
ZF-18(402.5)_15.1	4.30	546	626	1.18	12.3	159.6	±3.5	221	±420	0.175	18.0	0.0251	2.2	0.121
ZF-18(402.5)_14.1	0.42	491	323	0.68	10.6	159.8	±3.1	211	±120	0.174	5.7	0.0251	1.9	0.343
ZF-18(402.5)_10.1	0.08	863	871	1.04	18.9	161.8	±2.9	135	±62	0.171	3.2	0.0254	1.8	0.569
ZF-18(402.5)_5.2	0.98	447	452	1.05	9.9	162.8	±3.2	166	±210	0.174	9.0	0.0256	2.0	0.221
ZF-18(402.5)_8.2	0.22	681	641	0.97	15.0	163.1	±3.0	121	±81	0.171	3.9	0.0256	1.9	0.476
ZF-18(402.5)_5.1	0.01	380	816	2.22	8.4	164.3	±3.2	226	±80	0.180	4.0	0.0258	2.0	0.491
ZF-18(402.5)_4.1	0.79	811	783	1.00	18.2	165.1	±3.1	207	±170	0.180	7.8	0.0259	1.9	0.245
ZF-18(402.5)_3.1	0.30	1031	978	0.98	23.3	167.1	±3.0	130	±87	0.176	4.1	0.0263	1.8	0.446
ZF-18(402.5)_8.1	0.89	227	258	1.17	5.2	167.9	±3.8	146	±250	0.178	11.0	0.0264	2.3	0.213
ZF-18(402.5)_1.1	0.24	644	617	0.99	14.7	168.7	±3.1	122	±87	0.177	4.1	0.0265	1.9	0.454
ZF-18(402.5)_2.1	0.01	872	910	1.08	19.9	169.1	±3.1	121	±49	0.178	2.8	0.0266	1.8	0.660
ZF-18(402.5)_7.1	0.52	451	459	1.05	11.1	181.4	±3.7	172	±150	0.195	6.7	0.0285	2.1	0.309
ZF-18(402.5)_3.2	10.71	78	61	0.81	2.4	206.0	±15.0	920	±1600	0.310	79.0	0.0325	7.5	0.095
ZF-18(402.5)_11.1	0.01	1685	2329	1.43	59.3	258.7	±4.5	251	±30	0.289	2.2	0.0410	1.8	0.803

(continued)

Table 6 (continued)

Crater	$^{206}Pb_c$, %	U, ppm	Th, ppm	$^{232}Th/^{238}U$	$^{206}Pb^*$, ppm	$^{206}Pb^*/^{238}U$ Age		$^{207}Pb^*/^{206}Pb^*$ Age		$^{207}Pb^*/^{235}U$	±%	$^{206}Pb^*/^{238}U$	±%	KK
ZF-18(402.5)_11.2	0.01	2708	5023	1.92	95.3	258.9	±4.5	248	±29	0.289	2.2	0.0410	1.8	0.816
ZF-18(402.5)_9.1	0.09	2094	2108	1.04	74.0	259.5	±4.5	245	±43	0.289	2.6	0.0411	1.8	0.685
ZF-18(402.5)_13.1	0.15	3913	5874	1.55	148.0	277.3	±5.3	219	±55	0.306	3.1	0.0440	2.0	0.638
ZF 18 (409.2)														
ZF-18(409.2)_1.1	0.90	873	4397	5.21	30.9	257.8	±4.6	306	±120	0.295	5.7	0.0408	1.8	0.317
ZF-18(409.2)_2.1	0.01	636	1946	3.16	21.3	246.6	±4.4	242	±47	0.274	2.7	0.0390	1.8	0.665
ZF-18(409.2)_3.1	0.03	1893	5226	2.85	66.2	257.2	±4.4	303	±28	0.294	2.1	0.0407	1.8	0.821
ZF-18(409.2)_4.1	0.85	623	2946	4.89	21.9	256.3	±4.8	215	±140	0.282	6.3	0.0406	1.9	0.303
ZF-18(409.2)_5.1	0.06	2611	10,671	4.22	91.4	257.5	±4.4	205	±32	0.282	2.2	0.0408	1.7	0.784
ZF-18(409.2)_5.2	0.61	1443	4305	3.08	50.1	253.7	±4.8	277	±99	0.287	4.7	0.0402	1.9	0.405
ZF-18(409.2)_6.1	0.06	2811	8693	3.20	94.9	248.4	±4.4	238	±30	0.276	2.2	0.0393	1.8	0.812
ZF-18(409.2)_7.1	0.66	250	451	1.86	8.3	242.2	±5.0	123	±190	0.256	8.5	0.0383	2.1	0.248
ZF-18(409.2)_8.1	0.19	748	1503	2.08	25.7	252.2	±4.6	339	±67	0.293	3.5	0.0399	1.9	0.536
ZF-18(409.2)_9.1	0.51	1464	3334	2.35	51.7	258.2	±4.8	130	±100	0.274	4.7	0.0409	1.9	0.404
ZF-18(409.2)_10.1	0.34	1500	2937	2.02	52.3	255.5	±4.6	298	±61	0.292	3.2	0.0404	1.8	0.565
ZF-18(409.2)_11.1	0.01	838	1558	1.92	28.1	247.1	±4.5	216	±49	0.272	2.8	0.0391	1.9	0.657
ZF-18(409.2)_12.1	0.51	437	823	1.95	14.2	238.8	±4.7	153	±150	0.256	6.8	0.0377	2.0	0.296
ZF-18(409.2)_13.1	0.01	729	1000	1.42	24.8	249.9	±4.6	346	±49	0.291	2.9	0.0395	1.9	0.655
ZF 18 (419.1)														
ZF-18(419.1)_4.1	0.08	1992	3877	2.01	70.0	258.2	±2.4	268	±43	0.291	2.1	0.0409	0.9	0.443
ZF-18(419.1)_3.1	0.46	1678	2894	1.78	59.5	259.4	±2.5	234	±77	0.288	3.5	0.0411	1.0	0.279
ZF-18(419.1)_2.1	2.29	638	2057	3.33	23.5	264.4	±4.0	235	±370	0.294	16.0	0.0419	1.5	0.095
ZF-18(419.1)_5.1	0.06	3550	8189	2.38	128.0	265.7	±2.7	202	±49	0.291	2.3	0.0421	1.0	0.440
ZF-18(419.1)_6.1	0.26	920	498	0.56	37.9	301.2	±3.4	254	±76	0.338	3.5	0.0478	1.1	0.329
ZF-18(419.1)_1.1	0.01	2	1	0.03	0.5	2055	±98.0	2191	±140	7.100	9.8	0.3750	5.6	0.569
ZF 18 (437.2)														
ZF-18(437.2)_1.1	0.39	985	589	0.62	34.7	258.0	±2.9	239	±100	0.287	4.6	0.0408	1.1	0.248
ZF-18(437.2)_1.2	0.53	758	442	0.60	26.6	256.5	±3.0	370	±110	0.302	4.8	0.0406	1.2	0.246
ZF-18(437.2)_2.1	0.47	534	177	0.34	18.8	257.3	±3.1	310	±110	0.295	5.1	0.0407	1.2	0.240

(continued)

Table 6 (continued)

Crater	$^{206}Pb_c$, %	U, ppm	Th, ppm	$^{232}Th/^{238}U$	$^{206}Pb^*$, ppm	$^{206}Pb^*/^{238}U$ Age		$^{207}Pb^*/^{206}Pb^*$ Age		$^{207}Pb^*/^{235}U$	±%	$^{206}Pb^*/^{238}U$	±%	KK
ZF-18(437.2)_3.1	0.33	720	290	0.42	24.7	251.6	±2.8	227	±89	0.278	4.0	0.0398	1.2	0.286
ZF-18(437.2)_4.1	0.01	1704	1733	1.05	59.8	258.2	±2.6	227	±35	0.286	1.8	0.0409	1.0	0.554
ZF-18(437.2)_5.1	0.01	859	448	0.54	30.7	262.5	±2.8	232	±51	0.291	2.5	0.0416	1.1	0.448
ZF-18(437.2)_6.1	0.18	693	507	0.76	23.8	252.3	±2.8	295	±65	0.287	3.1	0.0399	1.1	0.372
ZF-18(437.2)_7.1	0.01	693	542	0.81	24.2	257.1	±2.9	276	±54	0.291	2.6	0.0407	1.1	0.433
ZF-18(437.2)_8.1	0.09	1523	1033	0.70	54.1	260.8	±2.6	246	±52	0.291	2.5	0.0413	1.0	0.418
ZF-18(437.2)_9.1	0.13	1539	754	0.51	53.7	256.2	±2.6	255	±46	0.287	2.3	0.0406	1.0	0.456
ZF-18(437.2)_10.1	0.31	1208	800	0.68	42.2	256.0	±2.7	154	±69	0.274	3.1	0.0405	1.1	0.337
ZF-18(437.2)_11.1	0.19	640	360	0.58	22.4	256.6	±2.9	237	±110	0.285	5.1	0.0406	1.2	0.226
ZF 18 (450)														
ZF18_450_1.1	0.32	288	383	1.37	9.8	249.4	±3.0	123	±110	0.264	5.0	0.0394	1.2	0.247
ZF18_450_2.1	0.01	26	21	0.85	0.8	235.7	±7.5	684	±230	0.320	11.0	0.0372	3.2	0.292
ZF18_450_3.1	0.01	147	230	1.62	4.8	242.9	±3.6	225	±110	0.268	5.1	0.0384	1.5	0.302
ZF18_450_4.1	1.58	351	344	1.01	11.8	244.9	±3.5	479	±200	0.303	9.3	0.0387	1.5	0.157
ZF18_450_4.2	0.85	2174	595	0.28	79.7	267.0	±2.2	160	±89	0.287	3.9	0.0423	0.9	0.220
ZF18_450_5.1	0.53	199	216	1.12	6.6	244.6	±3.4	201	±150	0.267	6.8	0.0387	1.4	0.208
ZF18_450_6.1	0.01	539	1455	2.79	17.6	240.9	±2.4	246	±71	0.268	3.3	0.0381	1.0	0.315
ZF18_450_7.1	0.15	1720	2436	1.46	60.5	258.4	±2.9	323	±51	0.298	2.5	0.0409	1.1	0.452
ZF18_450_8.1	3.58	1651	1122	0.70	62.7	269.1	±2.9	43	±200	0.276	8.5	0.0426	1.1	0.129
ZF18_450_8.2	0.91	947	755	0.82	32.7	252.2	±2.6	215	±110	0.277	5.0	0.0399	1.0	0.209
ZF18_450_9.1	0.31	390	236	0.63	13.0	244.1	±2.7	23	±110	0.247	4.8	0.0386	1.1	0.235
ZF18_450_10.1	0.25	223	114	0.53	36.8	1128.0	±11.0	1294	±35	2.217	2.1	0.1913	1.1	0.503
ZF18_450_11.1	0.22	244	293	1.24	18.9	554.0	±6.0	384	±75	0.672	3.5	0.0897	1.1	0.322
ZF18_450_12.1	0.01	127	140	1.14	4.3	247.6	±4.0	378	±110	0.292	5.4	0.0392	1.6	0.305
ZF18_450_13.1	0.01	103	116	1.17	3.4	243.4	±4.4	267	±140	0.274	6.2	0.0385	1.8	0.295
ZF18_450_14.1	0.57	268	422	1.63	8.7	238.3	±3.1	145	±160	0.254	6.9	0.0377	1.3	0.190
ZF18_450_15.1	0.01	262	258	1.02	8.7	246.6	±3.7	392	±150	0.293	6.7	0.0390	1.5	0.227

(continued)

Table 6 (continued)

Crater	$^{206}Pb_c$, %	U, ppm	Th, ppm	$^{232}Th/^{238}U$	$^{206}Pb^*$, ppm	$^{206}Pb^*/^{238}U$ Age		$^{207}Pb^*/^{206}Pb^*$ Age		$^{207}Pb^*/^{235}U$	±%	$^{206}Pb^*/^{238}U$	±%	KK
ZF 18 (460.9)														
ZF-18(460.3–460.9)_1.1	0.09	813	1091	1.39	29.6	267.4	±4.3	264	±46	0.301	2.6	0.0424	1.6	0.629
ZF-18(460.3–460.9)_2.1	0.01	978	1354	1.43	35.2	264.7	±4.4	264	±32	0.298	2.2	0.0419	1.7	0.772
ZF-18(460.3–460.9)_2.2	0.07	2799	6193	2.29	103.0	269.6	±4.4	251	±28	0.302	2.1	0.0427	1.7	0.813
ZF-18(460.3–460.9)_3.1	0.21	932	3220	3.57	32.7	257.5	±4.3	198	±57	0.281	3.0	0.0408	1.7	0.570
ZF-18(460.3–460.9)_4.1	0.14	1410	6076	4.45	50.4	262.3	±4.4	310	±55	0.301	3.0	0.0415	1.7	0.577
ZF-18(460.3–460.9)_5.1	0.08	1076	5623	5.40	37.7	257.8	±4.4	251	±37	0.288	2.4	0.0408	1.7	0.738
ZF-18(460.3–460.9)_6.1	0.09	1158	5285	4.71	40.6	257.3	±4.4	224	±46	0.284	2.6	0.0407	1.7	0.658
ZF-18(460.3–460.9)_6.2	0.10	1113	3105	2.88	39.5	260.5	±4.5	288	±47	0.296	2.7	0.0412	1.8	0.656
ZF-18(460.3–460.9)_7.1	0.18	865	3309	3.95	30.0	254.9	±4.6	263	±52	0.286	2.9	0.0403	1.9	0.636
ZF-18(460.3–460.9)_7.2	0.08	949	3889	4.23	33.2	256.8	±4.4	249	±46	0.287	2.7	0.0406	1.8	0.661
ZF-18(460.3–460.9)_8.1	0.05	1331	6528	5.07	47.4	261.7	±4.3	255	±33	0.293	2.2	0.0414	1.7	0.759
ZF-18(460.3–460.9)_9.1	0.12	1967	2388	1.25	72.3	269.8	±5.3	240	±47	0.301	2.9	0.0427	2.0	0.702
ZF 18 (471)														
ZF18_471_1.1	0.06	2777	4286	1.60	99.3	262.7	±2.1	211	±39	0.289	1.9	0.0416	0.8	0.445
ZF18_471_2.1	0.33	180	58	0.33	52.1	1868.0	±17.0	1859	±28	5.270	1.9	0.3361	1.1	0.565
ZF18_471_3.1	0.14	379	224	0.61	112.0	1906.0	±15.0	1903	±16	5.527	1.3	0.3441	0.9	0.731
ZF18_471_4.1	0.21	598	436	0.75	20.9	256.0	±2.9	207	±73	0.281	3.3	0.0405	1.2	0.351
ZF18_471_5.1	0.19	1039	1008	1.00	35.8	253.2	±2.6	203	±56	0.277	2.6	0.0401	1.1	0.402
ZF18_471_6.1	0.22	865	638	0.76	29.3	248.9	±2.3	228	±74	0.275	3.4	0.0394	1.0	0.284
ZF18_471_7.1	0.10	1028	1026	1.03	34.7	248.0	±2.6	238	±47	0.276	2.3	0.0392	1.1	0.464
ZF18_471_8.1	0.12	776	585	0.78	26.6	251.6	±2.4	219	±57	0.277	2.6	0.0398	1.0	0.372
ZF18_471_9.1	0.08	399	214	0.56	123.0	1974.0	±18.0	1894	±14	5.726	1.3	0.3583	1.1	0.806
ZF18_471_10.1	0.10	901	772	0.89	31.8	259.2	±2.5	244	±51	0.289	2.4	0.0410	1.0	0.401
ZF 19 (429.7)														
ZF-19-429.7_1.1	0.40	439	500	1.18	15.3	255.2	±3.2	177	±120	0.276	5.4	0.0404	1.3	0.239
ZF-19-429.7_2.1	0.45	576	902	1.62	20.4	259.7	±3.1	183	±120	0.282	5.2	0.0411	1.2	0.235
ZF-19-429.7_3.1	0.54	480	498	1.07	16.7	254.9	±3.2	238	±130	0.283	6.0	0.0403	1.3	0.216
ZF-19-429.7_4.1	0.39	526	563	1.10	18.0	250.2	±3.1	169	±120	0.270	5.3	0.0396	1.3	0.241

(continued)

Table 6 (continued)

Crater	$^{206}Pb_c$, %	U, ppm	Th, ppm	$^{232}Th/^{238}U$	$^{206}Pb^*$, ppm	$^{206}Pb^*/^{238}U$ Age		$^{207}Pb^*/^{206}Pb^*$ Age		$^{207}Pb^*/^{235}U$	±%	$^{206}Pb^*/^{238}U$	±%	KK
ZF-19-429.7_1.2	0.16	1984	1080	0.56	70.5	260.9	±2.6	249	±48	0.292	2.3	0.0413	1.0	0.440
ZF-19-429.7_1.3	0.14	2705	2730	1.04	96.4	261.8	±2.6	202	±40	0.287	2.0	0.0414	1.0	0.508
ZF-19-429.7_2.2	0.19	925	1719	1.92	31.7	251.7	±2.8	363	±63	0.295	3.0	0.0398	1.1	0.373
ZF 19 (445)														
ZF-19-445_1.1	0.81	1642	3642	2.29	60.9	270.2	±4.5	219	±110	0.298	4.9	0.0428	1.7	0.348
ZF-19-445_2.1	0.09	1175	3069	2.70	41.8	261.4	±4.3	303	±49	0.299	2.7	0.0414	1.7	0.616
ZF 19 (449)														
ZF-19-449_1.1	0.72	257	350	1.41	9.0	254.1	±4.9	338	±150	0.295	6.9	0.0402	2.0	0.286
ZF-19-449_2.1	0.61	403	793	2.03	14.5	262.8	±4.9	125	±190	0.278	8.3	0.0416	1.9	0.230
ZF-19-449_3.1	0.30	995	3253	3.38	36.8	271.0	±4.7	167	±82	0.292	3.9	0.0429	1.8	0.450
ZF-19-449_4.1	0.75	995	2163	2.24	36.5	267.1	±4.6	267	±110	0.301	4.9	0.0423	1.7	0.354
ZF-19-449_5.1	0.29	1093	1396	1.32	38.0	254.8	±4.3	190	±64	0.277	3.2	0.0403	1.7	0.527
ZF-19-449_6.1	0.16	1595	1660	1.08	57.4	264.2	±4.4	259	±39	0.297	2.4	0.0419	1.7	0.701
ZF-19-449_7.1	0.39	726	665	0.95	25.9	261.5	±4.4	169	±78	0.282	3.8	0.0414	1.7	0.459
ZF-19-449_8.1	0.31	1408	2144	1.57	49.8	259.3	±4.3	302	±58	0.297	3.1	0.0411	1.7	0.549
ZF 9 (479.1)														
ZF-19-479.1_9.1	0.11	966	1748	1.87	32.9	250.3	±2.7	223	±63	0.276	2.9	0.0396	1.1	0.373
ZF-19-479.1_10.2	0.06	2176	4151	1.97	75.0	253.4	±2.5	244	±34	0.282	1.8	0.0401	1.0	0.564
ZF-19-479.1_5.1	0.17	2224	2997	1.39	77.3	255.2	±2.8	245	±46	0.285	2.3	0.0404	1.1	0.487
ZF-19-479.1_4.1	0.22	2148	4350	2.09	74.9	255.9	±2.9	311	±52	0.294	2.6	0.0405	1.2	0.453
ZF-19-479.1_11.1	0.04	2243	2968	1.37	78.7	258.0	±2.6	269	±40	0.291	2.0	0.0408	1.0	0.507
ZF-19-479.1_10.1	0.08	1621	1497	0.95	57.0	258.5	±2.6	226	±41	0.286	2.1	0.0409	1.0	0.499
ZF-19-479.1_6.1	0.12	1294	3081	2.46	45.7	259.5	±2.5	293	±46	0.296	2.2	0.0411	1.0	0.446
ZF-19-479.1_2.1	0.09	5307	14,637	2.85	188.0	260.3	±3.0	228	±27	0.288	1.6	0.0412	1.2	0.707
ZF-19-479.1_11.2	0.06	2534	3523	1.44	90.1	261.3	±3.0	252	±33	0.292	1.9	0.0414	1.2	0.636
ZF-19-479.1_8.1	0.18	2280	2945	1.33	81.4	262.1	±3.0	228	±55	0.290	2.7	0.0415	1.2	0.445
ZF-19-479.1_1.1	0.01	2688	4865	1.87	96.2	263.0	±2.2	265	±25	0.296	1.4	0.0417	0.9	0.614
ZF-19-479.1_7.1	0.43	2619	5457	2.15	94.4	263.9	±2.5	272	±59	0.298	2.8	0.0418	1.0	0.351
ZF-19-479.1_12.1	0.01	5948	10,275	1.78	225.0	278.1	±3.2	277	±21	0.315	1.5	0.0441	1.2	0.798
ZF-19-479.1_3.1	0.23	2767	3452	1.29	107.0	281.9	±3.5	187	±61	0.307	2.9	0.0447	1.3	0.433

(continued)

Table 6 (continued)

Crater	$^{206}Pb_c$, %	U, ppm	Th, ppm	$^{232}Th/^{238}U$	$^{206}Pb^*$, ppm	$^{206}Pb^*/^{238}U$ Age		$^{207}Pb^*/^{206}Pb^*$ Age		$^{207}Pb^*/^{235}U$	±%	$^{206}Pb^*/^{238}U$	±%	KK
ZF 19 (488.6)														
ZF-19-488.6_3.1	0.08	631	2518	4.12	22.1	257.5	±2.8	251	±71	0.288	3.3	0.0408	1.1	0.335
ZF-19-488.6_3.2	0.08	1099	3866	3.63	41.2	275.2	±2.4	235	±48	0.306	2.3	0.0436	0.9	0.390
ZF-19-488.6_5.1	0.08	2759	5444	2.04	102.0	271.5	±2.7	276	±40	0.307	2.0	0.0430	1.0	0.502
ZF-19-488.6_6.1	0.11	2635	5801	2.28	98.9	275.5	±3.1	247	±39	0.308	2.1	0.0437	1.1	0.555
ZF-19-488.6_7.1	0.05	4110	12,133	3.05	155.0	277.2	±2.8	235	±30	0.308	1.6	0.0440	1.0	0.621
ZF-19-488.6_8.1	0.18	2988	8672	3.00	113.0	277.4	±2.8	240	±40	0.309	2.0	0.0440	1.0	0.515
ZF-19-488.6_10.1	0.01	2512	6262	2.58	94.5	276.4	±3.1	282	±39	0.314	2.0	0.0438	1.1	0.560
ZF-19-488.6_9.1	0.01	4905	10674	2.25	188.0	281.7	±2.8	241	±27	0.314	1.5	0.0447	1.0	0.663
ZF-19-488.6_2.1	4.07	1651	2961	1.85	64.6	275.5	±2.9	304	±230	0.316	10.0	0.0437	1.1	0.103
ZF-19-488.6_1.1	0.15	3770	6277	1.72	146.0	283.2	±2.2	251	±39	0.317	1.9	0.0449	0.8	0.424
ZF-19-488.6_4.1	0.09	5616	13,447	2.47	227.0	296.3	±2.5	236	±28	0.330	1.5	0.0470	0.9	0.584
ZF-19-488.6_7.2	0.14	5139	13,273	2.67	211.0	300.0	±3.2	211	±38	0.331	2.0	0.0476	1.1	0.564
ZF 21														
ZF-21_1.1	0.31	662	639	1.00	11.1	123.6	±1.8	75	±120	0.127	5.2	0.0194	1.5	0.283
ZF-21_3.1	0.06	1505	1846	1.27	52.8	257.9	±3.1	260	±35	0.289	2.0	0.0408	1.2	0.620
ZF-21_5.1	0.06	2917	4213	1.49	107.0	269.6	±3.0	266	±26	0.304	1.6	0.0427	1.1	0.708
ZF-21_4.2	0.08	2361	5864	2.57	88.3	274.5	±3.0	285	±39	0.312	2.0	0.0435	1.1	0.556
ZF-21_4.1	0.08	3600	7635	2.19	137.0	278.4	±3.0	269	±26	0.314	1.6	0.0441	1.1	0.703
ZF-21_2.1	0.10	4760	14,380	3.12	181.0	278.9	±3.0	265	±23	0.314	1.5	0.0442	1.1	0.747
ZF 21 (446.0–446.4)														
ZF-21-446.0–446.4_1.1	0.46	563	1134	2.08	20.4	265.0	±2.8	–	–	0.302	5.0	0.0450	1.1	0.215
ZF-21-446.0–446.4_1.2	5.01	519	1072	2.13	20.2	270.9	±4.2	–	–	0.365	14.0	0.0429	1.6	0.110
ZF 21 (457.6–457.9)														
ZF-21-457.6–457.9_4.1	0.77	2091	2516	1.24	78.0	272.0	±3.0	–	–	0.292	3.6	0.0431	1.1	0.294
ZF-21-457.6–457.9_3.2	0.24	2100	5438	2.68	79.2	276.2	±3.0	–	–	0.305	2.7	0.0438	1.1	0.403
ZF-21-457.6–457.9_2.1	0.32	2840	6670	2.43	106.0	273.3	±2.0	–	–	0.309	2.5	0.0433	0.8	0.305
ZF-21-457.6–457.9_3.1	0.05	3092	6480	2.17	119.0	283.1	±3.0	–	–	0.325	1.9	0.0449	1.1	0.580
ZF-21-457.6–457.9_1.1	1.57	1286	4226	3.39	49.4	277.4	±2.0	–	–	0.388	13.0	0.0440	0.9	0.074

(continued)

Table 6 (continued)

Crater	$^{206}Pb_c$, %	U, ppm	Th, ppm	$^{232}Th/^{238}U$	$^{206}Pb^*$, ppm	$^{206}Pb^*/^{238}U$ Age		$^{207}Pb^*/^{206}Pb^*$ Age		$^{207}Pb^*/^{235}U$	±%	$^{206}Pb^*/^{238}U$	±%	KK
ZF 21 (465.0–465.6)														
ZF-21-465.0–465.6_3.1	2.33	477	249	0.54	8.07	122.8	±1.8	–	–	0.13	18.0	0.0192	1.5	0.081
ZF-21-465.0–465.6_1.1	0.13	1957	3442	1.82	33.7	127.6	±1.4	–	–	0.132	2.8	0.020	1.1	0.391
ZF-21-465.0–465.6_2.1	0.12	6030	4458	0.76	230.0	279.8	±2.6	–	–	0.317	1.6	0.0444	1.0	0.600
ZF-21-465.0–465.6_2.2	0.05	8096	9100	1.16	322.0	291.6	±2.8	–	–	0.327	1.3	0.0463	1.0	0.758
ZF-21-465.0–465.6_4.1	0.02	1131	5	0.01	343.0	1950.0	±17.0	1939.7	±9	5.791	1.1	0.3533	1.0	0.892
ZF 21 (472.8–473.1)														
ZF-21-472.8–473.1_1.1	0.23	282	69	0.26	18.3	468.4	±4.5	–	–	0.576	3.3	0.0754	10.0	0.304
ZF-21-472.8-473.1_1.2	0.43	346	62	0.19	20.1	419.2	±3.9	–	–	0.518	3.9	0.0672	1.0	0.250
ZF 30 (578.7)														
ZF-30_1.1	0.19	859	1167	1.40	15.0	129.6	±1.7	154	±83	0.138	3.8	0.0203	1.3	0.355
ZF-30_3.1	0.37	459	317	0.71	20.6	326.5	±4.2	289	±97	0.373	4.4	0.0520	1.3	0.299
ZF-30_10.1	0.22	486	269	0.57	25.0	375.0	±4.7	345	±70	0.441	3.4	0.0599	1.3	0.389
ZF-30_5.1	0.46	276	108	0.40	17.1	447.6	±6.6	412	±93	0.545	4.4	0.0719	1.5	0.344
ZF-30_7.1	0.01	472	1333	2.92	53.9	805.4	±9.3	782	±29	1.197	1.9	0.1331	1.2	0.661
ZF-30_2.1	0.08	249	224	0.93	33.1	927.0	±11.0	901	±38	1.474	2.2	0.1547	1.3	0.581
ZF-30_4.1	0.22	95	58	0.63	12.7	929.0	±14.0	926	±67	1.494	3.6	0.1550	1.6	0.450
ZF-30_9.1	0.18	324	179	0.57	90.7	1815.0	±20.0	1827	±20	5.008	1.7	0.3253	1.3	0.752
ZF-30_8.1	0.01	53	81	1.58	15.0	1828.0	±29.0	1840	±39	5.090	2.8	0.3279	1.8	0.647
ZF-30_6.1	0.01	293	188	0.66	95.7	2075.0	±22.0	2095	±15	6.800	1.5	0.3798	1.3	0.835
ZF 31 (539.8)														
ZF-31_1.1	0.55	653	194	0.31	11.0	125.1	±1.7	8	±130	0.125	5.4	0.0196	1.3	0.250
ZF-31_2.1	0.93	467	323	0.71	8.16	128.5	±2.0	223	±310	0.140	13.0	0.0201	1.6	0.115
ZF-31_3.1	1.78	306	140	0.47	5.29	126.3	±2.3	–	–	0.116	16.0	0.0198	1.8	0.110
ZF-31_4.1	1.20	817	755	0.96	13.8	124.3	±1.8	298	±260	0.140	11.0	0.0195	1.4	0.128
ZF-31_5.1	0.98	714	387	0.56	12.2	125.9	±1.8	–	–	0.117	8.9	0.0197	1.4	0.161
ZF-31_6.1	0.69	443	205	0.48	7.6	127.1	±1.9	586	±190	0.163	9.0	0.0199	1.5	0.170
ZF-31_7.1	0.67	511	309	0.62	8.7	124.8	±1.8	34	±160	0.126	6.9	0.0196	1.4	0.209

(continued)

Table 6 (continued)

Crater	$^{206}Pb_c$, %	U, ppm	Th, ppm	$^{232}Th/^{238}U$	$^{206}Pb^*$, ppm	$^{206}Pb^*/^{238}U$ Age		$^{207}Pb^*/^{206}Pb^*$ Age		$^{207}Pb^*/^{235}U$	±%	$^{206}Pb^*/^{238}U$	±%	KK
ZF 43 (625.5)														
ZF43.1.1	0.01	5236	8072	1.59	181.0	254.2	±4.5	257	±23	0.285	2.1	0.0402	1.8	0.879
ZF43.2.1	0.57	5023	332	0.07	196.0	284.4	±5.0	412	±44	0.342	2.7	0.0451	1.8	0.676
ZF43.2.2	0.74	4683	214	0.05	183.0	285.1	±5.2	364	±53	0.336	3.0	0.0452	1.9	0.619
KZ 931 (616–622)														
KZ931_1.1	0.03	545	7	0.01	243.0	2691.0	±30.0	2693.2	±6.6	13.170	1.4	0.5180	1.4	0.960
KZ931_2.1	0.11	154	70	0.47	40.0	1701.0	±28.0	1724	±19	4.394	2.1	0.3019	1.9	0.871
KZ931_3.1	0.01	335	196	0.60	11.6	254.0	±3.8	307	±53	0.291	2.8	0.0402	1.5	0.549
KZ931_4.1	0.17	788	632	0.83	26.8	249.9	±3.4	231	±57	0.277	2.8	0.0395	1.4	0.492
KZ931_5.1	0.01	772	641	0.86	26.7	254.7	±3.6	266	±36	0.287	2.1	0.0403	1.4	0.676
KZ931_6.1	0.05	1082	979	0.94	37.1	252.3	±3.5	216	±34	0.278	2.0	0.0399	1.4	0.697
KZ931_7.1	0.26	428	278	0.67	14.5	248.1	±3.8	289	±66	0.282	3.3	0.0392	1.5	0.474
KZ931_8.1	0.01	449	224	0.51	15.0	245.4	±3.7	205	±49	0.269	2.6	0.0388	1.5	0.584
KZ931_9.1	0.13	1438	1285	0.92	50.2	256.5	±3.5	213	±39	0.282	2.2	0.0406	1.4	0.643
KZ931_10.1	0.06	325	1	0.01	46.0	983.0	±13.0	1009	±31	1.653	2.1	0.1646	1.4	0.691
KZ931_11.1	0.06	626	663	1.09	22.0	257.8	±3.8	215	±45	0.284	2.4	0.0408	1.5	0.610
KZ931_12.1	0.06	936	758	0.84	32.0	251.1	±3.6	254	±37	0.281	2.2	0.0397	1.4	0.670
KZ 1319 (626.8)														
KZ-13-19_12.1	0.29	325	111	0.35	28.0	615.0	±5.2	578	±66	0.818	3.2	0.1001	0.9	0.280
KZ-13-19_1.1	0.01	94	76	0.84	20.9	1491.0	±15.0	1455	±31	3.279	2.0	0.2602	1.1	0.572
KZ-13-19_7.1	0.74	43	29	0.71	10.0	1544.0	±22.0	1536	±79	3.560	4.5	0.2706	1.6	0.354
KZ-13-19_8.1	0.03	304	178	0.60	71.7	1562.0	±12.0	1540	±17	3.613	1.2	0.2741	0.0	0.682
KZ-13-19_7.2	0.01	49	42	0.88	11.6	1568.0	±20.0	1546	±40	3.643	2.6	0.2755	1.4	0.560
KZ-13-19_9.1	0.50	31	46	1.53	7.9	1659.0	±25.0	1588	±68	3.970	4.0	0.2935	1.7	0.425
KZ-13-19_2.1	0.01	12	17	1.56	2.9	1674.0	±39.0	1660	±75	4.170	4.9	0.2966	2.7	0.547
KZ-13-19_11.1	0.03	127	51	0.41	36.4	1852.0	±16.0	1863	±20	5.227	1.5	0.3328	1.1	0.662
KZ-13-19_11.2	0.18	132	55	0.43	38.0	1854.0	±16.0	1843	±26	5.176	1.8	0.3332	1.0	0.565
KZ-13-19_5.2	0.26	81	133	1.69	23.6	1871.0	±19.0	1860	±33	5.280	2.2	0.3367	1.2	0.544
KZ-13-19_5.1	0.01	48	56	1.22	14.0	1902.0	±23.0	1868	±32	5.410	2.3	0.3432	1.4	0.619

(continued)

Table 6 (continued)

Crater	$^{206}Pb_c$, %	U, ppm	Th, ppm	$^{232}Th/^{238}U$	$^{206}Pb^*$, ppm	$^{206}Pb^*/^{238}U$ Age		$^{207}Pb^*/^{206}Pb^*$ Age		$^{207}Pb^*/^{235}U$	±%	$^{206}Pb^*/^{238}U$	±%	KK
KZ-13-19_4.1	0.20	134	63	0.48	39.9	1920.0	±17.0	1921	±23	5.628	1.6	0.3470	1.1	0.611
KZ-13-19_6.1	0.09	157	47	0.31	47.2	1932.0	±18.0	1959	±18	5.792	1.5	0.3495	1.1	0.739
KZ-13-19_16.1	0.16	109	47	0.45	33.0	1935.0	±18.0	1977	±28	5.860	1.9	0.3501	1.1	0.555
KZ-13-19_13.1	0.20	31	48	1.60	9.5	1960.0	±28.0	1974	±40	5.940	2.8	0.3554	1.6	0.593
KZ-13-19_3.1	0.01	194	83	0.44	63.7	2090.0	±16.0	2054	±14	6.696	1.2	0.3830	0.9	0.738
KZ-13-19_10.1	0.38	425	300	0.73	181.0	2581.0	±17.0	2724	±9	12.760	1.0	0.4923	0.8	0.821
KZ-13-19_14.1	0.59	269	24	0.09	119.0	2659.0	±20.0	2644	±14	12.610	1.2	0.5107	0.9	0.745
KZ-13-19_15.2	0.14	59	12	0.21	26.0	2669.0	±28.0	2833	±19	14.200	1.7	0.5130	1.3	0.743
KZ-13-19_15.1	0.57	28	13	0.48	12.8	2755.0	±45.0	2782	±31	14.310	2.8	0.5330	2.0	0.720
PT 2														
PT-2_1.1	0.08	479	3543	7.64	16.5	253.4	±3.9	277	±45	0.287	2.5	0.0401	1.6	0.629
PT-2_1.2	0.01	268	289	1.11	9.2	252.3	±4.1	309	±51	0.289	2.8	0.0399	1.6	0.589
PT-2_2.1	0.09	352	691	2.03	12.1	251.6	±4.0	250	±51	0.281	2.7	0.0398	1.6	0.587
PT-2_3.1	0.01	328	1360	4.28	10.9	244.4	±3.9	291	±52	0.278	2.8	0.0386	1.6	0.586
PT-2_3.2	0.07	2253	10,671	4.89	78.1	254.9	±3.8	238	±23	0.283	1.8	0.0403	1.5	0.830
PT-2_4.1	0.01	570	931	1.69	19.3	249.5	±3.8	251	±36	0.279	2.2	0.0395	1.6	0.703
PT-2_5.1	0.01	918	1469	1.65	31.0	248.8	±3.7	247	±29	0.277	2.0	0.0393	1.5	0.768
PT-2_5.2	0.10	389	554	1.47	13.2	249.7	±3.9	306	±51	0.286	2.8	0.0395	1.6	0.580
PT-2_6.1	0.04	430	99	0.24	33.9	565.2	±8.4	576	±28	0.749	2.0	0.0916	1.6	0.773
PT-2_6.2	0.01	1420	657	0.48	114.0	574.1	±8.8	567	±18	0.758	1.8	0.0931	1.6	0.889
PT 2 (1368.4)														
PT2-1368.4_1.1	1.09	155	73	0.49	5.2	245.6	±4.8	76	±310	0.254	13	0.0388	2.0	0.149
PT2-1368.4_2.1	0.21	1370	2086	1.57	48.8	261.2	±3.5	260	±59	0.293	2.9	0.0414	1.4	0.468
PT2-1368.4_3.1	0.14	1176	1760	1.55	40.3	251.8	±3.4	139	±57	0.268	2.8	0.0398	1.4	0.493
PT2-1368.4_4.1	0.11	2587	6375	2.55	89.3	253.7	±3.3	282	±32	0.287	1.9	0.0401	1.3	0.679
PT2-1368.4_5.1	0.32	538	1325	2.55	18.4	251.5	±3.7	175	±100	0.272	4.6	0.0398	1.5	0.319
PT2-1368.4_6.1	0.50	251	171	0.71	8.5	249.1	±4.1	292	±140	0.283	6.3	0.0394	1.7	0.269
PT2-1368.4_7.1	0.08	1555	3384	2.25	53.6	253.2	±3.1	252	±65	0.283	3.1	0.0401	1.3	0.409
PT2-1368.4_8.1	0.11	1790	3340	1.93	60.4	248.2	±3.5	244	±42	0.276	2.3	0.0392	1.4	0.613

(continued)

Table 6 (continued)

Crater	$^{206}Pb_c$, %	U, ppm	Th, ppm	$^{232}Th/^{238}U$	$^{206}Pb^*$, ppm	$^{206}Pb^*/^{238}U$ Age		$^{207}Pb^*/^{206}Pb^*$ Age		$^{207}Pb^*/^{235}U$	±%	$^{206}Pb^*/^{238}U$	±%	KK
PT2-1368.4_9.1	0.30	2117	3260	1.59	72.5	251.3	±3.7	326	±72	0.290	3.5	0.0398	1.5	0.425
PT2-1368.4_10.1	0.20	704	585	0.86	23.1	241.0	±3.4	183	±74	0.261	3.5	0.0381	1.4	0.416
PT2-1368.4_11.1	0.13	3089	7168	2.40	106.0	251.4	±3.6	222	±44	0.277	2.4	0.0398	1.5	0.610
PT2-1368.4_12.1	0.01	597	743	1.29	21.2	260.8	±3.9	273	±69	0.294	3.4	0.0413	1.5	0.456
PT2-1368.4_13.1	0.18	887	1435	1.67	30.0	248.3	±3.4	244	±63	0.276	3.1	0.0393	1.4	0.459
PT2-1368.4_14.1	0.24	1475	2611	1.83	50.8	252.7	±3.4	323	±74	0.291	3.5	0.0340	1.4	0.385
PT 2 (1371.2–1371.8)														
PT-2 (1371.2–1371.8)_3.1	0.13	1052	1264	1.24	34.6	242.0	±2.3	311	±51	0.277	2.4	0.0382	1.0	0.395
PT-2 (1371.2–1371.8)_7.1	0.05	3608	7518	2.15	120.0	245.0	±2.1	254	±32	0.274	1.7	0.0388	0.9	0.535
PT-2 (1371.2–1371.8)_4.1	0.26	1395	2731	2.02	48.9	257.0	±2.3	267	±67	0.289	3.1	0.0407	0.9	0.299
PT-2 (1371.2–1371.8)_6.1	1.39	56	31	0.58	5.0	631.0	±26.0	426	±330	0.780	15.0	0.1028	4.3	0.278
PT-2 (1371.2–1371.8)_1.1	1.00	391	104	0.28	54.7	963.0	±9.1	1092	±67	1.685	3.5	0.1611	1.0	0.290
PT-2 (1371.2–1371.8)_5.1	0.22	137	49	0.37	24.4	1213.0	±14.0	1184	±49	2.269	2.8	0.2071	1.3	0.450
PT-2 (1371.2–1371.8)_2.1	0.32	143	118	0.86	37.2	1701.0	±18.0	1673	±44	4.270	2.7	0.3019	1.2	0.459
PT 2 (1415)														
PT2-1415_10.1	0.63	462	489	1.09	9.8	156.7	±2.8	39	±200	0.159	8.4	0.0246	1.8	0.217
PT2-1415_9.1	1.29	159	187	1.21	5.2	238.0	±4.9	105	±360	0.249	15.0	0.0376	2.1	0.137
PT2-1415_11.1	0.51	741	309	0.43	24.3	240.5	±3.6	238	±130	0.267	5.7	0.0380	1.5	0.266
PT2-1415_4.1	0.50	363	433	1.23	12.1	244.4	±3.8	316	±160	0.281	7.1	0.0386	1.6	0.226
PT2-1415_2.1	0.30	617	741	1.24	21.0	249.9	±3.7	265	±92	0.281	4.3	0.0395	1.5	0.348
PT2-1415_8.1	0.75	542	558	1.07	18.7	251.6	±3.8	224	±170	0.278	7.7	0.0398	1.6	0.203
PT2-1415_5.1	0.21	717	699	1.01	24.8	254.4	±3.6	346	±130	0.296	5.8	0.0403	1.5	0.251
PT2-1415_6.1	0.11	1540	1678	1.13	54.1	258.1	±3.5	226	±45	0.285	2.4	0.0409	1.4	0.576
PT2-1415_1.1	0.28	718	688	0.99	25.7	261.9	±3.7	228	±89	0.290	4.1	0.0415	1.5	0.351
PT2-1415_3.1	0.23	558	527	0.97	20.0	262.5	±3.9	196	±81	0.287	3.8	0.0416	1.5	0.393
PT2-1415_12.1	0.38	356	188	0.55	14.5	298.2	±4.5	241	±110	0.333	5.1	0.0474	1.5	0.299
PT2-1415_7.1	0.38	586	256	0.45	25.9	322.3	±4.7	203	±120	0.355	5.5	0.0513	1.5	0.269

(continued)

Table 6 (continued)

Crater	$^{206}Pb_c$, %	U, ppm	Th, ppm	$^{232}Th/^{238}U$	$^{206}Pb^*$, ppm	$^{206}Pb^*/^{238}U$ Age		$^{207}Pb^*/^{206}Pb^*$ Age		$^{207}Pb^*/^{235}U$	±%	$^{206}Pb^*/^{238}U$	±%	KK
PT 2 (1419.1)														
PT2 1419.1.1	0.21	1111	2172	2.02	38.0	251.0	±4.5	188	±64	0.273	3.3	0.0397	1.8	0.554
PT2 1419.2.1	0.03	198	178	0.93	53.4	1757.0	±28.0	1771	±19	4.678	2.1	0.3133	1.8	0.864
PT2 1419.3.1	0.17	448	2597	5.99	14.5	238.5	±4.3	203	±89	0.261	4.3	0.0377	1.8	0.431
PT2 1419.4.1	1.51	320	162	0.52	51.2	1085.0	±18.0	1442	±57	2.294	3.5	0.1833	1.8	0.515
PT2 1419.5.1	0.12	1639	1931	1.22	54.5	244.3	±4.1	274	±51	0.276	2.8	0.0386	1.7	0.617
PT2 1419.6.1	0.01	439	2972	7.00	14.5	242.9	±4.4	191	±65	0.264	3.3	0.0384	1.8	0.552
PT2 1419.7.1	0.01	1550	5797	3.86	51.1	243.1	±4.3	248	±45	0.271	2.7	0.0384	1.8	0.673
PT2 1419.8.1	0.20	837	2471	3.05	26.8	235.1	±4.1	236	±64	0.261	3.3	0.0371	1.8	0.541
PT2 1419.9.1	0.13	1174	4051	3.56	38.5	241.4	±4.1	176	±48	0.261	2.7	0.0382	1.7	0.647
PT2 1419.10.1	0.37	283	538	1.96	9.4	244.2	±4.7	243	±110	0.272	5.3	0.0386	2.0	0.367
PT2 1419.11.1	0.17	1002	3545	3.66	32.3	237.3	±4.1	226	±57	0.262	3.0	0.0375	1.8	0.582
PT2 1419.12.1	1.36	1637	8734	5.51	52.9	234.8	±4.0	265	±110	0.264	5.0	0.0371	1.8	0.349
PT2 1419.13.1	0.50	346	3127	9.34	11.0	232.9	±4.4	202	±130	0.254	6.0	0.0368	1.9	0.317
PT2 1419.14.1	0.11	1722	3899	2.34	56.4	240.9	±4.1	209	±42	0.264	2.5	0.0381	1.7	0.688
PT2 1419.15.1	0.30	579	1530	2.73	19.1	241.6	±4.3	177	±77	0.261	3.8	0.0382	1.8	0.478
PT2 1419.16.1	0.31	280	613	2.26	9.2	241.1	±4.6	119	±110	0.254	5.1	0.0381	1.9	0.379
PT 2 (1423)														
PT-2-1423_1.1	0.63	543	1417	2.69	19.4	261.1	±4.7	277	±130	0.295	5.9	0.0413	1.8	0.310
PT-2-1423_3.1	0.58	595	6765	11.75	113.0	260.7	±4.2	213	±42	0.287	2.4	0.0413	1.6	0.675
PT-2-1423_3.2	0.22	656	7882	12.42	59.2	264.1	±4.3	246	±50	0.295	2.7	0.0418	1.7	0.614
PT-2-1423_2.2	0.23	1644	5816	3.66	20.0	246.3	±4.3	265	±120	0.277	5.5	0.0390	1.8	0.325
PT-2-1423_4.1	0.09	1760	5725	3.36	22.3	249.3	±4.3	282	±77	0.282	3.8	0.0394	1.7	0.461
PT-2-1423_2.1	0.15	3172	24,899	8.11	62.4	260.5	±4.3	245	±40	0.291	2.4	0.0412	1.7	0.693
PT 2 (1425)														
PT-2-1425_1.1	0.09	1341	2988	2.30	46.0	252.1	±2.6	205	±45	0.276	2.2	0.0399	1.0	0.470
PT-2-1425_1.2	0.13	965	1913	2.05	32.6	248.0	±2.6	233	±55	0.275	2.6	0.0392	1.1	0.413
PT-2-1425_2.1	0.33	536	1271	2.45	18.6	254.7	±3.2	187	±110	0.277	4.7	0.0403	1.3	0.275
PT-2-1425_3.1	0.01	1555	2170	1.44	53.7	253.9	±2.5	271	±37	0.286	1.9	0.0402	1.0	0.536

(continued)

Table 6 (continued)

Crater	$^{206}Pb_c$, %	U, ppm	Th, ppm	$^{232}Th/^{238}U$	$^{206}Pb^*$, ppm	$^{206}Pb^*/^{238}U$ Age		$^{207}Pb^*/^{206}Pb^*$ Age		$^{207}Pb^*/^{235}U$	±%	$^{206}Pb^*/^{238}U$	±%	KK
PT-2-1425_4.1	0.20	861	2326	2.79	29.5	251.2	±2.7	296	±67	0.286	3.1	0.0397	1.1	0.350
PT-2-1425_5.1	0.66	257	703	2.82	9.1	257.2	±3.7	328	±170	0.297	7.7	0.0407	1.5	0.191
PT-2-1425_6.1	0.10	1747	5181	3.06	60.1	252.7	±2.5	226	±43	0.279	2.1	0.0400	1.0	0.480
PT-2-1425_7.1	0.17	2164	4111	1.96	75.9	257.5	±2.5	205	±48	0.282	2.3	0.0408	1.0	0.433
PT-2-1425_8.1	0.25	1045	3139	3.10	35.1	246.8	±2.6	347	±66	0.288	3.1	0.0390	1.1	0.346
PT-2-1425_9.1	0.32	925	2098	2.34	31.5	250.3	±2.7	168	±100	0.270	4.4	0.0396	1.1	0.247
PT 2 (1438.7–1439.3)														
PT-2 (1438.7–1439.3)_12.1	0.37	504	1190	2.44	17.2	250.0	±3.2	262	±110	0.281	4.9	0.0396	1.3	0.263
PT-2 (1438.7–1439.3)_10.1	0.11	1545	1584	1.06	52.9	252.0	±2.2	291	±44	0.286	2.1	0.0398	0.9	0.419
PT-2 (1438.7–1439.3)_13.2	0.01	2435	5597	2.38	83.2	252.0	±2.3	262	±32	0.282	1.7	0.0398	0.9	0.549
PT-2 (1438.7–1439.3)_13.1	0.08	2127	5538	2.69	73.1	253.0	±2.1	241	±38	0.281	1.9	0.0340	0.9	0.467
PT-2 (1438.7–1439.3)_5.1	0.16	1507	2726	1.87	51.9	253.0	±1.9	205	±49	0.277	2.2	0.0401	0.8	0.336
PT-2 (1438.7–1439.3)_8.1	2.06	438	192	0.45	80.7	1227.0	±13.0	1561	±75	2.790	4.2	0.2097	1.2	0.287
PT-2 (1438.7–1439.3)_1.1	0.73	478	36	0.08	95.0	1330.0	±12.0	1366	±56	2.760	3.1	0.2292	1.0	0.333
PT-2 (1438.7–1439.3)_4.1	0.37	73	73	1.02	15.8	1440.0	±18.0	1472	±62	3.180	3.5	0.2502	1.4	0.385
PT-2 (1438.7–1439.3)_6.1	0.29	115	49	0.44	26.8	1539.0	±16.0	1557	±49	3.590	2.9	0.2696	1.2	0.417
PT-2 (1438.7–1439.3)_9.1	0.36	101	59	0.61	24.5	1593.0	±19.0	1626	±50	3.870	3.0	0.2803	1.4	0.454
PT-2 (1438.7–1439.3)_11.1	0.10	150	100	0.69	36.4	1599.0	±17.0	1562	±29	3.760	2.0	0.2816	1.2	0.605
PT-2 (1438.7–1439.3)_14.2	0.66	111	147	1.37	31.4	1826.0	±18.0	1795	±54	4.950	3.2	0.3274	1.2	0.362
PT-2 (1438.7–1439.3)_7.1	0.16	182	90	0.51	53.4	1891.0	±18.0	1871	±25	5.380	1.8	0.3408	1.1	0.621
PT-2 (1438.7–1439.3)_2.1	0.53	101	59	0.60	30.7	1939.0	±24.0	1922	±46	5.700	2.9	0.3508	1.4	0.481
PT-2 (1438.7–1439.3)_3.1	0.01	81	97	1.25	25.4	2015.0	±25.0	2022	±29	6.300	2.2	0.3669	1.4	0.666
PT-2 (1438.7–1439.3)_14.1	0.02	396	282	0.74	165.0	2552.0	±20.0	2551.6	±9.9	11.340	1.1	0.4857	0.9	0.846
KZ 981														
KZ981_1.1	0.09	165	61	0.38	38.6	1548.0	±21.0	1541	±21	3.582	1.9	0.2715	1.5	0.800
KZ981_1.2	0.23	294	113	0.40	65.2	1476.0	±19.0	1487	±18	3.296	1.7	0.2572	1.4	0.833
KZ981_2.1	0.03	190	46	0.25	47.9	1662.0	±22.0	1651	±17	4.114	1.8	0.2940	1.5	0.843
KZ981_3.1	1.01	194	101	0.54	29.8	1050.0	±15.0	1572	±37	2.370	2.5	0.1768	1.5	0.613
KZ981_4.1	1.16	488	236	0.50	84.4	1168.0	±15.0	1931	±18	3.239	1.7	0.1986	1.4	0.821

(continued)

Table 6 (continued)

Crater	$^{206}Pb_c$, %	U, ppm	Th, ppm	$^{232}Th/^{238}U$	$^{206}Pb^*$, ppm	$^{206}Pb^*/^{238}U$ Age		$^{207}Pb^*/^{206}Pb^*$ Age		$^{207}Pb^*/^{235}U$	±%	$^{206}Pb^*/^{238}U$	±%	KK
KZ981_5.1	1.32	265	310	1.21	16.5	446.5	±6.8	415	±110	0.545	5.0	0.0717	1.6	0.316
KZ981_6.1	0.08	98	43	0.46	16.8	1167.0	±17.0	1195	±35	2.188	2.4	0.1985	1.6	0.674
KZ981_7.1	0.01	277	109	0.41	20.8	541.0	±7.8	559	±36	0.710	2.2	0.0875	1.5	0.671
KZ981_8.1	0.25	511	235	0.48	120.0	1556.0	±19.0	1943	±12	4.486	1.6	0.2731	1.4	0.902
KZ981_9.1	0.17	250	229	0.95	64.8	1700.0	±22.0	1716	±18	4.373	1.7	0.3017	1.4	0.834
KZ981_10.1	0.04	206	96	0.48	60.2	1884.0	±23.0	1873	±12	5.362	1.6	0.3394	1.4	0.899
KZ981_11.1	0.10	243	67	0.29	42.7	1198.0	±16.0	1219	±23	2.279	1.9	0.2043	1.5	0.786
KZ981_12.1	0.10	282	111	0.41	47.6	1155.0	±15.0	1227	±22	2.198	1.8	0.1963	1.4	0.792
KZ981_13.1	0.27	258	49	0.20	37.6	1006.0	±14.0	1046	±32	1.727	2.2	0.1689	1.5	0.675
KZ981_14.1	0.06	267	207	0.80	84.0	2009.0	±25.0	2026	±12	6.290	1.6	0.3656	1.4	0.902
KZ981_15.1	0.20	239	78	0.34	52.9	1477.0	±19.0	1502	±20	3.326	1.8	0.2575	1.5	0.816
KZ981_16.1	0.05	326	54	0.17	88.7	1774.0	±22.0	1828	±13	4.879	1.6	0.3167	1.4	0.896
KZ 981 (1126.3-1126.9)														
KZ-981(1126.3–1126.9)_1.1	0.34	447	968	2.24	14.8	243.6	±2.8	386	±95	0.289	4.4	0.0385	1.2	0.263
KZ-981(1126.3–1126.9)_4.1	0.36	1418	3244	2.36	47.3	244.7	±2.3	158	±80	0.263	3.5	0.0387	1.0	0.271
KZ-981(1126.3–1126.9)_2.1	0.54	1153	2166	1.94	38.6	245.3	±2.3	229	±88	0.271	3.9	0.0388	1.0	0.243
KZ-981(1126.3–1126.9)_3.1	0.54	1934	5357	2.86	68.0	257.1	±2.6	223	±83	0.284	3.7	0.0407	1.0	0.273
KZ-981(1126.3–1126.9)_5.1	0.31	4186	6226	1.54	148.0	258.4	±2.9	235	±54	0.287	2.6	0.0409	1.1	0.433
KZ 1112 (1098.4)														
KZ-1112-1098.4_8.1	0.07	4269	7097	1.72	156.0	269.0	±4.0	235	±30	0.299	2.1	0.0426	1.7	0.800
KZ-1112-1098.4_7.2	0.38	4052	6368	1.62	152.0	275.0	±5.0	205	±44	0.302	2.6	0.0436	1.8	0.681
KZ-1112-1098.4_7.1	0.37	2744	4016	1.51	105.0	280.0	±5.0	306	±56	0.321	3.1	0.0443	1.8	0.593
KZ-1112-1098.4_8.2	0.07	4156	4545	1.13	162.0	286.0	±5.0	289	±30	0.326	2.2	0.0454	1.8	0.802
KZ-1112-1098.4_4.1	0.60	120	111	0.96	5.3	319.0	±6.0	245	±170	0.357	7.5	0.0507	2.0	0.265
KZ-1112-1098.4_2.1	5.72	200	93	0.48	36.6	1174.0	±20.0	1263	±140	2.280	7.3	0.1997	1.9	0.259
KZ-1112-1098.4_3.1	0.47	389	158	0.42	73.1	1268.0	±19.0	1278	±28	2.500	2.2	0.2174	1.7	0.757
KZ-1112-1098.4_1.1	1.90	140	89	0.66	32.5	1510.0	±25.0	1481	±85	3.370	4.9	0.2640	1.8	0.379
KZ-1112-1098.4_5.1	0.31	114	48	0.43	26.0	1515.0	±24.0	1530	±36	3.470	2.6	0.2649	1.8	0.687
KZ-1112-1098.4_6.1	0.29	90	38	0.44	25.6	1842.0	±30.0	1811	±30	5.050	2.5	0.3307	1.8	0.740

(continued)

Table 6 (continued)

Crater	$^{206}Pb_c$, %	U, ppm	Th, ppm	$^{232}Th/^{238}U$	$^{206}Pb^*$, ppm	$^{206}Pb^*/^{238}U$ Age		$^{207}Pb^*/^{206}Pb^*$ Age		$^{207}Pb^*/^{235}U$	±%	$^{206}Pb^*/^{238}U$	±%	KK
KZ 361														
KZ361_1.1	2.13	789	320	0.42	84.6	742.2	±9.3	992	±37	1.215	2.2	0.1220	1.3	0.594
KZ361_2.1	0.07	224	55	0.25	61.4	1782.0	±21.0	1748	±15	4.695	1.6	0.3184	1.4	0.857
KZ361_2.2	0.06	222	52	0.24	54.6	1621.0	±20.0	1596	±18	3.883	1.7	0.2859	1.4	0.829
KZ361_3.1	0.08	3921	1025	0.27	149.0	279.5	±3.6	273	±18	0.316	1.5	0.0443	1.3	0.858
KZ361_3.2	0.05	3527	918	0.27	136	283.6	±3.6	274	±24	0.321	1.7	0.0450	1.3	0.780
KZ361_4.1	0.02	288	80	0.29	48.1	1146.0	±14.0	1145	±24	2.092	1.8	0.1946	1.4	0.752
KZ361_5.1	0.38	3479	1202	0.36	138.0	290.4	±3.8	267	±69	0.328	3.3	0.0461	1.3	0.404
KZ361_5.2	0.06	2973	761	0.26	116.0	286.7	±3.6	273	±18	0.324	1.5	0.0455	1.3	0.853
KZ 1084 (1150.1)														
KZ-1084_1.1	0.10	141	97	0.71	33.0	1548.0	±27.0	1558	±30	3.612	2.6	0.2714	2.0	0.772
KZ-1084_1.2	0.07	137	93	0.70	31.6	1526.0	±25.0	1574	±29	3.586	2.4	0.2672	1.8	0.768
KZ-1084_2.1	0.82	439	291	0.68	15.2	252.9	±6.3	40	±270	0.258	11.0	0.0400	2.5	0.223
KZ-1084_3.1	0.79	1983	1708	0.89	183.0	652.0	±10.0	671	±36	0.908	2.4	0.1063	1.7	0.704
KZ-1084_4.1	0.51	333	153	0.48	11.5	253.4	±5.2	175	±200	0.274	8.9	0.0401	2.1	0.235
KZ 774														
KZ774_1.1	0.02	2334	2866	1.27	78.8	248.5	±3.2	265	±23	0.279	1.7	0.0393	1.3	0.800
KZ774_1.2	0.04	2597	4574	1.82	89.0	252.0	±3.2	261	±20	0.283	1.6	0.0399	1.3	0.834
KZ774_2.1	0.04	2483	1296	0.54	89.8	265.8	±3.4	256	±19	0.298	1.6	0.0421	1.3	0.841
KZ774_3.1	0.34	7001	7219	1.07	256.0	267.9	±3.4	275	±26	0.303	1.7	0.0424	1.3	0.752
KZ774_4.1	0.09	333	181	0.56	75.9	1515.0	±19.0	1661	±13	3.725	1.5	0.2649	1.4	0.895
KZ774_5.1	0.04	2473	3374	1.41	83.4	248.0	±3.4	277	±23	0.280	1.7	0.0392	1.4	0.816
KZ774_6.1	0.54	1495	3608	2.49	53.4	261.3	±3.7	246	±48	0.292	2.5	0.0414	1.4	0.568
KZ774_7.1	24.30	702	1251	1.84	35.1	276.0	±13.0	1873	±820	0.690	46.0	0.0437	4.9	0.109
KZ774_8.1	1.34	483	553	1.18	93.3	1290.0	±17.0	1883	±19	3.518	1.8	0.2215	1.4	0.806
KZ774_9.1	0.01	4453	12,995	3.02	163.0	269.1	±3.8	271	±18	0.304	1.6	0.0426	1.4	0.879
KZ774_10.1	0.01	5991	6780	1.17	205.0	252.3	±3.4	246	±15	0.281	1.5	0.0399	1.4	0.909
KZ774_11.1	0.35	5499	7755	1.46	192.0	255.8	±3.5	280	±22	0.290	1.7	0.0405	1.4	0.819
KZ774_11.2	0.11	4267	5511	1.33	142.0	244.5	±3.3	253	±20	0.273	1.6	0.0387	1.4	0.846
KZ774_6.2	13.16	803	1324	1.70	33.1	262.2	±4.6	152	±550	0.281	24.0	0.0415	1.8	0.075

(continued)

Table 6 (continued)

Crater	$^{206}Pb_c$, %	U, ppm	Th, ppm	$^{232}Th/^{238}U$	$^{206}Pb^*$, ppm	$^{206}Pb^*/^{238}U$ Age		$^{207}Pb^*/^{206}Pb^*$ Age		$^{207}Pb^*/^{235}U$	±%	$^{206}Pb^*/^{238}U$	±%	KK
KZ 774 [744] (1023)														
KZ744 1023.1.1	0.24	1208	1385	1.18	42.7	259.1	±4.4	199	±77	0.283	3.8	0.0410	1.7	0.463
KZ744 1023.2.1	0.12	1200	876	0.75	42.6	260.8	±4.5	227	±44	0.289	2.6	0.0413	1.7	0.675
KZ744 1023.3.1	0.21	1314	1399	1.10	45.6	254.6	±4.4	218	±54	0.281	2.9	0.0403	1.8	0.601
KZ744 1023.4.1	0.13	603	363	0.62	21.1	257.0	±4.7	245	±82	0.287	4.0	0.0407	1.9	0.464
KZ744 1023.5.1	0.04	2034	2995	1.52	70.0	253.0	±4.3	247	±31	0.282	2.2	0.0400	1.7	0.782
KZ744 1023.6.1	0.03	2396	2027	0.87	84.2	258.4	±4.3	261	±28	0.290	2.1	0.0409	1.7	0.819
KZ744 1023.7.1	0.01	4040	7558	1.93	141.0	257.5	±4.3	258	±21	0.289	1.9	0.0408	1.7	0.878
KZ744 1023.8.1	0.13	1516	1576	1.07	50.8	246.3	±4.2	199	±51	0.269	2.8	0.0390	1.7	0.624
KZ744 1023.9.1	0.07	1249	2042	1.69	41.5	244.4	±4.1	216	±44	0.269	2.6	0.0386	1.7	0.671
KZ744 1023.10.1	0.01	675	506	0.78	22.4	244.9	±4.3	365	±63	0.288	3.3	0.0387	1.8	0.538
KZ744 1023.11.1	0.30	1845	3877	2.17	61.8	245.8	±4.2	189	±66	0.267	3.3	0.0389	1.7	0.523
KZ744 1023.12.1	0.15	760	554	0.75	25.2	244.1	±4.3	215	±61	0.268	3.2	0.0386	1.8	0.561
KZ 774 (1029)														
KZ774 1029.1.1	0.10	862	797	0.95	28.9	246.8	±4.3	187	±51	0.268	2.8	0.0390	1.8	0.627
KZ774 1029.2.1	0.03	2831	4164	1.52	95.2	247.4	±4.2	240	±27	0.275	2.1	0.0391	1.7	0.830
KZ774 1029.3.1	0.20	2117	2999	1.46	71.0	246.5	±4.2	220	±48	0.272	2.7	0.0390	1.7	0.643
KZ774 1029.4.1	0.57	406	350	0.89	13.0	233.9	±4.4	229	±130	0.259	5.8	0.0370	1.9	0.326
KZ774 1029.5.1	0.19	724	750	1.07	23.7	241.1	±4.2	166	±72	0.260	3.6	0.0381	1.8	0.498
KZ774 1029.6.1	0.07	4169	15,092	3.74	140.0	246.7	±4.1	229	±24	0.273	2.0	0.0390	1.7	0.854
KZ774 1029.7.1	0.01	674	1657	2.54	21.4	234.8	±4.1	316	±82	0.270	4.0	0.0371	1.8	0.445
KZ774 1029.8.1	0.16	1234	1262	1.06	40.2	239.8	±4.1	214	±49	0.263	2.7	0.0379	1.7	0.634
KZ774 1029.9.1	0.10	799	696	0.90	26.6	244.5	±4.2	206	±52	0.268	2.9	0.0387	1.8	0.617
KZ774 1029.10.1	0.01	3748	16,189	4.46	126.0	247.4	±4.1	280	±27	0.280	2.1	0.0391	1.7	0.821
KZ774 1029.11.1	0.13	2886	5443	1.95	95.2	242.7	±4.2	206	±34	0.266	2.3	0.0384	1.8	0.769
KZ774 1029.12.1	0.29	506	415	0.85	16.7	242.8	±4.4	233	±86	0.269	4.2	0.0384	1.8	0.443

(continued)

Table 6 (continued)

Crater	$^{206}Pb_c$, %	U, ppm	Th, ppm	$^{232}Th/^{238}U$	$^{206}Pb^*$, ppm	$^{206}Pb^*/^{238}U$ Age		$^{207}Pb^*/^{206}Pb^*$ Age		$^{207}Pb^*/^{235}U$	±%	$^{206}Pb^*/^{238}U$	±%	KK
KZ 774 (1032.4–1033.0)														
KZ-774 (1032.4–1033.0)_5.1	0.01	2624	5096	2.01	89.7	251.6	±2.2	286	±29	0.285	1.5	0.0398	0.9	0.577
KZ-774 (1032.4–1033.0)_1.1	1.32	2313	3924	1.75	81.7	256.2	±2.6	182	±110	0.278	5.0	0.0405	1.0	0.203
KZ-774 (1032.4–1033.0)_2.1	0.01	90	52	0.60	12.2	939.0	±14.0	1000	±96	1.568	5.0	0.1569	1.6	0.327
KZ-774 (1032.4–1033.0)_3.1	0.58	112	33	0.31	21.3	1278.0	±17.0	1148	±84	2.360	4.5	0.2193	1.5	0.326
KZ-774 (1032.4–1033.0)_4.1	0.81	55	52	0.97	24.7	2698.0	±42.0	2664	±40	12.980	3.0	0.5197	1.9	0.620
KZ-774_1.1	0.29	189	136	0.74	8.8	340.6	±6.3	359	±96	0.402	4.6	0.0543	1.9	0.411
OM 10 (1061)														
OM-10-1061_2.2	0.16	748	18	0.02	59.0	565.3	±6.2	589	±46	0.753	2.4	0.0917	1.2	0.479
OM-10-1061_4.2	0.29	368	6	0.02	29.6	575.6	±5.8	601	±72	0.772	3.5	0.0934	1.0	0.298
OM-10-1061_3.2	0.13	485	9	0.02	39.1	577.5	±5.8	505	±49	0.741	2.5	0.0937	1.0	0.423
OM-10-1061_2.1	0.77	345	24	0.07	30.1	620.0	±5.8	594	±91	0.831	4.3	0.1010	1.0	0.225
OM-10-1061_1.1	0.34	292	37	0.13	27.9	677.7	±6.4	640	±75	0.933	3.6	0.1109	1.0	0.272
OM-10-1061_4.1	0.16	266	68	0.26	29.3	775.9	±7.2	820	±43	1.172	2.3	0.1279	1.0	0.433
OM-10-1061_3.1	0.01	365	174	0.49	54.3	1030.0	±8.2	1015	±23	1.745	1.4	0.1732	0.9	0.614
OM 10 (1068)														
OM-10-1068_3.1	0.01	362	589	1.68	12.5	254.8	±4.6	255	±67	0.285	3.5	0.0403	1.8	0.530
OM-10-1068_12.1	0.01	401	751	1.93	13.9	255.1	±4.5	236	±64	0.283	3.3	0.0404	1.8	0.540
OM-10-1068_8.1	0.01	455	952	2.16	15.9	257.1	±4.5	258	±59	0.288	3.1	0.0407	1.8	0.572
OM-10-1068_6.1	0.19	413	801	2.00	14.5	257.2	±4.5	307	±74	0.295	3.7	0.0407	1.8	0.483
OM-10-1068_4.1	0.16	564	907	1.66	19.8	257.8	±4.4	160	±67	0.277	3.4	0.0408	1.8	0.524
OM-10-1068_2.1	0.25	1165	1704	1.51	41.3	260.2	±4.3	232	±59	0.288	3.1	0.0412	1.7	0.549
OM-10-1068_9.1	0.19	762	1586	2.15	27.1	261.5	±4.4	240	±71	0.291	3.5	0.0414	1.7	0.487
OM-10-1068_11.1	0.22	1102	3580	3.36	39.4	262.0	±4.4	277	±61	0.296	3.2	0.0415	1.7	0.538
OM-10-1068_10.1	0.10	1655	3026	1.89	59.1	262.1	±4.3	227	±38	0.290	2.4	0.0415	1.7	0.711
OM-10-1068_1.1	0.01	1641	4951	3.12	58.5	262.3	±4.3	317	±32	0.302	2.2	0.0415	1.7	0.769
OM-10-1068_7.1	0.14	1571	3309	2.18	56.7	264.8	±4.3	252	±38	0.296	2.4	0.0419	1.7	0.709
OM-10-1068_5.1	4.39	276	272	1.02	25.9	639.0	±12.0	860	±180	0.973	9.1	0.1042	2.0	0.223

(continued)

Table 6 (continued)

Crater	$^{206}Pb_c$, %	U, ppm	Th, ppm	$^{232}Th/^{238}U$	$^{206}Pb^*$, ppm	$^{206}Pb^*/^{238}U$ Age		$^{207}Pb^*/^{206}Pb^*$ Age		$^{207}Pb^*/^{235}U$	±%	$^{206}Pb^*/^{238}U$	±%	KK
OM 32 (1084.8)														
OM32_1.1	0.41	374	71	0.20	7.5	148.6	±2.1	28	±130	0.150	5.8	0.0233	1.4	0.245
OM32_2.1	1.19	797	245	0.32	16.8	154.7	±2.1	111	±200	0.162	8.7	0.0243	1.4	0.155
OM 123 (1005.6)														
OM123.1.1	0.01	2442	206	0.09	185.0	545.6	±9.4	592	±20	0.727	2.0	0.0883	1.8	0.893
OM123.2.1	0.83	4029	126	0.03	186.0	334.8	±6.2	451	±60	0.411	3.3	0.0533	1.9	0.571
OM123.3.1	3.53	119	88	0.76	1.8	105.9	±4.1	–	–	0.096	57.0	0.0166	3.9	0.068
OM123.4.1	0.01	2139	131	0.06	117.0	398.3	±7.3	482	±24	0.499	2.2	0.0637	1.9	0.864
OM123.5.1	0.09	1662	432	0.27	128.0	554.0	±10.0	564	±30	0.729	2.3	0.0898	1.9	0.806
OM123.6.1	0.20	1884	149	0.08	141.0	538.6	±9.7	498	±36	0.687	2.5	0.0871	1.9	0.751
OM 123 (1033.7)														
OM-123-1033.7_6.1	1.10	87	42	0.50	4.8	400.9	±9.4	325	±280	0.468	12.0	0.0642	2.4	0.196
OM-123-1033.7_6.2	0.55	141	67	0.49	7.9	406.6	±8.4	316	±140	0.473	6.6	0.0651	2.1	0.322
OM-123-1033.7_5.2	1.02	549	76	0.14	37.3	485.5	±8.8	517	±110	0.622	5.3	0.0782	1.9	0.353
OM-123-1033.7_5.1	0.11	885	127	0.15	60.3	491.5	±8.6	511	±58	0.628	3.2	0.0792	1.8	0.570
OM-123-1033.7_1.1	3.28	312	151	0.50	64.3	1339.0	±23.0	1885	±56	3.670	3.7	0.2308	1.9	0.525
OM-123-1033.7_2.1	0.03	519	30	0.06	191.0	2298.0	±35.0	2363	±10	8.950	1.9	0.4282	1.8	0.950
OM-123-1033.7_4.1	0.20	148	270	1.88	62.1	2557.0	±41.0	2702	±15	12.450	2.1	0.4867	1.9	0.902
OM-123-1033.7_3.1	0.33	113	108	0.99	48.0	2583.0	±42.0	2740	±17	12.900	2.2	0.4928	2.0	0.886
MP 25 KZ (37.8)														
MP25kz_1.1	0.10	142	73	0.53	44.2	1987.0	±29.0	1916	±21	5.840	2.0	0.3610	1.7	0.826
MP25kz_2.1	0.46	293	165	0.58	9.8	245.3	±4.3	–	–	0.268	5.0	0.0388	1.8	0.363
MP25kz_3.1	0.13	1434	1432	1.03	47.2	242.3	±3.8	–	–	0.272	2.3	0.0383	1.6	0.696
MP25kz_4.1	0.55	559	385	0.71	18.3	240.0	±4.0	–	–	0.257	5.8	0.0379	1.7	0.291
MP25kz_5.1	0.09	132	32	0.25	39.4	1919.0	±30.0	1905	±21	5.570	2.1	0.3467	1.8	0.843
MP25kz_6.1	0.27	489	68	0.14	47.8	693.0	±10.0	1709	±21	1.639	2.0	0.1135	1.6	0.807
MP25kz_7.1	0.17	264	14	0.05	78.8	1922.0	±27.0	1911	±25	5.600	2.1	0.3474	1.6	0.760
MP25kz_8.1	0.40	673	632	0.97	22.2	241.9	±4.0	–	–	0.266	3.8	0.0382	1.7	0.444
MP25kz_9.1	0.16	537	336	0.65	17.9	244.5	±4.0	–	–	0.270	3.4	0.0387	1.7	0.485

(continued)

Table 6 (continued)

Crater	$^{206}Pb_c$, %	U, ppm	Th, ppm	$^{232}Th/^{238}U$	$^{206}Pb^*$, ppm	$^{206}Pb^*/^{238}U$ Age		$^{207}Pb^*/^{206}Pb^*$ Age		$^{207}Pb^*/^{235}U$	±%	$^{206}Pb^*/^{238}U$	±%	KK
MP25kz_10.1	0.22	997	839	0.87	32.5	239.5	±3.8	–	–	0.264	2.6	0.0379	1.6	0.606
MP25kz_11.1	0.30	670	539	0.83	22.3	243.8	±4.0	–	–	0.268	3.3	0.0385	1.7	0.507
MP25kz_12.1	0.29	65	28	0.44	20.2	1975.0	±30.0	1932	±31	5.850	2.5	0.3584	1.8	0.715
MP25kz_13.1	0.04	96	35	0.37	28.4	1907.0	±36.0	1893	±23	5.500	2.6	0.3441	2.2	0.863
MP25kz_14.1	1.02	126	62	0.51	4.3	248.0	±5.6	–	–	0.265	15.0	0.0392	2.3	0.151
MP25kz_15.1	0.24	260	146	0.58	8.7	246.2	±4.5	–	–	0.271	4.6	0.0389	1.9	0.412
MP25kz_6.2	0.93	870	184	0.22	29.5	247.0	±4.5	–	–	0.302	6.2	0.0391	1.9	0.300
MP25kz_6.3	0.23	269	116	0.45	73.3	1775.0	±25.0	1922	±17	5.147	1.9	0.3171	1.6	0.869
MP25kz_6.4	0.04	648	199	0.32	88.1	947.0	±14.0	1780	±18	2.374	1.9	0.1582	1.6	0.860
12 N 18														
12N18_12.1	0.52	811	1199	1.53	27.0	243.4	±3.7	240	±92	0.270	4.3	0.0385	1.6	0.363
12N18_21.1	0.36	1420	1328	0.97	47.7	246.4	±3.6	253	±66	0.276	3.2	0.0390	1.5	0.464
12N18_5.1	0.16	1298	2823	2.25	44.6	252.5	±3.7	248	±49	0.282	2.6	0.0340	1.5	0.577
12N18_18.1	0.26	1212	2279	1.94	41.8	253.0	±3.7	236	±69	0.281	3.4	0.0400	1.5	0.449
12N18_1.1	0.08	1063	908	0.88	36.6	253.2	±3.8	261	±48	0.284	2.6	0.0401	1.5	0.596
12N18_8.1	0.41	1508	1518	1.04	52.5	254.9	±3.8	277	±65	0.288	3.2	0.0403	1.5	0.468
12N18_9.1	0.15	1609	1286	0.83	56.0	255.7	±3.7	245	±44	0.285	2.4	0.0405	1.5	0.612
12N18_16.1	0.10	1309	2660	2.10	45.7	256.3	±3.8	264	±44	0.288	2.4	0.0406	1.5	0.622
12N18_17.1	0.04	1937	5540	2.96	67.7	257.0	±3.7	251	±31	0.287	2.0	0.0407	1.5	0.741
12N18_19.1	0.07	2076	4672	2.33	72.9	257.9	±3.7	240	±36	0.287	2.2	0.0408	1.5	0.684
12N18_20.1	0.19	1643	3281	2.06	57.8	258.0	±3.8	278	±44	0.292	2.4	0.0408	1.5	0.611
12N18_6.1	0.54	1954	2735	1.45	69.1	258.6	±3.8	282	±66	0.293	3.3	0.0409	1.5	0.458
12N18_7.1	0.20	1814	1789	1.02	64.0	258.7	±3.9	229	±44	0.286	2.5	0.0410	1.5	0.623
12N18_10.1	0.17	1359	1256	0.95	48.2	260.5	±4.0	275	±43	0.294	2.4	0.0412	1.5	0.635
12N18_4.1	0.18	1784	3876	2.25	63.5	261.5	±3.8	299	±43	0.299	2.4	0.0414	1.5	0.617
12N18_3.1	0.03	3545	8174	2.38	127.0	262.4	±3.7	280	±25	0.297	1.8	0.0415	1.4	0.796
12N18_15.1	0.12	2576	4247	1.70	92.3	262.9	±3.9	242	±32	0.293	2.1	0.0416	1.5	0.734
12N18_13.1	0.00	2709	5132	1.96	98.0	265.9	±3.9	241	±24	0.296	1.8	0.0421	1.5	0.819
12N18_11.1	0.04	2522	13,173	5.40	92.3	268.8	±3.9	271	±26	0.303	1.9	0.0426	1.5	0.787

(continued)

Table 6 (continued)

Crater	$^{206}Pb_c$, %	U, ppm	Th, ppm	$^{232}Th/^{238}U$	$^{206}Pb^*$, ppm	$^{206}Pb^*/^{238}U$ Age		$^{207}Pb^*/^{206}Pb^*$ Age		$^{207}Pb^*/^{235}U$	±%	$^{206}Pb^*/^{238}U$	±%	KK
12N18_2.1	0.17	3002	4474	1.54	110.0	269.1	±3.8	266	±35	0.303	2.1	0.0426	1.5	0.692
12N18_14.1	0.02	4600	11,255	2.53	171.0	273.6	±3.9	262	±20	0.308	1.7	0.0434	1.4	0.862
12 N 19A														
12N19A_1.1	0.86	724	2829	4.04	24.9	250.9	±2.0	–	–	0.258	5.3	0.0397	0.8	0.156
12N19A_2.1	4.29	527	690	1.35	18.8	251.6	±2.6	–	–	0.297	9.5	0.0398	1.0	0.110
12N19A_3.1	4.29	1372	3531	2.66	48.8	250.4	±2.4	–	–	0.293	14.0	0.0396	1.0	0.070
12N19A_6.1	0.18	2683	6627	2.55	91.6	250.8	±1.6	–	–	0.276	1.6	0.0397	0.7	0.399
12N19A_7.1	0.01	325	806	2.56	11.1	250.7	±2.7	–	–	0.282	3.5	0.0397	1.1	0.318
12N19A_7.2	0.10	452	1306	2.99	15.5	251.6	±2.5	–	–	0.284	3.4	0.0398	10.0	0.295
12N19A_8.1	1.13	1745	2323	1.38	60.3	251.5	±1.8	–	–	0.265	4.7	0.0398	0.7	0.156
12N19A_9.1	0.08	1205	1788	1.53	41.2	251.4	±1.9	–	–	0.268	2.2	0.0398	0.8	0.338
12N19A_10.1	0.03	2987	6818	2.36	102.0	250.3	±1.7	–	–	0.277	1.5	0.0396	0.7	0.450
12N19A_11.1	0.07	763	2648	3.58	26.1	251.3	±2.3	–	–	0.274	2.7	0.0398	0.9	0.344
12N19A_12.1	0.11	2887	4894	1.75	98.9	251.6	±1.6	–	–	0.277	1.5	0.0398	0.7	0.439
12N19A_13.1	0.01	624	1942	3.22	21.2	249.8	±2.2	–	–	0.274	2.6	0.0395	0.9	0.341
12N19A_14.1	0.01	777	2317	3.08	26.5	250.6	±2.1	–	–	0.288	2.6	0.0397	0.9	0.329
12N19A_15.1	0.05	1084	2339	2.23	37.0	251.2	±1.9	–	–	0.284	2.2	0.0397	0.8	0.363
12N19A_16.1	0.44	589	1656	2.90	20.1	250.5	±2.3	–	–	0.241	5.6	0.0396	1.0	0.168
12N19A_17.1	0.04	759	1275	1.74	25.8	249.7	±2.1	–	–	0.264	2.5	0.0395	0.9	0.341
12N19A_18.1	0.01	1595	2995	1.94	54.1	249.4	±2.0	–	–	0.272	1.8	0.0395	0.8	0.461
12N19A_20.1	0.78	903	2410	2.76	31.1	251.1	±2.1	–	–	0.250	4.0	0.0397	0.9	0.209
12N19A_21.1	0.01	798	662	0.86	26.9	248.1	±1.9	–	–	0.256	2.2	0.0392	0.8	0.359
12N19A_22.1	0.27	1105	1731	1.62	37.8	251.1	±1.9	–	–	0.272	3.2	0.0397	0.8	0.239
12N19A_23.1	0.13	1422	2269	1.65	48.3	249.5	±1.7	–	–	0.265	1.9	0.0395	0.7	0.370
12N19A_25.1	0.01	1256	3939	3.24	42.6	249.5	±1.8	–	–	0.273	1.8	0.0395	0.7	0.408
12N19A_26.1	0.11	584	2451	4.33	19.8	248.9	±2.2	–	–	0.267	2.9	0.0395	0.9	0.314
12N19A_4.1	0.16	205	78	0.39	44.6	1456.0	±13.0	1421	±28	3.139	1.7	0.2535	1.0	0.566
12N19A_24.1	0.20	997	460	0.48	21.8	161.5	±1.3	–	–	0.165	4.4	0.0255	0.8	0.186

(continued)

Table 6 (continued)

Crater	$^{206}Pb_c$, %	U, ppm	Th, ppm	$^{232}Th/^{238}U$	$^{206}Pb^*$, ppm	$^{206}Pb^*/^{238}U$ Age		$^{207}Pb^*/^{206}Pb^*$ Age		$^{207}Pb^*/^{235}U$	±%	$^{206}Pb^*/^{238}U$	±%	KK
12 N 19V														
12N19V.1.1	0.59	431	493	1.18	14.8	251.1	±4.1	219	±180	0.277	7.9	0.0397	1.7	0.211
12N19V.2.1	0.01	3432	4419	1.33	120.0	257.7	±3.1	311	±28	0.296	1.7	0.0408	1.2	0.712
12N19V.3.1	0.20	1297	1353	1.08	44.6	252.6	±3.7	224	±84	0.279	3.9	0.0400	1.5	0.380
12N19V.4.1	0.06	1168	1554	1.37	39.1	246.4	±3.1	297	±40	0.281	2.2	0.0390	1.3	0.595
12N19V.5.1	0.80	786	1922	2.53	27.0	250.4	±3.4	11	±180	0.253	7.5	0.0396	1.4	0.187
12N19V.6.1	0.24	343	460	1.39	11.4	244.4	±3.6	192	±110	0.266	5.0	0.0386	1.5	0.296
12N19V.7.1	0.10	2039	3890	1.97	71.5	257.7	±3.2	222	±44	0.285	2.3	0.0408	1.3	0.556
12N19V.8.1	0.92	729	964	1.37	25.0	249.7	±3.4	–57	±200	0.245	8.2	0.0395	1.4	0.171
12N19V.9.1	0.93	1933	3817	2.04	67.8	255.6	±3.2	72	±94	0.265	4.2	0.0405	1.3	0.307
12N19V.10.1	0.12	1575	3094	2.03	53.6	250.2	±3.1	195	±50	0.273	2.5	0.0396	1.3	0.512
12N19V.11.1	2.62	2181	4285	2.03	78.3	256.9	±3.3	106	±130	0.270	5.6	0.0407	1.3	0.229
12N19V.12.1	0.38	1004	2294	2.36	35.3	257.7	±3.7	147	±100	0.275	4.5	0.0408	1.5	0.321
12N19V.13.1	2.90	198	657	3.42	7.0	250.4	±4.9	–	–	0.216	27.0	0.0396	2.0	0.072
12N19V.14.1	0.83	2299	4429	1.99	81.6	258.9	±3.2	137	±77	0.276	3.5	0.0410	1.2	0.355
12N19V .15.1	0.07	3171	5733	1.87	111.0	257.6	±3.1	284	±32	0.292	1.9	0.0408	1.2	0.661
12N19V.16.1	0.01	619	992	1.66	21.3	253.3	±3.6	360	±51	0.297	2.7	0.0401	1.4	0.533
12N19V.17.1	0.16	938	1324	1.46	31.8	249.2	±3.1	228	±53	0.276	2.6	0.0394	1.3	0.492
12N19V.18.1	0.98	938	2384	2.63	33.2	258.1	±3.7	–	–	0.258	7.6	0.0409	1.4	0.190
12N19V.19.1	0.07	812	1581	2.01	28.4	256.6	±3.4	238	±48	0.285	2.5	0.0406	1.4	0.545
12N19V.20.1	1.85	182	549	3.12	6.3	248.4	±4.9	–	–	0.191	28.0	0.0393	2.0	0.071
12N19V.21.1	4.33	507	1402	2.86	18.5	256.5	±4.1	130	±420	0.272	18.0	0.0406	1.6	0.090
12 N 21														
12N21_1.1	5.67	242	8	0.03	25.4	699.2	±6.9	658	±170	0.972	7.9	0.1146	1.0	0.133
12N21_3.1	0.05	243	221	0.94	27.2	791.8	±5.7	832	±30	1.204	1.6	0.1307	0.8	0.472
12N21_2.1	22.85	4179	7669	1.90	123.0	166.8	±1.6	–	–	0.179	16.0	0.0262	1.0	0.060
12N21_4.1	0.01	1273	478	0.39	28.4	165.3	±1.1	–	–	0.176	2.0	0.0260	0.7	0.330
12N21_5.1	0.22	1471	574	0.40	33.1	166.0	±1.1	–	–	0.179	2.1	0.0261	0.7	0.314

(continued)

Table 6 (continued)

Crater	$^{206}Pb_c$, %	U, ppm	Th, ppm	$^{232}Th/^{238}U$	$^{206}Pb^*$, ppm	$^{206}Pb^*/^{238}U$ Age		$^{207}Pb^*/^{206}Pb^*$ Age		$^{207}Pb^*/^{235}U$	±%	$^{206}Pb^*/^{238}U$	±%	KK
MP 2 206 mz														
MP2mz_1.1	0.06	5055	8696	1.78	167.0	243.1	±4.7	229	±20	0.269	2.2	0.0384	2.0	0.918
MP2mz_2.1	0.02	3582	6093	1.76	112.0	230.3	±4.5	234	±22	0.255	2.2	0.0364	2.0	0.900
MP2mz_3.1	0.05	1160	1603	1.43	36.5	231.7	±4.8	211	±40	0.254	2.7	0.0366	2.1	0.773
MP2mz_4.1	0.05	1765	1474	0.86	56.3	234.8	±4.7	234	±34	0.260	2.5	0.0371	2.0	0.810
MP2mz_5.1	0.04	203	62	0.32	60.0	1903.0	±34.0	1900	±17	5.500	2.3	0.3433	2.1	0.910
MP2mz_5.2	0.07	374	63	0.17	111.0	1916.0	±34.0	1909	±14	5.580	2.2	0.3462	2.0	0.936
MP2mz_6.1	0.01	152	51	0.35	44.9	1905.0	±35.0	1903	±19	5.520	2.4	0.3439	2.1	0.896
MP2mz_7.1	0.21	40	16	0.41	11.6	1862.0	±40.0	1938	±43	5.480	3.4	0.3349	2.5	0.716
MP2mz_8.1	0.07	1427	1895	1.37	45.1	232.7	±4.6	262	±40	0.261	2.7	0.0368	2.0	0.760
MP2mz_9.1	0.42	190	147	0.80	6.3	241.9	±5.5	–	–	0.263	6.8	0.0382	2.3	0.343
MP2mz_10.1	0.01	98	30	0.31	30.2	1966.0	±37.0	1920	±23	5.780	2.5	0.3565	2.2	0.865
12 N 05														
12N05.1.1	0.08	3450	11126	3.33	116.0	246.5	±4.4	291	±31	0.280	2.3	0.0390	1.8	0.797
12N05.2.1	0.26	2163	6851	3.27	72.2	245.1	±4.6	206	±87	0.268	4.2	0.0388	1.9	0.455
12N05.3.1	0.24	2965	7390	2.58	97.0	240.4	±4.6	235	±51	0.267	2.9	0.0380	1.9	0.663
12N05.4.1	0.25	1808	5218	2.98	59.1	240.0	±4.5	226	±87	0.265	4.2	0.0379	1.9	0.452
12N05.5.1	0.05	3190	10,088	3.27	107.0	245.8	±4.6	273	±36	0.277	2.5	0.0389	1.9	0.769
12N05.6.1	0.01	3641	11,288	3.20	120.0	241.9	±4.3	253	±33	0.270	2.3	0.0382	1.8	0.786
12N05.7.1	0.15	4231	13,769	3.36	141.0	245.0	±4.5	264	±42	0.275	2.6	0.0387	1.9	0.718
12N05.8.1	0.22	2342	7541	3.33	76.2	239.3	±4.6	232	±72	0.265	3.7	0.0378	2.0	0.536
12N05_1.1	0.08	3057	10,531	3.56	113.0	272.1	±3.3	271	±36	0.307	2.0	0.0431	1.2	0.621
12N05_2.1	0.01	2870	9355	3.37	101.0	257.7	±2.9	269	±27	0.290	1.6	0.0408	1.1	0.695
12N05_3.1	0.04	4517	18,269	4.18	164.0	266.2	±2.9	260	±23	0.299	1.5	0.0422	1.1	0.750
12N05_4.1	0.05	1935	6091	3.25	68.3	259.3	±3.0	245	±35	0.289	1.9	0.0411	1.2	0.605
12N05_5.1	0.05	1737	5487	3.26	59.7	252.8	±2.9	263	±37	0.284	2.0	0.0400	1.2	0.581

(continued)

Table 6 (continued)

Crater	$^{206}Pb_c$, %	U, ppm	Th, ppm	$^{232}Th/^{238}U$	$^{206}Pb^*$, ppm	$^{206}Pb^*/^{238}U$ Age		$^{207}Pb^*/^{206}Pb^*$ Age		$^{207}Pb^*/^{235}U$	±%	$^{206}Pb^*/^{238}U$	±%	KK
12 N 07														
12N07_1.1	1.78	880	972	1.14	18.6	154.1	±2.0	–	–	0.163	6.0	0.0242	1.3	0.224
12N07_2.1	0.12	565	325	0.59	17.8	231.2	±2.8	–	–	0.260	2.4	0.0365	1.2	0.504
12N07_3.1	0.16	477	342	0.74	9.7	150.9	±2.2	–	–	0.155	3.9	0.0237	1.4	0.370
12N07_4.1	0.75	111	113	1.05	3.5	228.5	±3.8	–	–	0.270	7.6	0.0361	1.7	0.220
12N07_4.2	0.32	1127	882	0.81	35.1	229.0	±2.8	–	–	0.252	2.3	0.0362	1.2	0.544
12N07_5.1	3.46	1814	2793	1.59	36.6	144.5	±1.8	–	–	0.146	6.3	0.0227	1.3	0.204
12N07_6.1	0.98	910	700	0.80	18.2	147.1	±1.9	–	–	0.159	4.2	0.0231	1.3	0.312
12N07_7.1	0.31	396	209	0.54	18.0	331.0	±4.3	–	–	0.381	3.2	0.0527	1.3	0.421
12N07_8.1	0.22	997	1153	1.19	20.7	153.7	±2.0	–	–	0.163	2.9	0.0241	1.3	0.440
12 N 08														
12N08_1.1	0.79	310	169	0.56	11.6	272.4	±4.1	225	±190	0.301	8.4	0.0432	1.5	0.182
N 2 N 2														
N-2_1.1	0.03	542	70	0.13	183.0	2136.0	±29.0	2016	±10	6.720	1.7	0.3927	1.6	0.942
N-2_1.2	0.10	352	4	0.01	94.1	1746.0	±26.0	1899	±15	4.986	1.9	0.3111	1.7	0.896
N-2_2.1	0.63	444	63	0.15	8.5	141.9	±2.7	–	–	0.147	7.0	0.0223	1.9	0.275
N-2_3.1	0.36	366	37	0.10	7.5	150.8	±2.8	–	–	0.155	6.1	0.0237	1.9	0.308
N-2_4.1	0.91	80	78	1.02	2.6	242.5	±5.8	–	–	0.275	12.0	0.0383	2.4	0.203
N-2_4.2	1.41	44	36	0.84	1.4	229.7	±6.3	–	–	0.271	16.0	0.0363	2.8	0.171
N-2_5.1	1.13	388	50	0.13	7.9	148.8	±2.7	–	–	0.137	11.0	0.0238	1.8	0.172
N-2_6.1	0.01	359	48	0.14	7.9	163.7	±3.0	–	–	0.176	4.3	0.0257	1.8	0.420
N-2_7.1	0.18	521	388	0.77	10.7	152.1	±3.3	–	–	0.165	4.1	0.0239	2.2	0.540
N-2_8.1	0.01	59	28	0.49	2.2	273.6	±6.4	–	–	0.321	7.0	0.0434	2.4	0.342
N-2_9.1	0.27	178	162	0.94	8.2	336.6	±6.0	–	–	0.400	4.7	0.0536	1.8	0.388
N-2_10.1	0.80	532	505	0.98	11.0	151.7	±2.7	–	–	0.153	8.9	0.0238	1.8	0.199
N-2_11.1	0.89	220	328	1.54	4.4	148.2	±2.9	–	–	0.156	11.0	0.0233	2.0	0.185
N-2_12.1	0.36	169	104	0.64	48.9	1863.0	±28.0	1904	±26	5.380	2.3	0.3351	1.7	0.775
N-2_13.1	0.16	322	245	0.78	12.5	284.0	±4.7	–	–	0.328	3.5	0.0450	1.7	0.484

(continued)

Table 6 (continued)

Crater	$^{206}Pb_c$, %	U, ppm	Th, ppm	$^{232}Th/^{238}U$	$^{206}Pb^*$, ppm	$^{206}Pb^*/^{238}U$ Age		$^{207}Pb^*/^{206}Pb^*$ Age		$^{207}Pb^*/^{235}U$	±%	$^{206}Pb^*/^{238}U$	±%	KK
N-2_14.1	0.33	77	55	0.74	25.2	2073.0	±33.0	1802	±39	5.760	2.8	0.3794	1.9	0.663
N-2_15.1	0.41	199	117	0.61	9.9	361.2	±6.3	–	–	0.435	6.8	0.0576	1.8	0.261
N-2_16.1	0.76	378	151	0.41	7.5	145.2	±2.6	–	–	0.163	6.7	0.0228	1.8	0.270
N-2_17.1	0.31	492	90	0.19	10.7	161.1	±2.8	–	–	0.174	5.2	0.0253	1.8	0.339
N-2_18.1	0.86	289	226	0.81	11.4	286.1	±5.1	–	–	0.322	7.5	0.0454	1.8	0.242
N-2_19.1	1.61	281	35	0.13	5.6	145.1	±2.8	–	–	0.139	12.0	0.0228	1.9	0.158
N-2_20.1	0.07	1095	115	0.11	351.0	2041.0	±27.0	2133	±13	6.810	1.7	0.3724	1.6	0.905
N-2_21.1	2.93	91	313	3.54	1.9	145.5	±4.9	–	–	0.148	38.0	0.0228	3.4	0.089
N-2_22.1	0.65	374	47	0.13	7.4	145.9	±2.9	–	–	0.149	15.0	0.0229	2.0	0.138
N-2_23.1	1.15	252	91	0.37	5.1	148.8	±2.9	–	–	0.135	12.0	0.0234	2.0	0.161
N 3														
N-3_2.2	4.28	38	20	0.56	1.1	209.8	±7.4	–	–	0.180	55.0	0.0331	3.6	0.065
N-3_2.1	2.53	86	57	0.69	2.5	210.0	±5.9	–	–	0.188	35.0	0.0331	2.9	0.081
N-3_1.2	0.42	436	83	0.20	16.1	270.4	±4.1	346	±120	0.315	5.4	0.0428	1.5	0.287
N-3_1.1	0.55	666	321	0.50	28.4	310.8	±4.4	211	±120	0.343	5.2	0.0494	1.4	0.277
OV 28 (703)														
OV-28-703_4.1	0.12	2351	2511	1.10	81.8	255.6	±3.3	220	±44	0.282	2.3	0.0404	1.3	0.568
OV-28-703_10.1	0.15	1519	2015	1.37	53.3	257.8	±2.2	223	±50	0.285	2.4	0.0408	0.9	0.377
OV-28-703_1.1	0.12	1582	1308	0.85	55.6	258.1	±2.0	221	±42	0.285	2.0	0.0409	0.8	0.402
OV-28-703_11.1	0.15	1381	3324	2.49	49.8	264.6	±2.2	223	±53	0.292	2.5	0.0419	0.9	0.344
OV-28-703_1.2	0.03	5248	10,982	2.16	194.0	271.4	±2.5	230	±23	0.301	1.4	0.0430	1.0	0.687
OV-28-703_3.1	0.04	5656	10,861	1.98	211.0	273.8	±2.4	221	±37	0.303	1.8	0.0434	0.9	0.484
OV-28-703_9.1	1.96	2194	3356	1.58	84.3	276.5	±2.5	415	±260	0.333	12.0	0.0438	0.9	0.082
OV-28-703_2.1	0.01	2708	4398	1.68	104.0	280.9	±3.6	268	±37	0.317	2.1	0.0445	1.3	0.632
OV-28-703_6.2	0.72	309	3	0.01	22.8	526.9	±6.5	378	±130	0.636	6.0	0.0852	1.3	0.213
OV-28-703_5.2	1.23	575	4	0.01	44.9	553.0	±5.2	579	±100	0.733	4.7	0.0896	1.0	0.208
OV-28-703_12.1	0.15	1868	68	0.04	147.0	565.1	±4.8	551	±43	0.740	2.2	0.0916	0.9	0.415
OV-28-703_5.1	0.26	257	41	0.16	25.2	695.6	±6.6	649	±60	0.963	3.0	0.1139	1.0	0.339

(continued)

Table 6 (continued)

Crater	$^{206}Pb_c$, %	U, ppm	Th, ppm	$^{232}Th/^{238}U$	$^{206}Pb^*$, ppm	$^{206}Pb^*/^{238}U$ Age		$^{207}Pb^*/^{206}Pb^*$ Age		$^{207}Pb^*/^{235}U$	±%	$^{206}Pb^*/^{238}U$	±%	KK
OV-28-703_13.1	0.50	428	115	0.28	43.0	709.0	±7.4	645	±82	0.981	4.0	0.1163	1.1	0.276
OV-28-703_7.1	0.68	264	59	0.23	30.6	810.4	±7.6	756	±84	1.190	4.1	0.1340	1.0	0.244
OV-28-703_6.1	0.37	210	58	0.29	26.9	891.4	±8.5	850	±59	1.378	3.0	0.1483	1.0	0.335
OV-28-703_8.1	0.20	341	143	0.43	46.0	937.7	±8.0	909	±38	1.497	2.0	0.1566	0.9	0.447
OV 28 (711)														
OV-28-711_8.1	0.23	1236	1716	1.43	42.2	250.8	±2.0	220	±92	0.276	4.1	0.0397	0.8	0.205
OV-28-711_10.1	0.33	881	1881	2.21	30.6	255.1	±2.2	242	±86	0.284	3.9	0.0404	0.9	0.231
OV-28-711_9.1	0.17	1257	1278	1.05	44.0	256.7	±2.1	213	±55	0.282	2.5	0.0406	0.8	0.329
OV-28-711_5.1	0.07	1356	1155	0.88	49.2	266.5	±2.2	180	±40	0.289	1.9	0.0422	0.8	0.442
OV-28-711_2.1	0.01	3086	2832	0.95	113.0	268.0	±2.7	278	±31	0.303	1.7	0.0425	1.0	0.597
OV-28-711_4.1	0.13	1907	2737	1.48	70.1	269.9	±2.6	284	±43	0.306	2.1	0.0428	1.0	0.466
OV-28-711_11.1	0.21	4002	5463	1.41	151	276.6	±3.5	239	±50	0.308	2.5	0.0438	1.3	0.507
OV-28-711_9.2	0.06	3397	5569	1.69	129.0	279.3	±2.4	231	±29	0.310	1.5	0.0443	0.9	0.567
OV-28-711_6.1	0.01	2284	5415	2.45	87.1	280.1	±2.6	227	±32	0.310	1.7	0.0444	1.0	0.561
OV-28-711_7.1	0.06	4952	8669	1.81	191.0	283.3	±3.3	247	±30	0.317	1.8	0.0449	1.2	0.680
OV-28-711_1.1	0.05	7259	10,064	1.43	286.0	288.7	±3.0	224	±35	0.320	1.9	0.0458	1.1	0.572
OV-28-711_6.2	0.05	7815	12,177	1.61	315.0	295.4	±2.7	210	±31	0.325	1.6	0.0469	0.9	0.566
OV-28-711_7.2	0.10	4925	8671	1.82	199.0	296.1	±3.4	265	±31	0.334	1.8	0.0470	1.2	0.657
OV-28-711_3.1	0.01	6023	7126	1.22	252.0	306.4	±3.5	254	±45	0.344	2.3	0.0487	1.2	0.508
OV 28 (827)														
OV-28-827_2.1	1.95	296	89	0.31	4.9	121.1	±2.5	308	±340	0.137	15.0	0.0190	2.1	0.137
OV-28-827_2.2	19.51	181	145	0.83	4.6	151.0	±11.0	2345	±620	0.490	37.0	0.0237	7.5	0.205
OV-28-827_1.1	0.21	1008	1618	1.66	35.1	255.3	±2.3	248	±60	0.285	2.7	0.0404	0.9	0.329
OV-28-827_11.1	0.01	569	444	0.81	19.8	256.3	±2.6	262	±57	0.288	2.7	0.0406	1.0	0.379
OV-28-827_10.1	0.18	1498	2545	1.76	52.5	257.4	±2.4	217	±58	0.284	2.7	0.0407	0.9	0.353
OV-28-827_8.1	4.16	1354	3383	2.58	49.7	258.7	±3.0	344	±310	0.301	14.0	0.0409	1.2	0.085
OV-28-827_7.1	0.14	2853	4536	1.64	101.0	259.1	±4.3	232	±59	0.287	3.1	0.0410	1.7	0.551
OV-28-827_4.2	0.19	1014	1419	1.45	35.8	259.4	±2.4	207	±61	0.285	2.8	0.0411	1.0	0.337

(continued)

Table 6 (continued)

Crater	$^{206}Pb_c$, %	U, ppm	Th, ppm	$^{232}Th/^{238}U$	$^{206}Pb^*$, ppm	$^{206}Pb^*/^{238}U$ Age		$^{207}Pb^*/^{206}Pb^*$ Age		$^{207}Pb^*/^{235}U$	±%	$^{206}Pb^*/^{238}U$	±%	KK
OV-28-827_3.1	0.01	1135	1219	1.11	40.3	261.0	±2.6	272	±44	0.295	2.2	0.0413	1.0	0.475
OV-28-827_7.2	0.10	4275	6267	1.51	152.0	262.0	±2.3	262	±30	0.294	1.6	0.0415	0.9	0.552
OV-28-827_5.1	0.17	2194	4684	2.21	78.4	262.4	±2.6	218	±53	0.289	2.5	0.0416	1.0	0.400
OV-28-827_9.1	0.20	2112	2633	1.29	75.7	262.9	±2.7	275	±52	0.297	2.5	0.0416	1.1	0.421
OV-28-827_6.1	0.18	2350	5517	2.43	84.4	263.4	±2.7	233	±48	0.292	2.3	0.0417	1.0	0.448
OV-28-827_12.1	0.27	2152	4779	2.29	77.9	265.3	±3.0	233	±61	0.294	2.9	0.0420	1.1	0.398
OV-28-827_4.1	0.28	1410	2157	1.58	51.3	266.5	±3.0	248	±72	0.298	3.3	0.0422	1.1	0.341
657 2b														
657-2b_1.1	0.70	267	22	0.08	4.8	131.9	±2.3	–	–	0.139	7.7	0.0207	1.7	0.225
657-2b_2.1	14.28	298	35	0.12	7.1	151.1	±3.5	–	–	0.161	44.0	0.0237	2.3	0.053
657-2b_3.1	0.12	1511	53	0.04	25.6	125.7	±1.6	–	–	0.132	3.1	0.0197	1.3	0.422
657-2b_4.1	0.30	602	21	0.04	10.2	126.0	±1.8	–	–	0.132	4.2	0.0197	1.5	0.355
657-2b_5.1	0.57	557	73	0.13	9.7	128.6	±1.9	–	–	0.134	4.9	0.0202	1.5	0.308
657-2b_6.1	0.63	309	39	0.13	5.2	124.7	±2.2	–	–	0.124	8.0	0.0195	1.8	0.224
657-2b_6.2	12.97	283	37	0.14	5.9	134.1	±4.2	–	–	0.180	76.0	0.0210	3.2	0.042
657-2b_7.1	0.85	435	15	0.04	7.4	124.5	±2.0	–	–	0.126	7.3	0.0195	1.6	0.224
657-2b_8.1	0.42	1137	255	0.23	19.7	127.9	±1.7	–	–	0.134	3.8	0.0200	1.3	0.354
657-2b_9.1	2.15	300	23	0.08	5.5	133.1	±2.5	–	–	0.134	15	0.0209	1.9	0.127
657-2b_10.1	0.51	580	62	0.11	9.6	122.3	±1.8	–	–	0.127	5.0	0.0192	1.5	0.301
657-2b_11.1	0.14	432	9	0.02	7.3	125.1	±2.0	–	–	0.127	4.2	0.0196	1.6	0.384
657-2b_11.2	0.06	989	33	0.03	17.1	128.6	±1.7	–	–	0.134	2.7	0.0201	1.3	0.491
657-2b_12.1	0.17	502	46	0.09	8.7	128.8	±2.0	–	–	0.134	4.0	0.0202	1.5	0.380
657-2b_12.2	0.33	263	18	0.07	4.5	126.0	±2.3	–	–	0.125	6.3	0.0197	1.8	0.292
I 1														
I-1_8.1	0.55	46	21	0.47	4.1	629.0	±9.8	526	±140	0.818	6.4	0.1025	1.6	0.255
I-1_13.2	0.50	72	22	0.32	6.4	629.9	±8.5	543	±130	0.826	5.9	0.1026	1.4	0.239
I-1_12.1	0.23	233	142	0.63	20.1	614.2	±5.5	561	±58	0.811	2.8	0.1000	1.0	0.337
I-1_7.1	0.34	127	53	0.43	10.9	613.9	±6.6	569	±82	0.814	3.9	0.1000	1.1	0.288

(continued)

Table 6 (continued)

Crater	$^{206}Pb_c$, %	U, ppm	Th, ppm	$^{232}Th/^{238}U$	$^{206}Pb^*$, ppm	$^{206}Pb^*/^{238}U$ Age		$^{207}Pb^*/^{206}Pb^*$ Age		$^{207}Pb^*/^{235}U$	±%	$^{206}Pb^*/^{238}U$	±%	KK
I-1_13.1	0.23	115	40	0.36	10.1	621.8	±6.9	600	±69	0.836	3.4	0.1013	1.2	0.341
I-1_5.1	0.01	71	38	0.55	6.5	644.6	±8.3	657	±68	0.892	3.5	0.1052	1.3	0.389
I-1_4.1	0.25	174	130	0.77	17.1	698.7	±6.6	663	±58	0.974	2.9	0.1145	1.0	0.342
I-1_4.2	0.43	56	30	0.55	5.5	690.0	±9.9	680	±100	0.969	5.1	0.1130	1.5	0.295
I-1_2.2	0.14	164	80	0.50	16.8	728.0	±8.2	702	±51	1.036	2.7	0.1195	1.2	0.445
I-1_2.1	0.01	126	60	0.49	12.7	714.2	±7.6	737	±52	1.032	2.7	0.1172	1.1	0.419
I-1_11.1	0.01	67	30	0.47	7.1	741.6	±9.4	746	±63	1.078	3.3	0.1219	1.3	0.413
I-1_1.1	0.02	394	176	0.46	51.2	907.1	±8.0	953	±24	1.476	1.5	0.1511	0.9	0.627
I-1_3.1	0.10	89	146	1.69	25.9	1872.0	±18.0	1843	±24	5.232	1.7	0.3369	1.1	0.640
I-1_6.1	0.41	293	506	1.78	69.3	1561.0	±12.0	2464	±16	6.073	1.3	0.2740	0.8	0.658
I-1_9.1	0.15	171	101	0.61	73.1	2597.0	±19.0	2561	±12	11.660	1.1	0.4962	0.9	0.792
I-1_10.1	0.17	69	45	0.68	30.4	2659.0	±30.0	2652	±17	12.670	1.7	0.5107	1.4	0.800
KYM 1														
KYM-1_1.1	0.04	302	185	0.63	140.0	2778.0	±21.0	2697	±10	13.730	1.1	0.5387	1.0	0.838
KYM-1_1.2	0.05	195	308	1.63	92.0	2819.0	±24.0	2716	±10	14.140	1.2	0.5486	1.1	0.860
N-1.1.1	0.27	849	927	1.13	27.2	235.8	±3.0	112	±90	0.248	4.0	0.0373	1.3	0.322
N-1.1.2	0.17	918	2004	2.26	32.8	262.6	±3.3	192	±69	0.286	3.2	0.0416	1.3	0.398
N-1.1.3	0.08	556	765	1.42	18.9	250.0	±3.3	239	±65	0.278	3.1	0.0396	1.3	0.428
N-1.1.4	0.01	231	217	0.97	8.0	256.6	±3.5	339	±55	0.298	2.8	0.0406	1.4	0.495
N-1.1.5	0.01	396	447	1.17	13.5	250.6	±3.1	227	±29	0.277	1.8	0.0396	1.3	0.708
N-1.2.1	0.22	589	727	1.28	19.1	238.9	±3.0	184	±62	0.259	3.0	0.0378	1.3	0.437
N-1.2.2	0.09	1892	3264	1.78	63.9	248.5	±3.0	191	±39	0.271	2.1	0.0393	1.2	0.595
N-1.2.3	0.11	3363	5956	1.83	104.0	228.1	±2.7	207	±13	0.250	1.3	0.0360	1.2	0.906
N-1.3.1	0.01	1280	2045	1.65	42.7	245.6	±3.0	269	±24	0.277	1.6	0.0388	1.3	0.762
N-1.3.2	0.14	2089	4030	1.99	67.8	238.9	±2.9	214	±29	0.263	1.8	0.0378	1.2	0.696
N-1.4.1	0.33	257	1108	4.46	9.0	255.9	±3.6	138	±180	0.272	7.7	0.0405	1.4	0.186
N-1.4.2	0.06	835	1232	1.52	28.6	251.9	±3.1	256	±60	0.282	2.9	0.0399	1.3	0.437
N-1.4.3	0.52	264	1869	7.31	9.2	254.6	±3.5	23	±150	0.258	6.6	0.0403	1.4	0.212

(continued)

Table 6 (continued)

Crater	$^{206}Pb_c$, %	U, ppm	Th, ppm	$^{232}Th/^{238}U$	$^{206}Pb^*$, ppm	$^{206}Pb^*/^{238}U$ Age		$^{207}Pb^*/^{206}Pb^*$ Age		$^{207}Pb^*/^{235}U$	±%	$^{206}Pb^*/^{238}U$	±%	KK
N-1.5.1	0.31	382	1282	3.47	13.1	251.6	±3.3	89	±130	0.262	5.6	0.0398	1.4	0.244
N-1.5.2	0.04	668	2630	4.07	23.1	254.3	±3.2	290	±87	0.289	4.0	0.0402	1.3	0.322
N-1.5.3	0.09	918	3974	4.47	30.9	247.4	±3.1	215	±40	0.272	2.2	0.0391	1.3	0.587
N-1.5.4	0.10	667	2185	3.39	23.4	258.1	±3.2	192	±39	0.281	2.1	0.0409	1.3	0.611
N-1.5.5	0.10	505	1718	3.51	16.6	242.3	±3.0	220	±38	0.267	2.1	0.0383	1.3	0.605
N-1.6.1	0.01	6002	13,790	2.37	194.0	238.1	±2.9	271	±11	0.268	1.3	0.0376	1.2	0.932
N-1.7.1	0.01	227	1173	5.33	7.9	255.2	±3.5	330	±130	0.295	6.0	0.0404	1.4	0.237
N-1.7.2	0.09	142	639	4.66	4.9	256.3	±3.7	312	±64	0.294	3.2	0.0406	1.5	0.460
N-1.7.3	0.40	172	818	4.91	5.9	250.4	±3.6	251	±140	0.280	6.2	0.0396	1.5	0.235
N-1.7.4	0.43	73	92	1.29	2.7	265.0	±4.3	125	±160	0.281	6.9	0.0420	1.6	0.239
N-1.7.5	0.59	64	204	3.28	2.2	244.6	±4.5	194	±290	0.266	13.0	0.0387	1.9	0.150
N-1.7.6	0.78	92	369	4.12	3.1	243.4	±4.5	22	±340	0.247	14.0	0.0385	1.9	0.131
N-1.7.7	0.16	54	152	2.90	1.8	243.4	±4.3	282	±160	0.276	7.1	0.0385	1.8	0.254
N-1.7.8	0.01	196	961	5.07	6.2	236.7	±3.1	284	±51	0.268	2.6	0.0374	1.3	0.511
N1-1.1.11.1	0.35	4100	9700	2.37	127.0	227.9	±8.0	133	±66	0.242	4.5	0.0360	3.6	0.787
N1-1.2.10.1	0.01	1083	2214	2.11	35.6	242.6	±8.6	395	±57	0.289	4.4	0.0383	3.6	0.818
N1-1.3.9.1	0.01	2255	7307	3.35	72.8	238.0	±8.3	305	±36	0.272	3.9	0.0376	3.6	0.914
N1-1.4.8.1	0.27	2225	5011	2.33	73.0	241.0	±8.5	166	±59	0.259	4.4	0.0381	3.6	0.819
N1-1.5.7.1	0.10	1674	2050	1.27	54.2	238.2	±8.4	265	±65	0.267	4.5	0.0376	3.6	0.785
N1-1.6.6.1	0.01	2935	5148	1.81	93.9	235.9	±8.2	263	±35	0.265	3.9	0.0373	3.6	0.918
N1-1.8.5.1	2.65	140	90	0.66	42.5	1911.0	±62.0	1933	±77	5.640	5.7	0.3450	3.7	0.652
N1-1.9.4.1	0.32	2060	8186	4.11	68.0	242.2	±8.5	143	±160	0.258	7.6	0.0383	3.6	0.473
N1-1.10.3.1	0.01	3518	4113	1.21	111.0	232.5	±8.1	293	±29	0.264	3.8	0.0367	3.6	0.942
N1-1.11.2.1	0.01	4753	7181	1.56	143.0	222.9	±7.8	329	±48	0.257	4.1	0.0352	3.6	0.858
N1-2.(15).6.1	0.19	917	3646	4.11	28.0	224.4	±6.0	340	±120	0.260	5.8	0.0354	2.7	0.466
N1-2.(15).6.2	0.69	926	3252	3.63	29.2	231.1	±6.7	98	±500	0.241	21.0	0.0365	2.9	0.137
N1-2.(16).5.1	1.61	1230	4354	3.66	36.8	217.5	±5.9	43	±360	0.222	15.0	0.0343	2.7	0.182
N1-2.(17).4.1	0.18	1210	3528	3.01	37.9	230.5	±5.9	284	±130	0.261	6.2	0.0364	2.6	0.418

(continued)

Table 6 (continued)

Crater	$^{206}Pb_c$, %	U, ppm	Th, ppm	$^{232}Th/^{238}U$	$^{206}Pb^*$, ppm	$^{206}Pb^*/^{238}U$ Age		$^{207}Pb^*/^{206}Pb^*$ Age		$^{207}Pb^*/^{235}U$	±%	$^{206}Pb^*/^{238}U$	±%	KK
N1-2.(17).4.2	0.01	1150	3864	3.47	35.2	225.8	±5.9	369	±120	0.265	6.0	0.0357	2.6	0.439
N1-2.(18).3.2	0.07	1380	4995	3.74	43.5	232.4	±5.9	196	±96	0.253	4.9	0.0367	2.6	0.531
N1-2.(19).2.1	0.59	1792	6531	3.76	54.7	223.7	±6.3	32	±160	0.227	7.3	0.0353	2.8	0.391
N1-2.(20).1.1	0.05	3242	9770	3.11	94.8	215.7	±5.4	277	±70	0.243	4.0	0.0340	2.5	0.637
N1-2.1.9.1	2.59	221	401	1.88	7.5	243.0	±10.0	–	–	0.171	56.0	0.0385	4.4	0.078
N1-2.11.4.1	0.75	2188	6226	2.94	64.7	216.6	±7.6	236	±120	0.240	6.5	0.0342	3.6	0.551
N1-2.12.3.1	0.02	1154	3576	3.20	36.8	234.9	±8.3	261	±58	0.263	4.4	0.0371	3.6	0.816
N1-2.13.2.1	0.06	2747	5941	2.23	85.8	230.0	±8.0	183	±42	0.249	4.0	0.0363	3.6	0.893
N1-2.14.1.1	0.01	1448	2572	1.83	48.1	245.2	±8.6	339	±75	0.285	4.9	0.0388	3.6	0.736
N1-2.16.1	0.01	1024	3350	3.38	33.9	243.9	±5.9	245	±40	0.272	3.0	0.0386	2.5	0.814
N1-2.19.1	0.01	2563	11,139	4.49	92.0	264.2	±7.2	252	±36	0.296	3.2	0.0418	2.8	0.871
N1-2.21.10.1	0.42	6305	25,938	4.25	166.0	193.3	±6.8	338	±77	0.223	4.9	0.0304	3.6	0.722
N1-2.3.8.1	0.01	2185	4733	2.24	68.7	231.9	±8.1	297	±43	0.264	4.0	0.0366	3.6	0.886
N1-2.4.7.1	0.01	2195	4174	1.96	74.4	250.2	±8.8	303	±96	0.286	5.5	0.0396	3.6	0.647
N1-2.6.6.1	1.29	2618	6486	2.56	82.3	228.8	±8.1	419	±130	0.275	6.8	0.0361	3.6	0.531
N1-2.7.5.1	0.13	4418	10,967	2.56	134.0	223.5	±7.8	212	±52	0.245	4.2	0.0353	3.5	0.846
N1-2.9(18).3.1	0.06	2001	4139	2.14	62.3	229.2	±6.0	227	±140	0.253	6.5	0.0362	2.7	0.409
N1-3.(1).5.1	2.95	422	1499	3.67	13.3	226.3	±6.7	–	–	0.172	33.0	0.0357	3.0	0.093
N1-3.(1).5.2	1.20	503	1877	3.86	17.1	247.5	±6.9	–	–	0.245	18.0	0.0391	2.8	0.154
N1-3.(2).4.1	0.01	3941	14,261	3.74	117.0	219.7	±5.9	201	±35	0.240	3.1	0.0347	2.7	0.873
N1-3.(2).4.2	0.32	2771	7216	2.69	81.8	217.2	±5.5	109	±140	0.228	6.6	0.0343	2.6	0.390
N1-3.(4).3.1	0.23	2012	7935	4.08	61.2	223.9	±5.7	201	±150	0.244	6.8	0.0354	2.6	0.379
N1-3.(5).2.1	0.11	1817	6036	3.43	56.4	228.7	±5.8	254	±62	0.255	3.7	0.0361	2.6	0.689
N1-3.(5).2.2	0.32	1913	5677	3.07	58.9	226.2	±5.8	81	±170	0.235	7.5	0.0357	2.6	0.349
N1-3.(6).1.1	0.48	943	2854	3.13	29.9	232.4	±6.3	126	±340	0.246	15.0	0.0367	2.8	0.190
N1-3.2.1	0.18	1784	4505	2.61	58.3	240.3	±5.8	243	±63	0.267	3.7	0.0380	2.5	0.669
N1-3.2.2	0.04	4686	14,806	3.26	148.0	232.8	±5.5	231	±25	0.257	2.6	0.0368	2.4	0.914
N1-3.3.1.1	0.01	204	450	2.28	6.8	250.4	±9.0	871	3±20	0.372	16.0	0.0396	3.7	0.235

(continued)

Table 6 (continued)

Crater	$^{206}Pb_c$, %	U, ppm	Th, ppm	$^{232}Th/^{238}U$	$^{206}Pb^*$, ppm	$^{206}Pb^*/^{238}U$ Age		$^{207}Pb^*/^{206}Pb^*$ Age		$^{207}Pb^*/^{235}U$	±%	$^{206}Pb^*/^{238}U$	±%	KK
N1-3.5.1	0.18	2084	7682	3.81	67.9	239.5	±5.8	231	±58	0.265	3.5	0.0379	2.5	0.698
N1-3.5.2	0.03	1715	5077	3.06	57.9	248.6	±5.9	227	±35	0.275	2.9	0.0393	2.4	0.852
N1-4.(12).3.1	0.23	3296	20768	6.51	98.3	219.6	±5.5	166	±82	0.236	4.3	0.0347	2.6	0.589
N1-4.(13).4.1	0.22	3121	12394	4.10	93.4	220.4	±5.8	248	±100	0.245	5.2	0.0348	2.7	0.513
N1-4.(13).4.2	0.03	6017	14354	2.46	182.0	222.6	±5.5	255	±32	0.249	2.9	0.0351	2.5	0.873
N1-4.(15).2.1	0.72	702	1138	1.68	21.7	226.8	±6.0	99	±200	0.237	9.0	0.0358	2.7	0.298
N1-4.(17).1.1	0.57	2357	4876	2.14	71.9	223.6	±5.7	115	±150	0.235	7.0	0.0353	2.6	0.372
N1-4.10.3.1	0.01	4042	10750	2.75	127.0	231.3	±7.2	343	±52	0.269	3.9	0.0365	3.2	0.809
N1-4.11.2.1	0.33	3908	15633	4.13	119.0	223.0	±7.0	137	±74	0.237	4.5	0.0352	3.2	0.708
N1-4.12.1	0.01	2846	16739	6.08	92.4	239.0	±5.7	234	±26	0.265	2.7	0.0378	2.4	0.908
N1-4:14.1.1	0.40	2502	20619	8.51	73.4	215.4	±6.8	111	±120	0.226	5.9	0.0340	3.2	0.544
N1-4.17.1	0.26	6384	16977	2.75	203.0	233.7	±5.5	218	±33	0.257	2.8	0.0369	2.4	0.864
N1-4.3.8.1	0.01	4768	8576	1.86	162.0	249.9	±7.8	266	±43	0.281	3.7	0.0395	3.2	0.859
N1-4.4.7.1	0.09	11,423	31,614	2.86	380.0	244.6	±7.6	194	±20	0.267	3.3	0.0387	3.2	0.965
N1-4.6.6.1	0.04	2127	5052	2.45	74.0	255.8	±8.0	241	±62	0.285	4.2	0.0405	3.2	0.767
N1-4.7.5.1	0.28	2632	4942	1.94	87.8	245.0	±7.7	164	±57	0.264	4.0	0.0387	3.2	0.795
N1-4.7.5.2	0.05	4579	13662	3.08	155.0	249.6	±7.8	219	±24	0.275	3.3	0.0395	3.2	0.949
N1-4.9.4.1	2.31	515	413	0.83	17.6	246.4	±8.3	–	–	0.181	25.0	0.0390	3.4	0.135
N1-5.(1).7.1	0.20	2266	8990	4.10	72.9	236.4	±6.0	208	±77	0.259	4.2	0.0374	2.6	0.612
N1-5.(10).5.1	0.08	7404	8450	1.18	230.0	229.0	±5.7	209	±33	0.251	2.9	0.0362	2.5	0.874
N1-5.(11).4.1	0.32	3361	10,655	3.28	102.0	222.5	±5.7	152	±150	0.238	7.0	0.0351	2.6	0.375
N1-5.(16).3.1	1.33	1279	4010	3.24	39.6	225.1	±6.0	–	–	0.213	16.0	0.0355	2.7	0.170
N1-5.(17).2.1	0.16	1577	3029	1.98	52.6	245.0	±6.3	257	±81	0.274	4.4	0.0387	2.6	0.597
N1-5.(17).2.2	0.65	527	537	1.05	17.5	242.8	±7.6	146	±290	0.259	13.0	0.0384	3.2	0.245
N1-5.(22).1.1	0.14	1553	5016	3.34	58.3	275.5	±7.1	281	±150	0.313	7.3	0.0437	2.6	0.361
N1-5.(22).1.2	0.50	1096	2416	2.28	34.0	227.7	±7.0	88	±280	0.237	12.0	0.0359	3.1	0.255
N1-5.(7).6.1	0.25	1297	4147	3.30	41.9	237.4	±6.1	230	±120	0.263	5.9	0.0375	2.6	0.443
N1-5.12.4.1	0.30	1695	3493	2.13	57.7	249.7	±8.8	160	±120	0.268	6.3	0.0395	3.6	0.569

(continued)

Table 6 (continued)

Crater	$^{206}Pb_c$, %	U, ppm	Th, ppm	$^{232}Th/^{238}U$	$^{206}Pb^*$, ppm	$^{206}Pb^*/^{238}U$ Age		$^{207}Pb^*/^{206}Pb^*$ Age		$^{207}Pb^*/^{235}U$	±%	$^{206}Pb^*/^{238}U$	±%	KK
N1-5.13.3.1	0.41	1141	3589	3.25	39.5	253.3	±9.1	68	±110	0.262	5.8	0.0401	3.6	0.633
N1-5.15.2.1	0.01	1712	4949	2.99	55.2	238.1	±8.3	354	±61	0.278	4.5	0.0376	3.6	0.797
N1-5.17.1	0.55	224	252	1.16	8.0	261.8	±7.0	290	±160	0.298	7.3	0.0415	2.7	0.371
N1-5.19.1.1	0.14	3002	10,453	3.60	91.4	224.2	±7.8	202	±56	0.245	4.3	0.0354	3.6	0.829
N1-5.2.9.1	0.26	1794	3522	2.03	59.1	242.1	±8.5	119	±67	0.255	4.6	0.0383	3.6	0.782
N1-5.4.8.1	0.01	957	3447	3.72	32.5	250.8	±8.9	331	±83	0.290	5.1	0.0397	3.6	0.704
N1-5.5.7.1	0.01	1835	7099	4.00	58.3	234.0	±8.2	224	±37	0.258	3.9	0.0370	3.6	0.911
N1-5.6.6.1	0.01	1573	5877	3.86	50.7	237.9	±8.3	360	±52	0.279	4.2	0.0376	3.6	0.841
N1-5.9.5.1	0.01	554	1494	2.79	17.9	238.2	±8.5	367	±62	0.280	4.6	0.0376	3.6	0.795
N1-6.(1).11.1	0.59	918	3093	3.48	35.3	280.9	±7.3	607	±95	0.369	5.1	0.0445	2.7	0.516
N1-6.(11).2.1	0.31	2107	7563	3.71	63.5	221.6	±5.8	198	±180	0.241	8.1	0.0350	2.6	0.327
N1-6.(18).1.1 (Prism)	0.01	4086	15,399	3.89	125.0	224.8	±5.6	248	±34	0.250	2.9	0.0355	2.5	0.868
N1-6.(18).1.2 (Pyramid)	0.19	2225	9573	4.44	70.7	233.6	±9.4	218	±80	0.257	5.4	0.0369	4.1	0.763
N1-6.(18).1.3	0.50	2373	6869	2.99	75.1	231.9	±6.9	280	±92	0.262	5.1	0.0366	3.0	0.602
N1-6.(18).1.4	0.01	1248	3343	2.77	39.4	232.6	±7.0	247	±40	0.259	3.5	0.0367	3.0	0.871
N1-6.(2).10.1	0.39	603	1390	2.38	22.2	269.1	±7.2	272	±170	0.304	8.1	0.0426	2.7	0.337
N1-6.(2).10.2	0.11	527	1264	2.48	17.6	245.0	±7.5	260	±160	0.275	7.5	0.0387	3.1	0.413
N1-6.(20).3.1	0.56	863	1586	1.90	25.4	216.3	±7.3	303	±190	0.247	8.9	0.0341	3.4	0.384
N1-6.(21).4.1	0.33	1136	5618	5.11	34.8	225.0	±5.9	312	±220	0.258	9.9	0.0355	2.7	0.270
N1-6.(26).5.1	1.26	937	1859	2.05	33.8	262.1	±8.4	–	–	0.242	17.0	0.0415	3.3	0.197
N1-6.(27).6.1	0.14	3108	6311	2.10	102.0	241.6	±6.1	183	±81	0.262	4.3	0.0382	2.6	0.592
N1-6.(29).7.1	0.25	607	2167	3.69	22.0	265.2	±10.0	429	±110	0.321	6.4	0.0420	3.8	0.603
N1-6.(4).9.1	0.07	1429	2921	2.11	49.8	256.1	±6.6	287	±180	0.291	8.3	0.0405	2.6	0.317
N1-6.(6).8.1	0.14	1294	3539	2.83	43.5	247.4	±8.5	323	±80	0.285	5.0	0.0391	3.5	0.705
N1-6.10.5.1	0.01	801	1988	2.56	26.5	244.3	±8.6	388	±65	0.290	4.6	0.0386	3.6	0.779
N1-6.11.1	0.39	494	813	1.70	16.9	250.7	±6.3	221	±94	0.276	4.8	0.0396	2.6	0.533
N1-6.11.2	0.37	644	1878	3.01	20.3	231.5	±5.7	202	±110	0.253	5.5	0.0366	2.5	0.455
N1-6.13.7.1	0.01	1520	2289	1.56	50.9	247.2	±8.7	332	±64	0.286	4.5	0.0391	3.6	0.787

(continued)

Table 6 (continued)

Crater	$^{206}Pb_c$, %	U, ppm	Th, ppm	$^{232}Th/^{238}U$	$^{206}Pb^*$, ppm	$^{206}Pb^*/^{238}U$ Age		$^{207}Pb^*/^{206}Pb^*$ Age		$^{207}Pb^*/^{235}U$	±%	$^{206}Pb^*/^{238}U$	±%	KK
N1-6.14.6.1	0.58	1276	4657	3.77	40.0	230.0	±8.1	129	±140	0.243	7.1	0.0363	3.6	0.505
N1-6.16.4.1	0.09	5821	10,899	1.93	175.0	221.7	±7.7	202	±40	0.242	3.9	0.0350	3.6	0.899
N1-6.23.3.1	0.01	1283	3495	2.81	42.8	245.8	±8.7	334	±140	0.285	7.2	0.0389	3.6	0.501
N1-6.26.1	0.54	1457	2196	1.56	46.4	233.2	±5.6	245	±76	0.259	4.1	0.0368	2.5	0.598
N1-6.32.2.1	0.01	1414	2689	1.97	46.4	242.2	±8.5	301	±71	0.276	4.7	0.0383	3.6	0.756
N1-6.33.1.1	0.67	1244	4130	3.43	40.2	236.5	±8.4	–	–	0.235	8.5	0.0374	3.6	0.424
N1-6.6.1	0.08	694	829	1.23	23.2	246.1	±6.0	236	±56	0.273	3.5	0.0389	2.5	0.712
N1-6.7.9.1	0.31	992	2231	2.32	33.2	245.9	±8.8	100	±160	0.257	7.5	0.0389	3.6	0.484
N1-6.9.8.1	0.01	1641	3284	2.07	55.0	246.9	±8.7	298	±36	0.281	3.9	0.0390	3.6	0.914
N1-7.(1).8.1	0.18	1530	2077	1.40	50.4	242.2	±7.3	223	±62	0.267	4.1	0.0383	3.1	0.753
N1-7.(13).7.1	0.79	825	1425	1.78	27.2	240.5	±7.3	–	–	0.241	7.6	0.0380	3.1	0.406
N1-7.(13).7.2	0.01	740	1506	2.10	24.3	241.5	±7.4	298	±46	0.275	3.7	0.0382	3.1	0.842
N1-7.(14).6.1	0.19	4046	9192	2.35	121.0	220.9	±6.6	174	±49	0.238	3.7	0.0349	3.0	0.825
N1-7.(14).6.2	0.13	918	1186	1.34	28.8	231.2	±7.0	249	±84	0.258	4.8	0.0365	3.1	0.646
N1-7.(18).5.1	0.10	1524	5367	3.64	53.1	256.0	±7.7	241	±52	0.285	3.8	0.0405	3.1	0.808
N1-7.(19).1.1 (4.3)	0.01	1228	1640	1.38	40.4	244.1	±6.4	550	±170	0.311	8.4	0.0386	2.7	0.318
N1-7.(19).4.1	0.15	993	1108	1.15	34.6	256.0	±11.0	204	±68	0.280	5.3	0.0405	4.4	0.831
N1-7.(19).4.2	0.10	1622	1927	1.23	56.8	257.2	±7.7	200	±85	0.281	4.8	0.0407	3.1	0.642
N1-7.(21).3.1	0.72	297	290	1.01	9.8	241.8	±7.6	68	±170	0.250	7.9	0.0382	3.2	0.408
N1-7.(21).3.2	0.18	1822	3033	1.72	57.6	232.4	±7.0	200	±49	0.254	3.7	0.0367	3.0	0.824
N1-7.(21).3.3	1.19	542	763	1.45	17.9	240.6	±7.8	–	–	0.220	18.0	0.0380	3.3	0.185
N1-7.(22).2.1	0.90	1673	3287	2.03	55.4	241.9	±7.3	–	–	0.233	7.6	0.0382	3.1	0.405
N1-7.(23).1.1	0.40	692	1029	1.54	24.4	257.8	±7.9	208	±140	0.283	6.9	0.0408	3.1	0.449
N1-7.11.7.1	0.01	541	1343	2.57	17.9	244.6	±7.8	403	±120	0.292	6.2	0.0387	3.2	0.524
N1-7.12.6.1	0.01	8220	26,516	3.33	239.0	214.7	±6.7	242	±25	0.238	3.3	0.0339	3.2	0.946
N1-7.15.5.1	0.86	607	1024	1.74	20.0	240.4	±7.8	90	±270	0.250	12.0	0.0380	3.3	0.274
N1-7.16.4.1	0.10	1048	3429	3.38	32.4	227.7	±7.3	202	±190	0.249	8.9	0.0360	3.2	0.363
N1-7.16.4.2	0.51	604	1286	2.20	20.7	251.0	±8.2	192	±310	0.273	14.0	0.0397	3.3	0.240

(continued)

Table 6 (continued)

Crater	$^{206}Pb_c$, %	U, ppm	Th, ppm	$^{232}Th/^{238}U$	$^{206}Pb^*$, ppm	$^{206}Pb^*/^{238}U$ Age		$^{207}Pb^*/^{206}Pb^*$ Age		$^{207}Pb^*/^{235}U$	±%	$^{206}Pb^*/^{238}U$	±%	KK
N1-7.17.3.1	0.01	184	403	2.27	6.8	274.0	±9.1	842	±64	0.402	4.6	0.0434	3.4	0.738
N1-7.17.3.2	2.06	147	394	2.76	5.2	253.0	±10.0	–	–	0.204	46.0	0.0401	4.0	0.087
N1-7.18.1	0.01	598	3016	5.21	19.5	240.8	±5.9	282	±62	0.273	3.7	0.0381	2.5	0.680
N1-7.19.1	0.01	718	825	1.19	24.4	250.0	±6.1	234	±45	0.277	3.2	0.0396	2.5	0.783
N1-7.20.2.1	0.01	2575	3953	1.59	80.8	231.4	±7.2	281	±25	0.262	3.4	0.0366	3.2	0.946
N1-7.22.1	0.07	1534	3656	2.46	50.8	243.6	±5.8	230	±42	0.270	3.0	0.0385	2.4	0.805
N1-7.24.1.1	0.14	2016	4806	2.46	59.3	216.8	±6.8	235	±50	0.240	3.9	0.0342	3.2	0.830
N1-7.3.14.1	0.01	625	1396	2.31	20.4	241.7	±7.9	362	±120	0.283	6.1	0.0382	3.3	0.547
N1-7.4.13.1	0.41	894	2073	2.40	28.0	229.7	±7.3	217	±140	0.252	7.0	0.0363	3.2	0.461
N1-7.4.13.2	0.12	2774	6944	2.59	88.0	233.5	±7.5	204	±58	0.255	4.1	0.0369	3.3	0.795
N1-7.5.12.1	0.01	530	849	1.65	18.2	252.9	±8.0	367	±80	0.297	4.8	0.0400	3.2	0.676
N1-7.6.11.1	0.32	2317	3003	1.34	73.3	232.3	±7.3	178	±75	0.251	4.5	0.0367	3.2	0.701
N1-7.7.10.1	0.01	1577	2862	1.88	52.0	242.9	±7.6	258	±37	0.272	3.6	0.0384	3.2	0.891
N1-7.9.8.1	0.01	536	1279	2.47	18.1	249.1	±7.9	332	±49	0.288	3.9	0.0394	3.2	0.832
N1-8.(1).12.1	0.07	3369	5978	1.83	98.3	215.1	±6.4	269	±35	0.242	3.4	0.0339	3.0	0.894
N1-8.(10).5.1	0.41	1530	4649	3.14	46.2	221.6	±6.7	176	±86	0.239	4.8	0.0350	3.1	0.637
N1-8.(16).4.1	0.21	512	1102	2.23	17.9	256.3	±7.8	171	±140	0.277	6.9	0.0406	3.1	0.451
N1-8.(17).3.1	1.18	194	289	1.54	5.9	222.9	±8.7	–	–	0.218	40.0	0.0352	4.0	0.099
N1-8.(18).2.1	0.26	1209	2381	2.04	39.7	241.3	±7.3	196	±110	0.263	5.5	0.0381	3.1	0.557
N1-8.(19).1.1	0.24	1671	4062	2.51	53.0	233.0	±7.3	134	±110	0.247	5.5	0.0368	3.2	0.577
N1-8.(19).1.2	0.07	504	505	1.03	16.6	242.1	±7.4	277	±81	0.273	4.7	0.0383	3.1	0.661
N1-8.(2).11.1	0.25	422	653	1.60	14.4	251.1	±7.7	160	±180	0.270	8.4	0.0397	3.1	0.375
N1-8.(3).10.1	0.22	510	567	1.15	17.3	249.4	±7.6	205	±86	0.273	4.9	0.0394	3.1	0.643
N1-8.(5).9.1	0.01	2946	7106	2.49	86.4	216.4	±6.5	229	±28	0.239	3.3	0.0341	3.0	0.929
N1-8.(6).8.1	0.40	598	1145	1.98	21.1	258.7	±7.9	146	±220	0.276	9.8	0.0409	3.1	0.319
N1-8.(7).7.1	0.33	677	687	1.05	21.9	237.3	±7.2	220	±110	0.261	5.7	0.0375	3.1	0.537
N1-8.(7).7.2	0.52	609	877	1.49	19.1	229.6	±7.0	126	±140	0.243	6.7	0.0363	3.1	0.464
N1-8.(7).7.3	0.01	739	1011	1.41	22.8	227.0	±6.9	333	±56	0.262	4.0	0.0358	3.1	0.781

(continued)

Table 6 (continued)

Crater	$^{206}Pb_c$, %	U, ppm	Th, ppm	$^{232}Th/^{238}U$	$^{206}Pb^*$, ppm	$^{206}Pb^*/^{238}U$ Age		$^{207}Pb^*/^{206}Pb^*$ Age		$^{207}Pb^*/^{235}U$	±%	$^{206}Pb^*/^{238}U$	±%	KK
N1-8.(8).6.1	0.09	1643	2262	1.42	51.6	231.2	±6.9	272	±59	0.260	4.0	0.0365	3.1	0.763
N1-8.10.1	0.01	809	1775	2.27	28.4	258.2	±6.4	242	±50	0.287	3.3	0.0409	2.5	0.757
N1-8.10.2	0.04	1277	3169	2.56	42.1	242.4	±5.8	230	±42	0.268	3.1	0.0383	2.5	0.802
N1-8.11.1.1	0.01	1442	4553	3.26	47.9	246.4	±8.7	452	±110	0.301	6.2	0.0390	3.6	0.580
N1-8.11.1.2	0.99	350	412	1.22	11.7	244.5	±9.2	85	±420	0.254	18.0	0.0387	3.8	0.210
N1-8.17.1	0.59	66	79	1.25	2.5	273.6	±8.7	316	±260	0.315	12.0	0.0434	3.3	0.270
N1-8.9.2.1	0.01	592	1132	1.97	19.8	246.5	±8.8	468	±70	0.303	4.8	0.0390	3.6	0.756
N1-9.(1).9.1	0.09	1930	2065	1.11	60.9	232.3	±6.9	218	±38	0.256	3.5	0.0367	3.0	0.881
N1-9.(17).6.1	0.22	1714	2448	1.48	53.1	228.0	±6.8	220	±69	0.251	4.3	0.0360	3.0	0.715
N1-9.(17).6.2	0.61	703	938	1.38	23.2	241.1	±7.4	83	±200	0.250	8.9	0.0381	3.1	0.347
N1-9.(17).6.3	0.19	1634	3233	2.04	50.9	229.2	±6.9	193	±90	0.249	4.9	0.0362	3.1	0.619
N1-9.(18).5.1	0.08	602	1116	1.91	18.3	223.7	±7.0	259	±83	0.250	4.8	0.0353	3.2	0.657
N1-9.(20).4.1	0.79	1084	2165	2.06	34.1	229.8	±7.0	–	–	0.223	10.0	0.0363	3.1	0.305
N1-9.(22).3.1	1.48	794	1364	1.78	24.8	226.8	±7.0	–	–	0.195	13.0	0.0358	3.1	0.245
N1-9.(23).2.1	1.55	1964	3280	1.73	58.0	214.4	±6.5	136	±140	0.227	6.6	0.0338	3.1	0.467
N1-9.(25).1.1	0.17	2669	3825	1.48	82.8	228.3	±6.8	275	±40	0.257	3.5	0.0360	3.1	0.870
N1-9.(5).8.1	0.02	7440	9661	1.34	219.0	217.2	±6.5	202	±27	0.237	3.2	0.0343	3.0	0.933
N1-9.(7).7.1	0.92	414	637	1.59	12.8	226.3	±7.0	–	–	0.223	11.0	0.0357	3.2	0.276
N1-9.12.6.1	0.94	1078	2097	2.01	36.7	248.4	±8.4	–	–	0.243	21.0	0.0393	3.4	0.166
N1-9.14.5.1	0.14	2644	3337	1.30	83.7	232.9	±7.3	198	±56	0.254	4.0	0.0368	3.2	0.799
N1-9.15.4.1	0.59	1475	4703	3.29	46.2	229.5	±7.3	–	–	0.225	7.2	0.0362	3.2	0.447
N1-9.16.3.1	0.01	921	1575	1.77	31.5	252.8	±8.2	412	±130	0.303	6.8	0.0400	3.3	0.489
N1-9.17.1	1.24	366	406	1.15	11.7	232.7	±6.1	206	±220	0.255	10.0	0.0368	2.7	0.264
N1-9.17.2	0.50	1029	1564	1.57	34.9	248.4	±6.0	231	±87	0.275	4.5	0.0393	2.5	0.549
N1-9.21.2.1	0.26	3878	5997	1.60	121.0	229.2	±7.1	136	±62	0.243	4.1	0.0362	3.2	0.771
N1-9.24.1.1	0.24	1634	3886	2.46	56.2	252.6	±8.0	203	±150	0.276	7.2	0.0400	3.2	0.445
N1-9.25.1	0.16	2406	4368	1.88	75.4	230.7	±5.5	231	±40	0.255	3.0	0.0364	2.4	0.818
N1-9.4.9.1	0.02	1571	4287	2.82	54.2	253.8	±7.9	307	±64	0.291	4.2	0.0402	3.2	0.752

(continued)

Table 6 (continued)

Crater	$^{206}Pb_c$, %	U, ppm	Th, ppm	$^{232}Th/^{238}U$	$^{206}Pb^*$, ppm	$^{206}Pb^*/^{238}U$ Age		$^{207}Pb^*/^{206}Pb^*$ Age		$^{207}Pb^*/^{235}U$	±%	$^{206}Pb^*/^{238}U$	±%	KK
N1-9.5.1	1.34	5974	7557	1.31	180.0	219.8	±5.2	222	±63	0.242	3.7	0.0347	2.4	0.663
N1-9.6.8.1	0.07	7775	11,684	1.55	241.0	228.7	±7.1	200	±30	0.250	3.4	0.0361	3.2	0.925
N1-9.8.7.1	0.01	1119	2074	1.92	37.4	247.7	±7.8	481	±33	0.306	3.5	0.0392	3.2	0.906
N1-10.2.6.1	0.12	2736	2566	0.97	98.7	264.9	±9.3	262	±54	0.298	4.3	0.0420	3.6	0.834
N1-10.3.5.1	0.01	940	2250	2.47	29.8	233.8	±8.3	546	±56	0.298	4.4	0.0369	3.6	0.813
N1-10.4.4.1	0.06	8404	6655	0.82	250.0	219.5	±7.6	213	±24	0.241	3.7	0.0346	3.5	0.961
N1-10.5.3.1	1.46	2922	4001	1.41	101.0	250.3	±8.8	9	±270	0.252	12.0	0.0396	3.6	0.307
N1-10.6.2.1	0.63	2163	1694	0.81	73.2	247.7	±8.7	146	±100	0.264	5.7	0.0392	3.6	0.627
N1-10.7.1.1	0.01	6819	6983	1.06	200.0	216.8	±7.6	285	±20	0.245	3.6	0.0342	3.5	0.971
N-11.1.1	0.01	3637	5608	1.59	120.0	244.0	±6.8	274	±22	0.275	3.0	0.0385	2.9	0.948
N-11.1.2	0.12	7344	17,886	2.52	271.0	271.0	±7.6	251	±22	0.303	3.0	0.0429	2.9	0.948
N-11.1.3	0.09	3763	9252	2.54	148.0	289.0	±8.1	210	±23	0.318	3.0	0.0458	2.9	0.945
N-11 3.1	0.05	3587	5866	1.69	129.0	263.5	±1.1	277	±29	0.298	1.4	0.0417	0.4	0.301
N-11 3.2	0.20	3952	10,149	2.65	128.0	238.8	±1.0	338	±44	0.277	2.0	0.0377	0.4	0.215
N-11 4.1	0.10	3515	4573	1.34	97.4	204.5	±0.9	358	±30	0.239	1.4	0.0322	0.4	0.316
N-11 4.2	0.05	4643	7026	1.56	157.0	248.5	±1.0	311	±24	0.285	1.1	0.0393	0.4	0.355
N-11 6.1	1.99	372	419	1.16	12.6	243.9	±2.6	1081	±150	0.401	7.7	0.0386	1.1	0.141
N-11 6.2	0.12	1154	1667	1.49	37.5	239.1	±1.5	269	±45	0.269	2.1	0.0378	0.6	0.300
N-11 7.1	0.04	1691	3115	1.90	52.3	227.8	±2.7	290	±32	0.258	1.8	0.0360	1.2	0.655
N1-1.12.1.1	0.01	994	1669	1.73	31.2	232.0	±8.3	320	±82	0.267	5.1	0.0366	3.6	0.707
N-12 1	0.25	478	852	1.84	16.6	254.7	±3.6	305	±70	0.292	3.4	0.0403	1.5	0.427
N-12 3	0.01	677	2238	3.42	22.1	241.0	±1.5	253	±45	0.269	2.1	0.0381	0.6	0.309
N-12 4	0.12	2477	9481	3.96	84.3	250.1	±1.1	244	±40	0.279	1.8	0.0396	0.5	0.258
N-12 5	0.30	462	1666	3.73	14.7	233.5	±1.9	419	±110	0.281	5.0	0.0369	0.8	0.168
N-12 6	0.27	471	1666	3.65	14.7	229.0	±2.8	149	±84	0.245	3.8	0.0362	1.2	0.326
T-1/T-22 (n = 220) Talnakh intrusive														
T1-6.1	0.08	6228	9366	1.55	225.0	265.0	±5.4	251	±31	0.297	2.5	0.0420	2.1	0.842
T1 9.1	0.02	6365	8077	1.31	191.0	221.0	±5.8	240	±19	0.245	2.8	0.0349	2.7	0.955

(continued)

Table 6 (continued)

Crater	$^{206}Pb_c$, %	U, ppm	Th, ppm	$^{232}Th/^{238}U$	$^{206}Pb^*$, ppm	$^{206}Pb^*/^{238}U$ Age		$^{207}Pb^*/^{206}Pb^*$ Age		$^{207}Pb^*/^{235}U$	±%	$^{206}Pb^*/^{238}U$	±%	KK
T1_15.1	0.52	1577	3218	2.11	44.4	207.0	±5.2	164	±77	0.222	4.2	0.0326	2.5	0.608
T1_15.2	0.06	2442	4919	2.08	75.9	229.0	±5.6	266	±36	0.257	2.9	0.0362	2.5	0.846
T1 17.1	0.13	1774	4008	2.33	71.0	294.0	±8.3	246	±41	0.329	3.4	0.0467	2.9	0.849
T1 19.1	0.01	10,290	17,425	1.75	285.0	205.0	±5.8	283	±14	0.231	2.9	0.0323	2.9	0.979
T1 23.1	0.02	6122	8273	1.40	162.0	196.0	±5.2	214	±35	0.214	3.1	0.0308	2.7	0.871
T1 24.1	1.06	829	1808	2.25	24.0	212.0	±6.1	–	–	0.208	7.6	0.0334	2.9	0.389
T1 24.2	0.67	489	811	1.71	18.0	268.0	±7.7	185	±140	0.291	6.6	0.0424	2.9	0.446
T1 25.1	0.17	6634	7346	1.14	170.0	189.0	±5.0	285	±42	0.213	3.3	0.0297	2.7	0.824
T1 26.1	0.01	2626	1896	0.75	87.5	245.0	±6.4	228	±22	0.271	2.8	0.0388	2.7	0.940
T1 26.2	0.92	1411	3730	2.73	45.0	233.0	±6.1	295	±83	0.265	4.5	0.0368	2.7	0.593
T1 27.1	0.16	1471	1041	0.73	44.0	221.0	±6.3	213	±43	0.242	3.4	0.0348	2.9	0.843
T1 29.1	0.22	6090	8521	1.45	214.0	258.0	±5.3	163	±51	0.278	3.0	0.0409	2.1	0.691
T1 30.1	0.02	4674	14,895	3.29	167.0	262.0	±5.4	322	±38	0.302	2.7	0.0415	2.1	0.784
T1 30.2	0.03	1516	1154	0.79	52.9	257.0	±5.4	367	±45	0.302	2.9	0.0406	2.1	0.734
T1 33.1	0.08	3834	4091	1.10	115.0	221.0	±5.8	235	±31	0.244	3.0	0.0348	2.7	0.895
T1 35.1	0.06	3867	7095	1.90	115.0	219.0	±5.7	228	±27	0.241	2.9	0.0345	2.7	0.916
T1 36.1	3.05	1509	2088	1.43	48.1	228.0	±6.0	320	±100	0.262	5.2	0.0360	2.7	0.517
T1 37.1	0.04	3029	2884	0.98	94.6	230.0	±6.0	252	±23	0.257	2.9	0.0363	2.7	0.934
T1 37.2	0.09	4797	5480	1.18	139.0	214.0	±5.6	237	±26	0.236	2.9	0.0337	2.7	0.923
T1 38.1	0.19	2187	6325	2.99	69.9	235.0	±6.2	254	±42	0.263	3.2	0.0372	2.7	0.823
T2-11.1	0.03	8600	16,624	2.00	301.0	258.0	±7.2	249	±28	0.288	3.1	0.0408	2.9	0.921
T2-11.2	0.10	3778	5966	1.63	134.0	261.0	±7.3	237	±28	0.290	3.1	0.0413	2.9	0.919
T2-12.1	0.58	1655	1713	1.07	57.1	253.0	±5.4	112	±160	0.266	6.9	0.0399	2.2	0.313
T2-2.1	0.29	628	1077	1.77	20.6	240.0	±1.8	338	±84	0.279	3.8	0.0380	0.8	0.203
T2-2.2	0.44	1868	6302	3.49	49.6	195.0	±4.9	470	±90	0.239	4.8	0.0308	2.6	0.533
T2-21.1	0.07	6328	26,377	4.31	215.0	250.0	±7.0	255	±22	0.280	3.0	0.0396	2.9	0.948
T2-4.1	1.18	1964	2231	1.17	66.0	246.0	±7.0	278	±87	0.278	4.8	0.0389	2.9	0.605
T2-8.1	0.01	3516	5165	1.52	125.0	261.0	±7.3	244	±20	0.291	3.0	0.0414	2.9	0.956

(continued)

Table 6 (continued)

Crater	$^{206}Pb_c$, %	U, ppm	Th, ppm	$^{232}Th/^{238}U$	$^{206}Pb^*$, ppm	$^{206}Pb^*/^{238}U$ Age		$^{207}Pb^*/^{206}Pb^*$ Age		$^{207}Pb^*/^{235}U$	±%	$^{206}Pb^*/^{238}U$	±%	KK
T2-8.2	0.06	2683	9038	3.48	92.0	251.0	±7.0	236	±25	0.279	3.1	0.0397	2.9	0.934
T2-9.1	3.04	1778	2014	1.17	67.0	267.0	±7.8	279	±250	0.302	11.0	0.0423	3.0	0.263
T2 17.1	0.15	8142	23,014	2.92	259.0	235.0	±6.2	279	±42	0.265	3.2	0.0370	2.7	0.822
T2 17.2	0.33	4476	12,556	2.90	134.0	220.0	±5.8	224	±31	0.242	3.0	0.0347	2.7	0.897
T2 27.1	0.06	4951	10,472	2.19	151.0	225.0	±5.9	220	±18	0.248	2.8	0.0355	2.7	0.958
T2 32.1	0.07	2131	3370	1.63	68.5	237.0	±6.2	248	±36	0.264	3.1	0.0374	2.7	0.865
T2 32.2	0.06	3614	11,551	3.30	100.0	205.0	±5.4	228	±32	0.226	3.0	0.0323	2.7	0.890
T2 36.1	0.18	863	875	1.05	28.4	242.0	±6.4	229	±80	0.267	4.4	0.0382	2.7	0.618
T2 36.2	0.18	1770	5695	3.33	52.5	219.0	±5.7	202	±46	0.239	3.3	0.0345	2.7	0.800
T2 43.1	1.22	1945	2280	1.21	63.1	236.0	±6.2	319	±110	0.271	5.4	0.0373	2.7	0.498
T2-12.2	0.12	2483	3891	1.62	86.4	256.0	±5.3	321	±51	0.295	3.1	0.0405	2.1	0.689
T2-38.1	0.24	1975	3380	1.77	70.8	263.0	±5.6	200	±97	0.287	4.7	0.0416	2.2	0.459
T2-38.2	0.44	1758	1673	0.98	62.2	259.0	±5.4	174	±95	0.280	4.6	0.0410	2.1	0.466
T2-41.1	0.10	2535	5301	2.16	89.9	260.0	±5.4	234	±51	0.289	3.1	0.0412	2.1	0.693
T2-41.2	2.77	2467	3561	1.49	91.1	264.0	±5.6	96	±280	0.276	12.0	0.0418	2.2	0.178
T2-42.1	0.74	1396	1297	0.96	47.3	248.0	±5.3	250	±120	0.276	5.8	0.0391	2.2	0.377
T3 12,1	0.14	3331	5150	1.60	97.8	216.0	±5.7	236	±29	0.240	2.9	0.0341	2.7	0.906
T3 12.2	0.07	3584	6033	1.74	110.0	226.0	±5.9	228	±31	0.249	3.0	0.0356	2.7	0.894
T3-10.1	0.01	2906	4717	1.68	106.0	269.0	±6.7	301	±32	0.307	2.9	0.0426	2.5	0.877
T3-10.2	0.49	1518	1502	1.02	56.1	270.0	±6.8	220	±94	0.298	4.8	0.0428	2.6	0.535
T3-11.1	0.37	3281	5846	1.84	125.0	278.0	±6.8	196	±68	0.304	3.8	0.0441	2.5	0.652
T3-13.1	0.02	4941	10,580	2.21	192.0	285.0	±6.9	285	±22	0.324	2.7	0.0451	2.5	0.933
T3-20.1	0.44	2052	3674	1.85	76.5	273.0	±6.8	317	±93	0.314	4.8	0.0432	2.5	0.527
T3-22.1	2.29	1854	2842	1.58	68.6	266.0	±6.7	354	±150	0.311	7.2	0.0421	2.6	0.360
T3-22.2	0.18	2651	4925	1.92	93.9	260.0	±6.4	215	±52	0.286	3.4	0.0411	2.5	0.747
T3-26.1	0.13	2043	2431	1.23	76.6	275.0	±6.8	219	±47	0.304	3.2	0.0436	2.5	0.780
T3-27.1	0.19	3680	6810	1.91	143.0	285.0	±7.0	238	±47	0.318	3.2	0.0453	2.5	0.775
T3-27.2	0.01	5008	10,689	2.21	190.0	278.0	±6.8	278	±22	0.315	2.7	0.0441	2.5	0.934

(continued)

Table 6 (continued)

Crater	$^{206}Pb_c$, %	U, ppm	Th, ppm	$^{232}Th/^{238}U$	$^{206}Pb^*$, ppm	$^{206}Pb^*/^{238}U$ Age		$^{207}Pb^*/^{206}Pb^*$ Age		$^{207}Pb^*/^{235}U$	±%	$^{206}Pb^*/^{238}U$	±%	KK
T3-28.1	0.13	2199	5516	2.59	84.8	283.0	±7.0	226	±44	0.313	3.2	0.0448	2.5	0.797
T3-28.2	0.17	1935	4569	2.44	71.2	270.0	±6.7	273	±49	0.305	3.3	0.0428	2.5	0.766
T3-28.3	0.77	1717	5378	3.24	64.6	274.0	±6.9	391	±150	0.326	7.0	0.0435	2.6	0.368
T3-33.1	0.29	817	734	0.93	27.0	246.0	±5.0	312	±110	0.280	5.2	0.0390	2.1	0.401
T3-33.2	0.07	4020	5473	1.41	139.0	254.0	±5.0	222	±32	0.280	2.4	0.0400	2.0	0.819
T3-36.1	1.65	316	209	0.68	10.0	236.0	±6.4	–	–	0.230	18.0	0.0370	2.7	0.150
T3-36.2	0.19	2804	4354	1.60	90.0	236.0	±4.7	170	±46	0.250	2.8	0.0370	2.0	0.716
T3-38.1	1.47	159	58	0.38	5.0	230.0	±6.2	–	–	0.230	27.0	0.0360	2.7	0.101
T3-40.1	0.27	537	328	0.63	18.0	247.0	±5.3	364	±120	0.290	5.9	0.0390	2.2	0.367
T3-3.1	0.01	606	444	0.76	20.6	251.0	±7.2	206	±49	0.275	3.6	0.0397	2.9	0.811
T3-3.2	0.01	3477	1570	0.47	122.0	258.0	±7.2	254	±19	0.289	3.0	0.0408	2.9	0.960
T3-4.1	0.07	3868	5753	1.54	139.0	264.0	±6.5	283	±27	0.300	2.8	0.0418	2.5	0.904
T3-5.1	0.27	1453	1691	1.20	51.1	258.0	±6.4	252	±71	0.288	4.0	0.0408	2.5	0.634
T3-5.2	0.77	552	394	0.74	19.3	255.0	±6.9	213	±230	0.280	10.0	0.0404	2.7	0.262
T3-6.1	0.15	3520	4348	1.28	125.0	262.0	±6.5	198	±35	0.286	2.9	0.0414	2.5	0.857
T3-6.2	0.13	1826	2568	1.45	68.3	275.0	±6.8	224	±51	0.304	3.4	0.0435	2.5	0.752
T3-7.1	0.24	2857	4035	1.46	101.0	260.0	±6.4	201	±58	0.285	3.6	0.0412	2.5	0.706
T3-8.1	0.11	982	1381	1.45	32.4	242.0	±6.9	285	±43	0.275	3.5	0.0383	2.9	0.840
T3-8.2	0.23	2540	3041	1.24	88.0	254.0	±7.1	193	±50	0.277	3.6	0.0402	2.9	0.797
T5-10.1	0.02	18,348	34,670	1.95	635.0	255.0	±5.2	266	±16	0.286	2.2	0.0403	2.1	0.948
T5-10.2	0.20	1456	2243	1.59	48.9	247.0	±5.2	232	±65	0.273	3.5	0.0390	2.2	0.609
T5-11.1	0.28	5226	8258	1.63	192.0	269.0	±5.5	250	±55	0.301	3.2	0.0427	2.1	0.657
T5-4.1	0.35	5637	9717	1.78	204.0	266.0	±5.5	298	±44	0.303	2.8	0.0421	2.1	0.736
T5-8.1	0.01	3958	6463	1.69	143.0	265.0	±5.4	271	±28	0.299	2.4	0.0420	2.1	0.867
T5-9.1	0.43	1993	4014	2.08	70.2	258.0	±5.4	262	±74	0.290	3.9	0.0408	2.1	0.552
T6-11.1	0.21	2483	2462	1.02	85.0	251.0	±3.4	213	±53	0.276	2.7	0.0397	1.4	0.514
T6-11.2	0.01	4372	4234	1.00	158.0	265.0	±3.5	248	±22	0.297	1.7	0.0420	1.4	0.815
T6-12.1	0.09	2898	8124	2.90	100.0	254.0	±3.4	234	±40	0.282	2.2	0.0402	1.4	0.622

(continued)

Table 6 (continued)

Crater	$^{206}Pb_c$, %	U, ppm	Th, ppm	$^{232}Th/^{238}U$	$^{206}Pb^*$, ppm	$^{206}Pb^*/^{238}U$ Age		$^{207}Pb^*/^{206}Pb^*$ Age		$^{207}Pb^*/^{235}U$	±%	$^{206}Pb^*/^{238}U$	±%	KK
T6-14.1	0.01	2617	4193	1.66	90.6	255.0	±3.4	227	±31	0.282	1.9	0.0403	1.4	0.718
T6-2.1	0.01	7839	9236	1.22	287.0	269.0	±3.5	254	±17	0.301	1.5	0.0426	1.3	0.872
T6-5.1	0.01	10,300	8954	0.90	379.0	271.0	±3.5	263	±15	0.304	1.5	0.0429	1.3	0.897
T6-5.2	0.64	702	855	1.26	22.9	239.0	±3.7	158	±180	0.257	8.1	0.0378	1.6	0.196
T6-6.1	0.04	3788	4371	1.19	135.0	263.0	±3.5	259	±28	0.295	1.8	0.0416	1.4	0.749
T6-8.1	0.06	4711	11,564	2.54	160.0	250.0	±3.3	236	±32	0.277	1.9	0.0395	1.3	0.697
T6-9.1	0.02	8278	13,919	1.74	291.0	258.0	±3.4	242	±17	0.287	1.5	0.0409	1.3	0.869
T8-1.1	0.03	5784	9092	1.62	204.0	260.0	±3.4	255	±24	0.291	1.7	0.0411	1.3	0.786
T8-20.1	0.04	5365	8638	1.66	186.0	255.0	±3.4	255	±22	0.286	1.7	0.0404	1.3	0.813
T8-21.1	0.01	13,600	29,346	2.23	504.0	272.0	±3.5	252	±13	0.305	1.4	0.0432	1.3	0.922
T8-21.2	0.01	11,144	25,377	2.35	407.0	269.0	±3.5	238	±15	0.299	1.5	0.0425	1.3	0.895
T8-24.1	0.16	4267	5275	1.28	153.0	263.0	±3.5	225	±49	0.290	2.5	0.0416	1.4	0.545
T8-26.1	0.10	2501	4682	1.93	88.7	261.0	±7.3	239	±37	0.290	3.3	0.0413	2.9	0.872
T8-31.1	0.09	2994	4500	1.55	101.0	247.0	±3.3	228	±39	0.273	2.2	0.0391	1.4	0.632
T8-32.1	0.03	3436	5301	1.59	119.0	255.0	±3.4	253	±27	0.285	1.8	0.0404	1.4	0.756
T8-33.1	0.08	2472	3282	1.37	86.8	258.0	±3.5	226	±36	0.285	2.1	0.0408	1.4	0.670
T8-37.1	0.05	5071	7902	1.61	174.0	253.0	±3.4	275	±28	0.286	1.8	0.0400	1.4	0.746
T8-6.1	0.01	6390	11691	1.89	226.0	261.0	±3.4	270	±19	0.294	1.6	0.0412	1.3	0.853
T8-6.2	0.01	4447	6947	1.61	152.0	252.0	±3.3	272	±26	0.285	1.8	0.0399	1.4	0.764
T8-9.1	0.18	3071	4197	1.41	106.0	254.0	±3.4	213	±54	0.279	2.7	0.0402	1.4	0.504
T8 14.1	0.01	745	764	1.06	24.5	242.0	±6.4	283	±41	0.274	3.2	0.0383	2.7	0.835
T8 14.2	0.25	1553	2215	1.47	47.6	226.0	±6.0	191	±59	0.245	3.7	0.0356	2.7	0.730
T8 16.1	0.04	1320	1614	1.26	43.0	240.0	±6.3	255	±32	0.268	3.0	0.0379	2.7	0.886
T8 17.1	0.15	2169	5089	2.42	64.7	220.0	±5.8	215	±35	0.241	3.1	0.0347	2.7	0.869
T8-29.1	0.11	1750	1838	1.09	62.0	260.0	±3.6	272	±42	0.294	2.3	0.0412	1.4	0.604
T8_15.1	0.07	3000	5488	1.89	94.8	233.0	±5.6	221	±38	0.256	3.0	0.0368	2.5	0.831
T8_2.1	0.20	2556	4216	1.70	80.8	233.0	±5.6	176	±49	0.251	3.2	0.0367	2.5	0.761
T8_34.1	0.15	1652	4213	2.63	50.0	223.0	±5.5	262	±64	0.249	3.8	0.0351	2.5	0.667

(continued)

Table 6 (continued)

Crater	$^{206}Pb_c$, %	U, ppm	Th, ppm	$^{232}Th/^{238}U$	$^{206}Pb^*$, ppm	$^{206}Pb^*/^{238}U$ Age		$^{207}Pb^*/^{206}Pb^*$ Age		$^{207}Pb^*/^{235}U$	±%	$^{206}Pb^*/^{238}U$	±%	KK
T8_34.2	0.17	2564	2591	1.04	80.9	232.0	±5.6	159	±52	0.249	3.3	0.0367	2.5	0.745
T10-11.1	0.11	1928	3289	1.76	67.6	258.0	±5.4	226	±80	0.285	4.0	0.0408	2.1	0.526
T10-12.1	0.01	1192	2505	2.17	36.5	226.0	±1.3	285	±42	0.256	1.9	0.0357	0.6	0.307
T10-16.1	0.97	399	554	1.43	13.9	253.0	±6.1	278	±240	0.286	11.0	0.0401	2.4	0.223
T10-16.2	0.48	2363	8804	3.85	82.3	255.0	±5.3	172	±96	0.276	4.6	0.0404	2.1	0.458
T10-18.1	0.39	2647	6073	2.37	93.7	259.0	±5.4	240	±72	0.288	3.8	0.0410	2.1	0.559
T10-20.1	0.06	1092	1215	1.15	38.9	262.0	±5.6	319	±74	0.302	3.9	0.0415	2.2	0.553
T10-20.2	1.39	471	398	0.87	16.8	258.0	±6.4	141	±310	0.275	13.0	0.0409	2.5	0.191
T10-21.1	0.64	1124	2123	1.95	40.2	261.0	±5.7	414	±210	0.314	9.6	0.0414	2.2	0.232
T10-25.1	0.58	2260	3509	1.60	80.4	260.0	±5.6	181	±150	0.282	6.6	0.0412	2.2	0.328
T10-26.1	0.51	1370	2065	1.56	48.8	261.0	±5.6	108	±140	0.274	6.2	0.0412	2.2	0.351
T10-26.2	0.23	1404	2543	1.87	49.0	256.0	±5.4	339	±59	0.297	3.4	0.0405	2.2	0.638
T10-27.1	3.44	581	1391	2.47	20.0	245.0	±6.1	–	–	0.240	21.0	0.0387	2.5	0.123
T10-4.1	1.23	814	1317	1.67	27.7	248.0	±5.6	265	±230	0.278	10.0	0.0391	2.3	0.226
T10-5.1	0.68	1160	3739	3.33	41.2	259.0	±5.7	62	±170	0.267	7.4	0.0410	2.2	0.305
T10-5.2	0.56	1123	1816	1.67	39.0	254.0	±5.5	159	±130	0.273	5.9	0.0402	2.2	0.374
T10-6.1	0.17	1203	2596	2.23	41.4	253.0	±5.4	177	±84	0.274	4.2	0.0400	2.2	0.516
T10-6.2	0.16	1184	2720	2.37	40.5	251.0	±5.3	370	±62	0.296	3.5	0.0397	2.2	0.616
T10_19.1	0.29	1204	3659	3.14	37.0	226.0	±5.6	126	±94	0.239	4.7	0.0357	2.5	0.528
T10_24.1	0.09	889	2379	2.77	28.9	239.0	±5.9	190	±52	0.260	3.3	0.0378	2.5	0.748
T10_24.2	0.01	706	1542	2.26	22.7	237.0	±6.0	332	±51	0.274	3.4	0.0374	2.6	0.749
T10_29.1	0.01	357	463	1.34	11.1	231.0	±6.1	633	±150	0.306	7.5	0.0364	2.7	0.358
T10_3.1	0.43	1338	3802	2.94	39.1	215.0	±5.5	163	±90	0.231	4.7	0.0339	2.6	0.555
T10_3.2	0.01	1298	3341	2.66	44.3	251.0	±6.2	420	±51	0.302	3.4	0.0398	2.5	0.739
T10_31.1	0.09	650	1336	2.12	20.3	230.0	±5.7	277	±66	0.259	3.8	0.0363	2.5	0.663
T10-7.1	0.10	1200	4962	4.27	40.2	246.0	±8.6	263	±49	0.276	4.1	0.0389	3.5	0.858
T10 2.1	0.01	1381	2387	1.79	47.6	253.0	±6.6	258	±27	0.284	2.9	0.0401	2.7	0.913
T10 2.2	1.47	518	1300	2.59	17.5	246.0	±6.9	80	±230	0.255	10.0	0.0388	2.9	0.286

(continued)

Table 6 (continued)

Crater	$^{206}Pb_c$, %	U, ppm	Th, ppm	$^{232}Th/^{238}U$	$^{206}Pb^*$, ppm	$^{206}Pb^*/^{238}U$ Age		$^{207}Pb^*/^{206}Pb^*$ Age		$^{207}Pb^*/^{235}U$	±%	$^{206}Pb^*/^{238}U$	±%	KK
T12 6.1	1.95	222	326	1.52	8.2	265.0	±7.6	7	±300	0.267	13.0	0.0420	2.9	0.226
T12 6.2	0.04	785	1784	2.35	26.8	251.0	±6.6	235	±39	0.279	3.2	0.0397	2.7	0.848
T12-4.2	0.27	2007	4753	2.45	66.5	243.0	±3.5	195	±110	0.265	5.0	0.0385	1.5	0.299
T12-4.3	1.55	1583	4880	3.18	56.3	257.0	±4.2	267	±200	0.290	8.8	0.0407	1.7	0.190
T12-4.4	0.06	4704	17,426	3.83	167.0	260.0	±3.5	230	±26	0.288	1.8	0.0412	1.4	0.763
T12-4.5	0.66	4594	14,775	3.32	159.0	253.0	±3.4	273	±64	0.285	3.1	0.0400	1.4	0.443
T12-5.1	0.01	2102	5020	2.47	73.6	258.0	±3.5	286	±32	0.292	2.0	0.0408	1.4	0.697
T12-7.1	0.09	1565	2532	1.67	54.7	257.0	±3.5	269	±45	0.289	2.4	0.0407	1.4	0.585
T12-8(2).1	0.90	1082	3841	3.67	37.9	256.0	±3.7	138	±150	0.272	6.5	0.0405	1.5	0.228
T12-8.1	0.01	15,208	28,250	1.92	565.0	273.0	±3.5	263	±12	0.307	1.4	0.0433	1.3	0.933
T12-9.1	0.45	1989	2827	1.47	74.4	274.0	±4.8	142	±79	0.292	3.8	0.0434	1.8	0.470
T12-9.2	0.01	5328	14732	2.86	189.0	261.0	±3.4	289	±31	0.297	1.9	0.0414	1.3	0.700
T12-1.1	0.32	985	3454	3.62	33.8	252.0	±3.7	278	±97	0.285	4.5	0.0398	1.5	0.335
T12-1.2	0.01	3215	11,771	3.78	109.0	249.0	±3.4	230	±27	0.276	1.8	0.0394	1.4	0.759
T12-10.1	0.13	4276	8694	2.10	151.0	259.0	±3.4	268	±53	0.291	2.7	0.0409	1.3	0.504
T12-13.1	0.16	1952	4252	2.25	68.8	259.0	±3.5	238	±56	0.288	2.8	0.0410	1.4	0.498
T12-17.1	0.55	1603	8592	5.54	55.6	254.0	±3.5	211	±92	0.279	4.2	0.0402	1.4	0.337
T12-2.1	0.01	2471	6234	2.61	85.8	256.0	±3.4	270	±30	0.288	1.9	0.0404	1.4	0.722
T12-2.2	0.15	1523	4774	3.24	53.2	257.0	±3.6	258	±65	0.288	3.2	0.0406	1.4	0.445
T12-23.1	0.14	3844	8253	2.22	135.0	259.0	±3.4	279	±42	0.293	2.3	0.0409	1.4	0.599
T12-4.1	0.01	5418	20744	3.96	197.0	267.0	±3.5	248	±22	0.298	1.6	0.0422	1.3	0.819
T12_11.1	0.01	1526	3603	2.44	47.0	227.0	±5.6	336	±60	0.263	3.6	0.0359	2.5	0.690
T12_12.1	0.34	1226	3169	2.67	41.6	249.0	±6.1	156	±80	0.267	4.2	0.0394	2.5	0.591
T12_12.2	0.04	2627	3797	1.49	89.8	252.0	±6.5	222	±32	0.277	3.0	0.0398	2.6	0.883
T12_19.1	0.10	1313	3554	2.80	44.6	250.0	±6.1	262	±51	0.280	3.3	0.0395	2.5	0.746
T12_19.2	0.14	1152	1945	1.74	35.7	228.0	±5.6	184	±57	0.247	3.5	0.0360	2.5	0.714
T12_24.1	0.11	2036	4026	2.04	67.0	242.0	±5.9	183	±53	0.263	3.4	0.0383	2.5	0.737
T12_25.1	0.15	1200	2980	2.57	36.9	227.0	±5.6	224	±69	0.250	3.9	0.0358	2.5	0.643

(continued)

Table 6 (continued)

Crater	$^{206}Pb_c$, %	U, ppm	Th, ppm	$^{232}Th/^{238}U$	$^{206}Pb^*$, ppm	$^{206}Pb^*/^{238}U$ Age		$^{207}Pb^*/^{206}Pb^*$ Age		$^{207}Pb^*/^{235}U$	±%	$^{206}Pb^*/^{238}U$	±%	KK
T13-1.1	0.09	2914	3613	1.28	102.0	258.0	±1.3	267	±36	0.290	1.6	0.0408	0.5	0.326
T13-1.2	0.02	12,888	37,065	2.97	488.0	278.0	±1.0	283	±14	0.316	0.7	0.0441	0.4	0.512
T13-11.1	0.05	2528	4276	1.75	89.8	261.0	±1.2	266	±35	0.295	1.6	0.0413	0.5	0.300
T13-11.2	0.13	1738	2566	1.53	60.7	257.0	±1.4	219	±44	0.284	2.0	0.0406	0.6	0.283
T13-12.1	0.26	1644	3611	2.27	58.7	262.0	±1.5	254	±57	0.294	2.5	0.0414	0.6	0.228
T13-16.1	0.12	2593	7002	2.79	92.0	261.0	±1.2	283	±35	0.296	1.6	0.0413	0.5	0.299
T13-2.1	0.19	1179	3104	2.72	43.1	268.0	±1.7	269	±61	0.302	2.8	0.0425	0.7	0.237
T13-2.2	0.16	968	2189	2.34	34.5	262.0	±1.8	193	±77	0.285	3.4	0.0414	0.7	0.210
T13-3.1	0.03	5668	17,041	3.11	204.0	264.0	±1.0	265	±23	0.297	1.1	0.0418	0.4	0.356
T13-4.1	0.01	2529	5725	2.34	90.8	264.0	±1.3	300	±57	0.302	2.6	0.0419	0.5	0.195
T13-6.1	0.04	7439	16,000	2.22	276.0	272.0	±1.0	262	±21	0.306	1.0	0.0431	0.4	0.362
T13-6.2	0.63	2915	7089	2.51	104.0	261.0	±5.4	282	±110	0.296	5.4	0.0414	2.1	0.391
T13-6.3	0.22	2725	4854	1.84	97.1	261.0	±1.2	222	±44	0.289	1.9	0.0414	0.5	0.248
T13-7.1	0.68	266	315	1.22	8.7	240.0	±7.1	267	±160	0.269	7.5	0.0379	3.0	0.401
T13-10.1	0.16	1252	5462	4.51	42.3	248.0	±6.2	254	±62	0.277	3.7	0.0392	2.5	0.687
T13-10.2	0.43	683	1487	2.25	24.8	266.0	±7.0	326	±120	0.307	6.1	0.0421	2.7	0.441
T13-5.1	2.16	619	1757	2.93	23.5	272.0	±7.2	719	±180	0.377	8.7	0.0431	2.7	0.309
T13-5.2	1.04	682	1673	2.53	24.0	256.0	±6.8	39	±190	0.262	8.5	0.0406	2.7	0.317
T14-1.2	0.38	1501	345	0.24	31.3	154.0	±4.0	152	±110	0.164	5.3	0.0242	2.6	0.494
T15-1.1	0.02	2453	10,347	4.36	87.2	261.0	±1.2	272	±31	0.295	1.4	0.0414	0.5	0.342
T15-4.1	0.05	3565	14,086	4.08	127.0	262.0	±1.1	262	±32	0.294	1.4	0.0414	0.4	0.305
T16-1.1	0.01	1071	1813	1.75	37.6	258.0	±2.0	271	±47	0.291	2.2	0.0409	0.8	0.357
T16-3.1	0.01	8588	19,953	2.40	320.0	274.0	±0.9	238	±21	0.305	1.0	0.0434	0.4	0.368
T16-6.1	0.26	1391	2667	1.98	49.3	260.0	±1.7	269	±82	0.293	3.6	0.0412	0.7	0.179
T16-6.2	2.61	1667	4163	2.58	60.7	261.0	±2.0	407	±190	0.312	8.4	0.0413	0.8	0.094
T17-4.1	0.60	3733	6472	1.79	136.0	265.0	±1.2	294	±63	0.302	2.8	0.0420	0.5	0.164
T17-1.1	4.85	75	89	1.22	2.6	238.0	±11.0	–	–	0.210	62.0	0.0376	4.6	0.074
T17-1.2	4.06	119	187	1.63	4.2	251.0	±9.7	–	–	0.230	50.0	0.0397	3.9	0.078

(continued)

Table 6 (continued)

Crater	$^{206}Pb_c$, %	U, ppm	Th, ppm	$^{232}Th/^{238}U$	$^{206}Pb^*$, ppm	$^{206}Pb^*/^{238}U$ Age		$^{207}Pb^*/^{206}Pb^*$ Age		$^{207}Pb^*/^{235}U$	±%	$^{206}Pb^*/^{238}U$	±%	KK
T17-2.1	0.01	804	1156	1.49	28.4	260.0	±6.6	431	±78	0.315	4.4	0.0412	2.6	0.591
T17-2.2	0.34	1607	3013	1.94	57.8	263.0	±6.5	260	±62	0.296	3.7	0.0417	2.5	0.683
T17-2.3	12.03	931	4172	4.63	42.0	291.0	±8.7	1487	±470	0.590	25.0	0.0461	3.1	0.122
T18-7.1	0.01	4783	6857	1.48	184.0	282.0	±1.3	251	±21	0.316	1.0	0.0447	0.5	0.442
T18-7.2	0.27	2272	1582	0.72	79.2	256.0	±1.3	243	±72	0.285	3.2	0.0405	0.5	0.167
T18-7.3	0.18	4707	5807	1.27	179.0	279.0	±1.3	286	±41	0.317	1.9	0.0442	0.5	0.254
T18-9.1	0.11	6457	8295	1.33	226.0	257.0	±1.0	277	±24	0.291	1.1	0.0408	0.4	0.345
T18-11.1	0.18	837	751	0.93	29.0	255.0	±1.6	251	±77	0.285	3.4	0.0403	0.7	0.193
T18-11.2	0.36	2292	3350	1.51	73.7	236.0	±1.1	211	±58	0.259	2.6	0.0373	0.5	0.185
T18-22.1	0.03	4946	8098	1.69	151.0	225.0	±0.8	281	±17	0.254	0.8	0.0355	0.4	0.415
T18-22.2	0.10	1848	1433	0.80	59.6	237.0	±1.1	265	±38	0.267	1.7	0.0375	0.5	0.270
T18-8	0.02	1800	1415	0.81	51.2	210.0	±1.0	332	±29	0.242	1.4	0.0331	0.5	0.364
T18-3.1	0.05	9653	49,226	5.27	358.0	272.0	±6.6	270	±25	0.307	2.7	0.0431	2.5	0.919
T18-3.2	0.02	4684	15,931	3.51	179.0	281.0	±6.9	269	±24	0.317	2.7	0.0446	2.5	0.925
T18-3.3	0.13	3153	4410	1.45	124.0	289.0	±7.2	305	±57	0.331	3.6	0.0458	2.5	0.712
T18-4.1	0.07	3947	5869	1.54	142.0	265.0	±6.5	338	±32	0.308	2.9	0.0420	2.5	0.869
T18-5.1	0.01	4270	8858	2.14	166.0	285.0	±7.0	337	±22	0.332	2.7	0.0452	2.5	0.929
T18-6.1	0.33	6362	16,283	2.64	252.0	290.0	±7.1	301	±54	0.332	3.4	0.0460	2.5	0.729
T22-1 1	0.19	687	1588	2.39	22.0	237.0	±4.9	250	±90	0.260	4.5	0.0370	2.1	0.474
T22-2 1	0.87	868	2471	2.94	29.0	241.0	±5.0	–	–	0.240	7.8	0.0380	2.1	0.273
T22-4 1	0.01	1285	2498	2.01	43.0	245.0	±5.0	293	±49	0.280	3.0	0.0390	2.1	0.698
844 (n = 45) Kharaelakh intrusive														
84410,11Z42.1	8.82	507	1665	3.39	20.0	264.0	±9.5	1390	±520	0.500	27.0	0.0420	3.7	0.135
84410,11Z42.2	1.22	359	935	2.69	11.7	237.0	±7.7	200	±180	0.300	8.2	0.0380	3.3	0.400
8441Z1.1	0.04	1244	3810	3.16	41.4	245.0	±7.7	296	±26	0.300	3.4	0.0390	3.2	0.940
8441Z10.1	0.09	393	411	1.08	18.2	338.0	±11.0	413	±66	0.400	4.4	0.0540	3.3	0.747
8441Z10.2	0.03	1552	3519	2.34	56.0	265.0	±8.3	222	±25	0.300	3.4	0.0420	3.2	0.949
8441Z12.1	0.06	1570	4651	3.06	52.4	246.0	±7.7	260	±30	0.300	3.5	0.0390	3.2	0.924

(continued)

Table 6 (continued)

Crater	$^{206}Pb_c$, %	U, ppm	Th, ppm	$^{232}Th/^{238}U$	$^{206}Pb^*$, ppm	$^{206}Pb^*/^{238}U$ Age		$^{207}Pb^*/^{206}Pb^*$ Age		$^{207}Pb^*/^{235}U$	±%	$^{206}Pb^*/^{238}U$	±%	KK
8441Z14.1	0.01	1025	3237	3.26	34.6	249.0	±7.8	259	±27	0.300	3.4	0.0390	3.2	0.940
8441Z16.1	0.11	952	2765	3.00	33.1	256.0	±8.0	247	±44	0.300	3.7	0.0410	3.2	0.857
8441Z19.1	0.33	805	1758	2.26	29.4	267.0	±8.4	229	±61	0.300	4.2	0.0420	3.2	0.775
8441Z2.1	0.01	1681	4580	2.82	57.0	250.0	±8.0	256	±21	0.300	3.4	0.0400	3.3	0.962
8441Z2.2	0.12	2023	9510	4.86	69.5	252.0	±7.9	233	±32	0.300	3.5	0.0400	3.2	0.917
8441Z20.1	0.11	1230	3684	3.09	41.7	249.0	±7.8	258	±42	0.300	3.7	0.0390	3.2	0.870
8441Z22.1	0.04	3571	14,600	4.09	114.0	235.0	±7.3	266	±20	0.300	3.3	0.0370	3.2	0.965
8441Z24.1	0.19	870	2679	3.18	28.4	240.0	±7.5	201	±46	0.300	3.8	0.0380	3.2	0.850
8441Z26.1	0.10	856	2849	3.44	27.6	237.0	±7.5	273	±36	0.300	3.6	0.0380	3.2	0.898
8441Z3.1	0.04	1780	3424	1.99	59.5	246.0	±7.7	238	±26	0.300	3.4	0.0390	3.2	0.944
8441Z3.2	0.05	1670	3586	2.22	55.5	245.0	±7.7	272	±26	0.300	3.4	0.0390	3.2	0.942
8441Z6.1	0.03	3525	7134	2.09	116.0	242.0	±7.6	269	±18	0.300	3.3	0.0380	3.2	0.971
8441Z7.1	0.01	1834	2306	1.30	57.1	229.0	±7.2	259	±22	0.300	3.3	0.0360	3.2	0.958
8441Z9.1	0.12	3092	4740	1.58	106.0	251.0	±7.9	250	±25	0.300	3.4	0.0400	3.2	0.948
8446Z31.1	0.01	370	147	0.41	12.4	247.0	±7.8	224	±49	0.300	3.9	0.0390	3.2	0.834
8446Z31.2	0.52	6389	1588	0.26	240.0	274.0	±8.7	224	±46	0.300	3.8	0.0440	3.2	0.851
8446Z31.3	0.09	9409	4039	0.44	334.0	261.0	±8.1	263	±15	0.300	3.3	0.0410	3.2	0.979
8446Z33.1	0.01	2719	3512	1.33	89.7	243.0	±7.6	232	±18	0.300	3.3	0.0380	3.2	0.971
8446Z33.2	0.03	1613	1675	1.07	51.7	236.0	±8.0	289	±24	0.300	3.6	0.0370	3.5	0.958
8446Z35.1	0.12	358	523	1.51	12.1	248.0	±8.0	286	±63	0.300	4.3	0.0390	3.3	0.764
8447Z37.1	0.01	689	1252	1.88	25.3	270.0	±8.5	271	±37	0.300	3.6	0.0430	3.2	0.894
8447Z37.2	1.67	427	509	1.23	21.1	355.0	±12.0	434	±210	0.400	10.0	0.0570	3.5	0.345
8447Z37.3	0.62	713	857	1.24	29.8	304.0	±9.6	93	±92	0.300	5.0	0.0480	3.2	0.640
8447Z37.4	2.11	773	978	1.31	30.6	284.0	±9.3	–	–	0.300	12.0	0.0450	3.4	0.281
844-1.13.1	0.08	1710	7804	4.72	59.4	255.0	±1.4	237	±43	0.300	1.9	0.0400	0.6	0.292
844-1.15.1	0.17	858	3671	4.42	29.6	253.0	±1.9	307	±80	0.300	3.6	0.0400	0.8	0.212
844-6.9.1	0.30	1885	1775	0.97	65.5	255.0	±1.5	239	±60	0.300	2.7	0.0400	0.6	0.222
844-6.9.2	0.05	3906	3261	0.86	135.0	254.0	±1.1	263	±31	0.300	1.4	0.0400	0.4	0.304

(continued)

Table 6 (continued)

Crater	$^{206}Pb_c$, %	U, ppm	Th, ppm	$^{232}Th/^{238}U$	$^{206}Pb^*$, ppm	$^{206}Pb^*/^{238}U$ Age		$^{207}Pb^*/^{206}Pb^*$ Age		$^{207}Pb^*/^{235}U$	±%	$^{206}Pb^*/^{238}U$	±%	KK
844-10,11.40.1	0.42	470	663	1.46	16.0	249.0	±6.6	291	±76	0.300	4.3	0.0390	2.7	0.630
844-10,11.40.2	0.43	622	2086	3.46	21.2	250.0	±6.7	203	±90	0.300	4.7	0.0400	2.7	0.573
844-10,11.41.1	0.01	260	503	2.00	8.8	249.0	±6.9	292	±69	0.300	4.1	0.0390	2.8	0.682
844-6.28.1	0.21	491	707	1.49	17.2	257.0	±6.8	299	±69	0.300	4.1	0.0410	2.7	0.667
844-6.28.2	0.01	1651	5434	3.40	52.3	233.0	±6.1	226	±28	0.300	2.9	0.0370	2.7	0.911
844-6.28.3	1.31	187	124	0.69	6.6	256.0	±7.4	357	±280	0.300	13.0	0.0410	2.9	0.231
844-6.30.1	0.62	195	246	1.30	6.7	250.0	±7.1	93	±210	0.300	9.3	0.0400	2.8	0.304
844-6.30.2	0.92	129	131	1.05	4.4	246.0	±7.1	119	±240	0.300	11.0	0.0390	2.9	0.278
844-6.30.3	0.06	534	964	1.87	17.3	238.0	±6.5	229	±69	0.300	4.1	0.0380	2.8	0.684
844-6.32.1	0.14	6774	5711	0.87	207.0	225.0	±5.9	223	±24	0.200	2.9	0.0360	2.7	0.930
844-6.32.2	0.16	3458	4908	1.47	104.0	221.0	±5.8	248	±28	0.200	2.9	0.0350	2.7	0.908
31 (n = 112) Lower Talnakh intrusive														
31-1 11.1	0.17	5094	3197	0.65	155.0	223.3	±5.4	222	±37	0.246	2.9	0.0350	2.5	0.842
31-1 17.1	0.24	3375	4508	1.38	102.0	223.1	±5.4	152	±47	0.238	3.2	0.0350	2.5	0.777
31-1 2.1	0.23	7016	8851	1.30	188.0	197.4	±4.8	211	±32	0.216	2.8	0.0310	2.4	0.874
31-1 2.2	0.16	4018	4357	1.12	114.0	209.4	±5.1	240	±42	0.232	3.1	0.0330	2.5	0.805
31-1 20.1	0.12	4095	10,407	2.63	118.0	211.9	±5.2	270	±43	0.238	3.1	0.0330	2.5	0.800
31-1 4.1	0.46	4179	5640	1.39	115.0	202.3	±4.9	367	±89	0.237	4.7	0.0320	2.5	0.527
31-1 8.1	0.12	6374	10,719	1.74	190.0	220.1	±5.3	196	±27	0.240	2.7	0.0350	2.5	0.903
31-1 9.1	0.25	4341	5528	1.32	140.0	236.2	±5.7	166	±47	0.254	3.2	0.0370	2.5	0.773
31-1 9.2	0.21	5036	7362	1.51	146.0	213.6	±5.2	215	±47	0.234	3.2	0.0340	2.5	0.772
31-3 1.1	0.22	1825	2918	1.65	55.6	224.1	±1.0	220	±51	0.246	2.3	0.0350	2.5	0.205
31-3 1.2	0.01	3328	6527	2.03	111.0	245.3	±0.9	297	±21	0.280	1.0	0.0390	2.4	0.373
31-3 2.1	0.01	1621	2171	1.38	49.8	226.5	±5.5	291	±35	0.257	2.9	0.0360	2.5	0.850
31-7 1.1	0.01	9673	4887	0.52	386.0	293.0	±8.1	268	±11	0.330	2.9	0.0460	2.8	0.987
31-7 2.1	0.13	3858	4608	1.23	130.0	248.0	±7.0	246	±31	0.276	3.2	0.0390	2.9	0.903
31-7 4.1	0.13	1818	1367	0.78	60.2	243.3	±5.9	303	±48	0.278	3.2	0.0380	2.5	0.763
31-7 5.1	0.07	7244	3986	0.57	227.0	230.5	±5.6	242	±22	0.256	2.6	0.0360	2.5	0.930

(continued)

Table 6 (continued)

Crater	$^{206}Pb_c$, %	U, ppm	Th, ppm	$^{232}Th/^{238}U$	$^{206}Pb^*$, ppm	$^{206}Pb^*/^{238}U$ Age		$^{207}Pb^*/^{206}Pb^*$ Age		$^{207}Pb^*/^{235}U$	±%	$^{206}Pb^*/^{238}U$	±%	KK
31-7 5.2	0.08	3558	3956	1.15	111.0	229.2	±5.6	233	±30	0.254	2.8	0.0360	2.5	0.884
31-7 6.1	0.36	1669	1151	0.71	53.1	233.7	±5.7	173	±68	0.252	3.8	0.0370	2.5	0.646
31-7 9.1	0.33	3292	2198	0.69	103.0	228.8	±5.6	116	±64	0.241	3.7	0.0360	2.5	0.672
31-7 9.2	0.07	4323	2147	0.51	139.0	237.3	±5.7	207	±29	0.260	2.8	0.0370	2.5	0.891
31-9 1.1	0.26	379	285	0.78	11.4	221.3	±5.9	464	±87	0.271	4.8	0.0350	2.7	0.564
31-9 1.2	0.02	4188	4148	1.02	123.0	217.0	±5.3	260	±25	0.243	2.7	0.0340	2.5	0.915
31-9 2.1	0.24	3679	3018	0.85	109.0	217.6	±5.3	183	±47	0.236	3.2	0.0340	2.5	0.775
31-9 3.1	0.29	1487	2265	1.57	45.3	223.9	±5.5	185	±69	0.243	3.9	0.0350	2.5	0.640
31-9 4.1	0.44	1305	2593	2.05	39.5	222.5	±5.4	145	±77	0.237	4.1	0.0350	2.5	0.605
31-10 1.1	0.12	1460	1195	0.85	47.2	238.0	±5.8	289	±54	0.270	3.4	0.0380	2.5	0.726
31-10 4.1	0.45	1819	3558	2.02	54.9	221.8	±5.4	116	±70	0.233	3.9	0.0350	2.5	0.645
31-11 1.1	0.11	2140	3430	1.66	63.2	217.6	±5.3	218	±46	0.239	3.2	0.0340	2.5	0.780
31-11 2.1	0.31	1303	2547	2.02	41.1	232.0	±5.7	150	±79	0.248	4.2	0.0370	2.5	0.599
31-13 10.1	0.26	1926	4991	2.68	64.0	245.0	±6.9	204	±50	0.268	3.6	0.0390	2.9	0.800
31-13 11.1	0.09	1817	1838	1.05	58.0	234.0	±6.6	231	±40	0.259	3.4	0.0370	2.9	0.859
31-13 11.2	0.09	1422	1058	0.77	46.0	239.0	±6.7	217	±40	0.263	3.4	0.0380	2.9	0.856
31-13 12.1	0.18	1214	2958	2.52	37.0	225.0	±6.4	307	±54	0.257	3.7	0.0360	2.9	0.771
31-13 14.1	0.09	2107	4457	2.19	71.1	248.0	±7.5	298	±36	0.283	3.5	0.0390	3.1	0.891
31-13 14.2	0.16	2539	4042	1.65	86.4	250.1	±7.6	237	±38	0.278	3.5	0.0400	3.1	0.881
31-13 19.1	0.21	4562	6712	1.52	160.0	256.7	±7.8	211	±32	0.282	3.4	0.0410	3.1	0.912
31-13 2.1	0.51	1718	4351	2.62	62.6	266.5	±8.1	247	±71	0.298	4.4	0.0420	3.1	0.710
31-13 20.1	0.44	968	1510	1.61	34.5	260.8	±8.0	159	±89	0.280	4.9	0.0410	3.1	0.635
31-13 21.1	0.15	2520	7789	3.19	75.0	220.0	±6.2	250	±36	0.245	3.3	0.0350	2.9	0.880
31-13 23.1	0.36	1726	2513	1.51	60.9	258.8	±7.9	126	±75	0.274	4.4	0.0410	3.1	0.699
31-13 25.1	0.21	3901	5640	1.49	130.0	244.8	±7.5	238	±50	0.272	3.8	0.0390	3.1	0.821
31-13 27.1	0.60	1976	4367	2.28	60.5	224.3	±6.9	157	±95	0.240	5.1	0.0350	3.1	0.610
31-13 27.2	0.01	10,722	38,408	3.70	380.0	260.4	±7.9	256	±16	0.292	3.2	0.0410	3.1	0.976
31-13 28.1	0.01	1891	1969	1.08	69.0	269.0	±7.6	254	±27	0.302	3.1	0.0430	2.9	0.926

(continued)

Table 6 (continued)

Crater	$^{206}Pb_c$, %	U, ppm	Th, ppm	$^{232}Th/^{238}U$	$^{206}Pb^*$, ppm	$^{206}Pb^*/^{238}U$ Age		$^{207}Pb^*/^{206}Pb^*$ Age		$^{207}Pb^*/^{235}U$	±%	$^{206}Pb^*/^{238}U$	±%	KK
31-13 28.2	0.66	823	1213	1.52	28.0	245.0	±7.0	290	±110	0.278	5.6	0.0390	2.9	0.519
31-13 34.1	0.06	4435	4266	0.99	145.0	239.9	±7.3	239	±28	0.266	3.3	0.0380	3.1	0.929
31-13 35.1	0.38	2143	3538	1.71	73.4	250.9	±7.6	212	±60	0.276	4.0	0.0400	3.1	0.768
31-13 38.1	0.25	2778	6044	2.25	95.4	252.0	±7.6	231	±51	0.279	3.8	0.0400	3.1	0.815
31-13 38.2	0.01	2164	4132	1.97	76.2	259.3	±7.9	311	±30	0.298	3.4	0.0410	3.1	0.923
31-13 4.2	0.17	616	1726	2.90	22.9	273.1	±8.4	232	±79	0.303	4.7	0.0430	3.1	0.675
31-13 42.1	0.03	1613	2276	1.46	53.0	242.0	±6.8	272	±31	0.272	3.2	0.0380	2.9	0.907
31-13 44.1	0.11	3196	4413	1.43	111.0	255.7	±7.7	266	±30	0.288	3.3	0.0410	3.1	0.922
31-13 46.1	0.01	5965	8346	1.45	208.0	256.7	±7.7	275	±19	0.290	3.2	0.0410	3.1	0.965
31-13 46.2	0.17	4047	5881	1.50	140.0	254.0	±7.7	225	±32	0.281	3.4	0.0400	3.1	0.913
31-13 48.1	0.26	1923	1347	0.72	72.4	275.9	±8.4	223	±58	0.305	4.0	0.0440	3.1	0.779
31-13 5.1	0.04	1635	4407	2.79	53.0	240.0	±7.0	246	±30	0.267	3.2	0.0380	3.0	0.915
31-13 50.1	0.01	2601	3884	1.54	90.2	255.0	±7.7	302	±27	0.292	3.3	0.0400	3.1	0.933
31-13 50.2	0.04	1826	2712	1.53	65.2	262.3	±8.0	301	±35	0.300	3.5	0.0420	3.1	0.896
31-13 53.1	0.01	3008	4469	1.53	100.0	245.4	±7.4	313	±25	0.282	3.3	0.0390	3.1	0.942
31-13 54.1	1.44	1367	1377	1.04	49.9	264.6	±8.2	164	±140	0.285	6.9	0.0420	3.2	0.457
31-13 54.2	0.16	3521	4281	1.26	121.0	252.3	±7.6	248	±32	0.282	3.4	0.0400	3.1	0.912
31-13 54.3	0.13	2875	6268	2.25	104.0	266.3	±8.1	279	±43	0.302	3.6	0.0420	3.1	0.854
31-13 56.1	0.12	3780	4256	1.16	130.0	253.3	±7.7	240	±36	0.282	3.5	0.0400	3.1	0.895
31-13 6.1	0.66	794	1393	1.81	28.6	263.2	±8.1	158	±110	0.283	5.8	0.0420	3.1	0.543
31-13 7.1	0.13	1988	8244	4.29	67.9	250.9	±7.6	247	±51	0.280	3.8	0.0400	3.1	0.815
31-13 8.1	0.23	3104	7453	2.48	109.0	257.8	±7.8	126	±49	0.273	3.7	0.0410	3.1	0.828
31-13 8.2	1.02	887	1725	2.01	37.3	304.7	±9.5	93	±220	0.320	9.7	0.0480	3.2	0.331
31-13 9.1	0.01	1174	984	0.87	40.5	254.2	±8.1	359	±39	0.298	3.7	0.0400	3.2	0.883
31-13 9.2	0.55	418	753	1.86	16.6	288.9	±9.0	460	±94	0.355	5.3	0.0460	3.2	0.600
31-16.13.1	0.11	1799	3222	1.85	65.2	265.9	±6.1	279	±36	0.301	2.8	0.0420	2.3	0.830
31-16.17.1	0.15	970	1668	1.78	32.5	246.0	±6.5	362	±64	0.288	3.9	0.0390	2.7	0.685
31-16.18.1	0.64	1103	1651	1.55	37.5	248.5	±5.8	34	±150	0.253	6.8	0.0390	2.4	0.349

(continued)

Table 6 (continued)

Crater	$^{206}Pb_c$, %	U, ppm	Th, ppm	$^{232}Th/^{238}U$	$^{206}Pb^*$, ppm	$^{206}Pb^*/^{238}U$ Age		$^{207}Pb^*/^{206}Pb^*$ Age		$^{207}Pb^*/^{235}U$	±%	$^{206}Pb^*/^{238}U$	±%	KK
31-16.19.1	0.31	1016	2775	2.82	33.8	244.3	±5.7	206	±73	0.268	3.9	0.0390	2.4	0.607
31-16.19.2	0.46	791	2630	3.44	26.0	240.8	±5.8	200	±120	0.263	5.7	0.0380	2.4	0.430
31-16.2.1	0.14	1117	2413	2.23	38.9	255.7	±5.9	224	±55	0.283	3.3	0.0400	2.4	0.708
31-16.29.1	0.23	678	1599	2.44	21.4	231.7	±5.5	183	±84	0.251	4.3	0.0370	2.4	0.554
31-16.39.1	4.35	750	1244	1.71	27.1	254.2	±6.2	–	–	0.229	14.0	0.0400	2.5	0.180
31-16.39.2	0.85	311	402	1.34	9.1	213.2	±5.6	125	±340	0.225	15.0	0.0340	2.7	0.180
31-16.43.1	0.08	1790	2992	1.73	61.7	253.4	±5.8	213	±41	0.279	2.9	0.0400	2.3	0.796
31-16.44.1	0.13	1507	3369	2.31	53.9	262.5	±6.1	311	±47	0.301	3.1	0.0420	2.4	0.753
31-16.5.1	0.07	838	896	1.10	31.2	273.0	±6.4	383	±49	0.324	3.2	0.0430	2.4	0.742
31-16.51.1	0.10	1419	3479	2.53	44.8	232.2	±5.4	258	±43	0.260	3.0	0.0370	2.4	0.787
31-16.52.1	0.09	1399	4110	3.04	47.6	250.4	±5.8	295	±47	0.285	3.1	0.0400	2.4	0.753
31-16.55.1	1.29	569	1890	3.43	19.0	243.2	±5.9	91	±230	0.254	10.0	0.0380	2.5	0.245
31-16.6.1	0.01	590	2230	3.90	21.5	268.0	±6.4	492	±62	0.334	3.7	0.0420	2.5	0.657
31-16.61.1	0.09	3848	9738	2.61	119.0	227.8	±5.2	259	±31	0.255	2.7	0.0360	2.3	0.864
31-16.65.1	0.21	287	984	3.55	9.87	252.8	±6.2	431	±93	0.306	4.9	0.0400	2.5	0.512
31-16.71.1	0.48	1505	3331	2.29	48.0	233.7	±5.4	156	±68	0.250	3.7	0.0370	2.4	0.633
31-16.72.1	0.66	1004	1140	1.17	34.3	249.6	±5.9	154	±130	0.267	6.1	0.0390	2.4	0.395
31-16.72.2	0.06	3345	3892	1.20	106.0	232.8	±5.3	304	±29	0.266	2.7	0.0370	2.3	0.876
31-16.73.1	3.78	1244	1739	1.44	44.8	254.7	±6.2	349	±430	0.297	19.0	0.0400	2.5	0.130
31-16.74.1	0.07	6192	2920	0.49	194.0	230.4	±5.3	269	±22	0.259	2.5	0.0360	2.3	0.926
31-16.81.1	1.18	745	7380	10.24	24.1	235.6	±5.7	–	–	0.235	9.6	0.0370	2.4	0.254
31-16.81.2	1.07	770	8746	11.74	22.9	217.3	±5.2	–	–	0.215	7.4	0.0340	2.5	0.331
31-16.9.1	1.09	879	1185	1.39	32.7	270.6	±6.4	–	–	0.260	7.0	0.0430	2.4	0.346
31-16 10.1	0.02	3918	5539	1.5	138.0	260.0	±14.0	238	±18	0.289	5.7	0.0410	5.6	0.991
31-16 20.1	0.15	1808	2915	1.7	56.6	230.0	±13.0	242	±37	0.256	5.9	0.0360	5.7	0.963
31-16 20.2	0.07	393	359	0.9	10.6	198.0	±11.0	273	±60	0.223	6.3	0.0310	5.7	0.908
31-16 22.1	0.22	323	355	1.1	11.6	264.0	±15.0	331	±93	0.305	7.0	0.0420	5.7	0.811
31-16 22.2	0.59	145	130	0.9	5.3	267.0	±15.0	274	±210	0.301	11.0	0.0420	5.8	0.534

(continued)

Table 6 (continued)

Crater	$^{206}Pb_c$, %	U, ppm	Th, ppm	$^{232}Th/^{238}U$	$^{206}Pb^*$, ppm	$^{206}Pb^*/^{238}U$ Age		$^{207}Pb^*/^{206}Pb^*$ Age		$^{207}Pb^*/^{235}U$	±%	$^{206}Pb^*/^{238}U$	±%	KK
31-16 25.1	0.01	316	338	1.1	12.7	295.0	±16.0	357	±66	0.347	6.4	0.0470	5.7	0.889
31-16 26.1	0.07	1871	7127	3.9	58.6	231.0	±13.0	197	±36	0.252	5.9	0.0370	5.7	0.965
31-16 26.2	0.16	1430	1890	1.4	52.5	270.0	±15.0	223	±34	0.298	5.8	0.0430	5.7	0.968
31-16 31.1	0.10	1239	2737	2.3	44.2	262.0	±15.0	226	±50	0.290	6.1	0.0420	5.7	0.934
31-16 37.1	0.06	2477	4390	1.8	91.3	271.0	±15.0	238	±28	0.301	5.8	0.0430	5.7	0.977
31-16 38.1	0.01	1375	1902	1.4	48.8	261.0	±14.0	261	±37	0.293	5.9	0.0410	5.7	0.963
31-16 41.1	0.12	2231	4377	2.0	72.5	239.0	±13.0	243	±31	0.266	5.8	0.0380	5.7	0.972
31-16 47.1	0.06	1192	1339	1.2	40.0	247.0	±14.0	265	±36	0.277	5.9	0.0390	5.7	0.964
31-16 49.1	0.01	4094	5370	1.4	141.0	253.0	±14.0	264	±16	0.285	5.7	0.0400	5.6	0.992
31-16 54.1	0.01	288	265	1.0	10.0	256.0	±14.0	347	±64	0.298	6.4	0.0410	5.7	0.896
31-16 67.1	0.21	1262	3248	2.7	40.1	233.0	±13.0	222	±58	0.257	6.2	0.0370	5.7	0.915
31-16 78.1	0.01	803	747	1.0	27.8	255.0	±14.0	235	±56	0.283	6.2	0.0400	5.7	0.919
27 (n = 60) Zub-Marksheidersky intrusive														
27-1.1.1	0.26	205	85	0.43	5.5	198.5	±4.0	362	±120	0.232	5.8	0.0313	2.0	0.351
27-1.2.1	0.24	501	249	0.51	16.0	234.9	±4.2	295	±83	0.267	4.1	0.0371	1.8	0.447
27-1.3.1	0.04	793	697	0.91	24.3	225.3	±4.0	288	±46	0.255	2.7	0.0356	1.8	0.665
27-1.3.2	0.15	1522	1549	1.05	42.8	207.3	±3.5	238	±54	0.230	2.9	0.0327	1.7	0.588
27-1.4.1	0.24	506	357	0.73	14.4	209.7	±3.8	312	±100	0.240	4.8	0.0331	1.8	0.381
27-1.4.2	0.27	453	349	0.80	12.2	198.9	±3.8	280	±110	0.224	5.3	0.0313	1.9	0.363
27-1.5.1	0.02	707	517	0.76	20.7	216.2	±3.9	319	±48	0.248	2.8	0.0341	1.8	0.652
27-1.6.1	0.11	326	224	0.71	10.0	225.0	±4.2	208	±81	0.246	4.0	0.0355	1.9	0.473
27-1.7.1	0.03	1124	1015	0.93	35.3	231.1	±3.9	274	±38	0.261	2.4	0.0365	1.7	0.724
27-3-1.1	0.01	4928	1797	0.38	175.0	260.5	±4.3	277	±21	0.295	1.9	0.0412	1.7	0.880
27-3-1.2	0.03	2665	508	0.20	80.1	221.7	±3.7	357	±27	0.259	2.1	0.0350	1.7	0.816
27-3-2.1	0.11	3541	795	0.23	121.0	251.4	±4.2	214	±32	0.276	2.2	0.0398	1.7	0.775
27-3-3.1	0.03	3437	1119	0.34	108.0	232.5	±3.9	431	±22	0.281	2.0	0.0367	1.7	0.860
27-4.10.1	0.18	4374	5775	1.36	121.0	203.2	±3.4	248	±35	0.226	2.3	0.0320	1.7	0.744
27-4.11.1	0.06	1372	1039	0.78	36.0	193.9	±3.3	294	±39	0.220	2.4	0.0305	1.7	0.710

(continued)

Table 6 (continued)

Crater	$^{206}Pb_c$, %	U, ppm	Th, ppm	$^{232}Th/^{238}U$	$^{206}Pb^*$, ppm	$^{206}Pb^*/^{238}U$ Age		$^{207}Pb^*/^{206}Pb^*$ Age		$^{207}Pb^*/^{235}U$	±%	$^{206}Pb^*/^{238}U$	±%	KK
27-4.12.1	0.11	10,319	12,064	1.21	323.0	230.2	±4.6	265	±35	0.258	2.5	0.0364	2.0	0.800
27-4.12.2	0.11	9171	9623	1.08	272.0	218.7	±4.0	267	±21	0.245	2.1	0.0345	1.9	0.899
27-4.14.1	0.89	1416	1299	0.95	36.3	187.8	±4.0	340	±92	0.217	4.6	0.0296	2.2	0.471
27-4.2.1	0.04	7673	10,403	1.40	261.0	250.1	±4.1	248	±17	0.279	1.8	0.0396	1.7	0.912
27-4.4.1	0.15	1566	1213	0.80	45.1	212.1	±4.5	262	±44	0.237	2.9	0.0335	2.2	0.744
27-4.4.2	0.10	8745	13,825	1.63	252.0	212.5	±3.7	247	±22	0.236	2.0	0.0335	1.8	0.879
27-4.5.1	0.01	11,403	26,926	2.44	380.0	245.4	±4.0	260	±12	0.275	1.7	0.0388	1.7	0.957
27-4.6.1	0.15	11,095	22,871	2.13	348.0	230.5	±3.8	274	±16	0.260	1.8	0.0364	1.7	0.921
27-4.7.1	0.13	8889	18,458	2.15	276.0	228.8	±3.8	279	±23	0.258	1.9	0.0361	1.7	0.857
27-4.8.1	0.02	12,249	14,654	1.24	433.0	259.6	±4.2	254	±12	0.291	1.7	0.0411	1.7	0.956
27-5-1.1	0.01	4180	7248	1.79	127.0	224.4	±3.7	238	±20	0.249	1.9	0.0354	1.7	0.888
27-5-3.1.1	0.08	6770	5690	0.87	189.0	205.8	±4.0	243	±23	0.228	2.2	0.0324	2.0	0.893
27-5-3.1.2	0.01	24,685	22,110	0.93	899.0	267.7	±4.5	236	±19	0.298	1.9	0.0424	1.7	0.900
27-5.2.1	0.06	6382	12,937	2.09	203.0	234.0	±6.6	231	±22	0.259	3.0	0.0370	2.9	0.950
27-5.2.2	0.01	11,409	28,326	2.57	458.0	294.5	±8.2	260	±11	0.332	2.9	0.0467	2.9	0.987
27-6 3.1	0.14	1446	1472	1.05	47.0	238.9	±4.9	123	±69	0.252	3.6	0.0378	2.1	0.579
27-6 3.2	0.08	496	426	0.89	17.4	258.0	±5.9	316	±100	0.297	5.1	0.0408	2.3	0.460
27-6 6	0.52	373	546	1.51	12.7	249.1	±5.6	233	±210	0.276	9.4	0.0394	2.3	0.246
27-6 7	0.09	1353	2843	2.17	46.8	254.0	±5.1	216	±56	0.280	3.2	0.0402	2.0	0.648
27-6 9	0.93	1335	2321	1.80	45.8	250.0	±5.2	–	–	0.243	10.0	0.0395	2.1	0.203
27-6 10	0.91	244	96	0.40	8.4	249.4	±6.1	262	±330	0.280	15.0	0.0395	2.5	0.171
27-7.10.1	0.13	1205	2276	1.95	43.2	263.0	±7.5	287	±53	0.299	3.7	0.0416	2.9	0.780
27-7.2.1	0.25	556	347	0.64	19.1	252.3	±7.2	244	±66	0.281	4.1	0.0399	2.9	0.714
27-7.2.2	0.10	395	240	0.63	12.9	239.8	±7.0	275	±67	0.271	4.2	0.0379	3.0	0.715
27-7.7.1	0.09	1797	1358	0.78	61.0	249.6	±7.0	268	±51	0.281	3.6	0.0395	2.9	0.790
27-7.9.1	0.05	3401	5209	1.58	126.0	272.2	±7.6	264	±21	0.306	3.0	0.0431	2.9	0.951
27-7.9.2	0.25	3233	7326	2.34	104.0	237.4	±6.7	206	±47	0.260	3.5	0.0375	2.9	0.817
27-7-11.1	0.88	1085	3356	3.20	32.4	218.6	±3.8	184	±110	0.237	4.9	0.0345	1.8	0.361

(continued)

Table 6 (continued)

Crater	$^{206}Pb_c$, %	U, ppm	Th, ppm	$^{232}Th/^{238}U$	$^{206}Pb^*$, ppm	$^{206}Pb^*/^{238}U$ Age		$^{207}Pb^*/^{206}Pb^*$ Age		$^{207}Pb^*/^{235}U$	±%	$^{206}Pb^*/^{238}U$	±%	KK
27-7-11.2	0.61	569	872	1.58	16.0	206.4	±3.7	111	±110	0.216	5.1	0.0325	1.8	0.356
27-7-3.1.1	0.06	444	605	1.41	15.9	262.5	±4.9	227	±65	0.291	3.4	0.0416	1.9	0.559
27-7-3.1.2	0.19	1661	5592	3.48	51.9	229.8	±3.9	168	±60	0.247	3.1	0.0363	1.7	0.558
27-7-4.1	0.60	667	988	1.53	19.4	213.6	±3.8	135	±170	0.226	7.4	0.0337	1.8	0.246
27-7-5.1	0.01	420	337	0.83	12.7	221.9	±4.0	193	±64	0.241	3.3	0.0350	1.8	0.556
27-7-6.1	0.05	2729	10,299	3.90	79.1	213.9	±3.6	230	±31	0.236	2.2	0.0337	1.7	0.787
27-7-8.1	0.19	1232	1236	1.04	30.5	182.6	±3.2	261	±51	0.204	2.8	0.0287	1.8	0.622
27-7-8.2	0.45	2630	10,596	4.16	70.9	198.4	±3.3	274	±53	0.223	2.9	0.0313	1.7	0.590
27-13.1.1	0.01	3317	5575	1.74	113.0	250.9	±4.9	316	±42	0.288	2.7	0.0397	2.0	0.729
27-13.1.2	0.02	5300	12,937	2.52	151.0	209.9	±4.5	248	±24	0.234	2.4	0.0331	2.2	0.903
27-13.1.3	0.03	5282	12,717	2.49	169.0	235.6	±3.9	274	±21	0.266	1.9	0.0372	1.7	0.878
27-13.2.1	0.07	8232	14,073	1.77	253.0	226.0	±4.0	272	±22	0.254	2.0	0.0357	1.8	0.886
27-14.1.1	0.13	272	132	0.50	7.8	210.7	±4.0	243	±93	0.234	4.5	0.0332	2.0	0.436
27-14.1.2	0.24	413	231	0.58	10.9	194.1	±3.6	239	±95	0.215	4.5	0.0306	1.9	0.413
27-14.2.1	0.09	496	264	0.55	15.9	235.2	±4.2	116	±68	0.248	3.4	0.0372	1.8	0.532
27-14.3.1	0.19	1154	889	0.80	35.8	228.3	±3.9	183	±62	0.247	3.2	0.0361	1.8	0.552
27-14.3.2	0.38	640	427	0.69	18.4	211.9	±3.7	63	±97	0.218	4.4	0.0334	1.8	0.405
4 Imangda intrusive (n = 18)														
4-3 1	0.08	5147	7892	1.58	155.0	222.3	±2.6	211	±30	0.244	1.8	0.0351	1.2	0.679
4-6 1	0.68	2377	7425	3.23	77.3	237.8	±3.0	363	±74	0.279	3.5	0.0376	1.3	0.358
4-6 2	0.01	3891	7574	2.01	120.0	227.0	±2.7	218	±28	0.250	1.7	0.0359	1.2	0.708
4-6 6	0.11	1619	3931	2.51	51.4	233.8	±2.9	198	±59	0.255	2.8	0.0369	1.3	0.449
4-6 7	0.16	7653	24,090	3.25	228.0	219.0	±2.6	191	±52	0.238	2.5	0.0346	1.2	0.471
4-8 1	0.23	1273	1127	0.91	42.0	242.4	±3.2	192	±73	0.264	3.4	0.0383	1.3	0.390
4-9 1	0.15	2260	217	0.10	80.8	262.4	±3.2	255	±64	0.294	3.1	0.0416	1.3	0.414
4-9 3	0.20	1695	2961	1.81	53.6	232.7	±3.0	230	±84	0.257	3.9	0.0368	1.3	0.3360
4-9 8.1	0.51	2884	3236	1.16	92.1	234.1	±2.9	214	±74	0.257	3.4	0.0370	1.3	0.3722
4-9 8.2	0.58	1859	7604	4.23	62.0	244.1	±3.1	264	±100	0.274	4.7	0.0386	1.3	0.2749

(continued)

Table 6 (continued)

Crater	$^{206}Pb_c$, %	U, ppm	Th, ppm	$^{232}Th/^{238}U$	$^{206}Pb^*$, ppm	$^{206}Pb^*/^{238}U$ Age		$^{207}Pb^*/^{206}Pb^*$ Age		$^{207}Pb^*/^{235}U$	±%	$^{206}Pb^*/^{238}U$	±%	KK
4-9 9	0.03	5590	10 957	2.03	183.0	240.8	±2.9	215	±25	0.265	1.6	0.0381	1.2	0.7517
4-9 10	0.27	3370	10 161	3.12	110.0	240.0	±2.9	209	±69	0.263	3.2	0.0379	1.2	0.3870
4-9 11	0.09	2679	4480	1.73	87.9	241.6	±3.0	254	±50	0.270	2.5	0.0382	1.2	0.4991
4-9 13	0.45	1819	151	0.09	64.7	260.3	±3.3	332	±79	0.302	3.7	0.0412	1.3	0.3477
4-10 2	0.32	2291	5089	2.30	70.1	224.8	±3.5	255	±84	0.251	4.0	0.0355	1.6	0.4031
4-10 4	0.08	2457	5433	2.28	75.2	225.5	±2.8	185	±45	0.244	2.3	0.0356	1.3	0.5493
4-10 7	0.54	952	1713	1.86	31.7	243.6	±3.4	177	±200	0.263	8.7	0.0385	1.4	0.1651
4-10 8	0.26	6348	8914	1.45	199.0	230.8	±2.8	238	±68	0.256	3.2	0.0365	1.2	0.3915
Chernogorsk intrusive (n = 21)														
CH-9_2	0.21	1179	2198	1.93	39.1	243.7	±3.2	274	±82	0.275	3.8	0.0385	1.3	0.347
CH-9_3	0.14	735	1109	1.56	24.8	248.3	±3.3	274	±65	0.280	3.1	0.0393	1.4	0.435
CH-9_4	0.25	3412	4480	1.36	109.0	234.9	±7.2	222	±96	0.259	5.2	0.0371	3.1	0.600
CH-9-5	0.40	1557	3276	2.17	51.2	241.3	±3.1	222	±110	0.266	4.9	0.0381	1.3	0.264
CH-9-6	0.19	487	319	0.68	16.3	246.6	±3.5	254	±98	0.276	4.5	0.0390	1.5	0.323
CH-9-7	0.14	1918	3409	1.84	63.5	243.5	±3.1	218	±63	0.268	3.0	0.0385	1.3	0.431
CH-9-8	0.01	1616	4212	2.69	49.7	227.0	±2.9	246	±58	0.253	2.8	0.0358	1.3	0.455
CH-10_7	1.68	727	1504	2.14	24.1	240.1	±5.4	–	–	0.204	19.0	0.0380	2.3	0.122
CH-10_8	0.12	1732	3106	1.85	56.6	240.1	±4.9	207	±61	0.263	3.4	0.0380	2.1	0.614
CH-11_1.1	0.36	481	762	1.64	15.9	242.4	±2.0	157	±110	0.260	4.6	0.0383	0.9	0.184
CH-11_1.2	0.05	498	1627	3.37	16.6	245.8	±1.9	230	±52	0.272	2.4	0.0389	0.8	0.332
CH-11_2.1	0.01	461	794	1.78	15.1	241.2	±1.8	303	±53	0.275	2.4	0.0381	0.7	0.305
CH-11_3.1	0.08	715	2003	2.89	23.9	246.2	±1.7	256	±51	0.276	2.3	0.0389	0.7	0.304
CH-11_3.2	7.54	434	1749	4.16	18.8	292.5	±5.4	1280	±1400	0.530	70.0	0.0464	1.9	0.027
CH-11_4.1	0.12	387	877	2.34	12.9	245.0	±2.3	223	±75	0.270	3.4	0.0387	0.9	0.278
CH-11_4.2	0.14	798	2667	3.45	27.0	248.3	±1.5	239	±60	0.276	2.7	0.0393	0.6	0.229
CH-11_5.1	0.17	846	1251	1.53	28.4	247.0	±1.7	273	±66	0.279	3.0	0.0391	0.7	0.233
CH-11_5.2	0.03	2859	6668	2.41	96.6	248.7	±0.9	281	±24	0.282	1.1	0.0393	0.4	0.341
CH-11_6.1	0.01	363	384	1.09	12.0	242.6	±2.4	315	±63	0.279	3.0	0.0383	1.0	0.342

(continued)

Table 6 (continued)

Crater	$^{206}Pb_c$, %	U, ppm	Th, ppm	$^{232}Th/^{238}U$	$^{206}Pb^*$, ppm	$^{206}Pb^*/^{238}U$ Age		$^{207}Pb^*/^{206}Pb^*$ Age		$^{207}Pb^*/^{235}U$	±%	$^{206}Pb^*/^{238}U$	±%	KK
CH-11_7.1	0.52	807	1131	1.45	27.8	251.8	±1.7	298	±130	0.287	5.8	0.0398	0.7	0.121
CH-11_8.1	0.02	2766	5616	2.10	93.8	249.6	±1.1	246	±23	0.278	1.1	0.0395	0.4	0.395
25 (n = 19) South Pyasina intrusive														
25-20 1.1	0.10	1231	2632	2.21	42.7	254.8	±5.2	257	±58	0.286	3.3	0.0403	2.1	0.633
25-31 2	0.15	745	2968	4.11	25.5	251.6	±3.4	301	±68	0.287	3.3	0.0398	1.4	0.418
25-31 3.1	0.17	1841	5523	3.10	60.8	242.5	±4.9	216	±71	0.267	3.7	0.0383	2.1	0.563
25-35 1.1	2.96	397	1417	3.68	13.8	247.9	±9.0	–	–	0.180	77.0	0.0392	3.7	0.048
25-35 2	0.71	311	858	2.85	10.3	241.8	±3.9	364	±170	0.284	7.6	0.0382	1.6	0.216
25-4 1.1	1.45	206	90	0.45	7.1	250.6	±6.3	124	±450	0.265	19.0	0.0396	2.6	0.132
25-4 2.1	0.78	1155	2023	1.81	39.1	247.1	±5.3	83	±300	0.257	13.0	0.0391	2.2	0.167
25-4 4.1	0.76	1436	3130	2.25	47.5	241.9	±3.2	213	±140	0.266	6.3	0.0382	1.3	0.214
25-4 5.1	0.38	1367	2137	1.62	45.2	242.7	±3.3	267	±94	0.273	4.3	0.0384	1.4	0.316
25-4 5.2	0.21	1088	2288	2.17	35.6	240.3	±3.3	290	±79	0.273	3.7	0.0380	1.4	0.378
25-46	0.18	2009	5247	2.70	62.8	229.9	±2.9	204	±76	0.251	3.5	0.0363	1.3	0.363
25-4 7.1	0.60	770	1729	2.32	25.4	242.0	±5.2	223	±230	0.267	10.0	0.0383	2.2	0.219
25-4 9.1	0.30	2078	4683	2.33	70.5	248.8	±5.0	222	±81	0.274	4.1	0.0394	2.0	0.503
25-44 2	0.09	1071	886	0.85	30.8	211.9	±2.8	232	±65	0.234	3.1	0.0334	1.4	0.431
25-44 4	0.21	2025	5682	2.90	64.8	235.1	±3.0	200	±86	0.257	3.9	0.0371	1.3	0.329
25-44 5.1	0.01	1342	2695	2.07	46.6	256.5	±5.3	447	±130	0.313	6.3	0.0406	2.1	0.337
25-44 6.1	0.21	3315	6531	2.04	113.0	249.7	±5.0	184	±85	0.271	4.2	0.0395	2.0	0.488
25-9-10 1.1	0.01	3732	4789	1.33	134.0	264.3	±5.2	275	±32	0.299	2.4	0.0419	2.0	0.821
25-9-10 2.1	0.87	1298	1314	1.05	44.3	248.9	±5.4	–	–	0.249	14.0	0.0394	2.2	0.161
29 Vologochan intrusive (n = 19)														
29-5,6 1	0.02	9495	14,553	1.58	351.0	271.6	±0.8	259	±14	0.305	0.7	0.0430	0.3	0.427
29-9 1.1	0.08	3152	2302	0.75	111.0	260.0	±14.0	237	±31	0.289	5.8	0.0411	5.7	0.972
29-9 2.1	0.27	3180	5180	1.68	98.1	227.0	±13.0	261	±49	0.254	6.0	0.0358	5.7	0.936
29-9 2.2	0.01	1770	2160	1.26	62.5	260.0	±14.0	218	±28	0.286	5.8	0.0411	5.7	0.978
29-9 2.3	0.22	1783	2178	1.26	58.0	239.0	±13.0	214	±55	0.263	6.1	0.0378	5.7	0.921

(continued)

Table 6 (continued)

Crater	$^{206}Pb_c$, %	U, ppm	Th, ppm	$^{232}Th/^{238}U$	$^{206}Pb^*$, ppm	$^{206}Pb^*/^{238}U$ Age		$^{207}Pb^*/^{206}Pb^*$ Age		$^{207}Pb^*/^{235}U$	±%	$^{206}Pb^*/^{238}U$	±%	KK
29-9 7	0.73	3723	6314	1.75	133.0	259.8	±1.5	217	±190	0.286	8.3	0.0411	0.6	0.071
29-9 8.1	0.01	3878	6698	1.78	135.0	256.0	±14.0	234	±20	0.285	5.7	0.0406	5.6	0.989
29-9 10.1	0.63	484	964	2.06	15.2	230.0	±13.0	142	±160	0.245	9.0	0.0364	5.7	0.636
29-9 10.2	1.48	160	382	2.47	4.9	223.0	±13.0	115	±310	0.234	15.0	0.0351	5.8	0.400
29-9 16.1	0.36	1480	3150	2.20	50.8	252.0	±14.0	229	±62	0.278	6.3	0.0398	5.7	0.904
29-9 19.1	0.06	1726	2289	1.37	62.3	265.0	±15.0	242	±32	0.296	5.8	0.0420	5.7	0.972
29-9 21.1	0.07	3416	3307	1.00	123.0	264.0	±15.0	279	±20	0.299	5.7	0.0419	5.6	0.989
29-16.1.1	1.12	1665	7373	4.57	56.4	246.6	±7.0	119	±110	0.260	5.5	0.0390	2.9	0.528
29-16 2	0.04	1242	2783	2.31	41.5	246.0	±1.3	211	±34	0.270	1.6	0.0389	0.5	0.337
29-16 4	0.01	1659	1963	1.22	51.4	228.6	±0.9	193	±29	0.248	1.3	0.0361	0.4	0.322
29-16 6	0.04	996	797	0.83	45.2	331.6	±2.1	347	±38	0.389	1.8	0.0528	0.6	0.355
29-17 1.1	0.34	573	1615	2.91	17.3	222.1	±1.6	176	±100	0.240	4.3	0.0351	0.7	0.167
29-17 1.2	0.69	769	1940	2.61	25.8	245.6	±1.9	327	±170	0.284	7.4	0.0388	0.8	0.104
29-17 2.1	0.02	2200	2595	1.22	67.6	226.5	±1.0	238	±27	0.251	1.2	0.0358	0.5	0.373
MD-48, S-1, TP-43 Mikchangda, Binyuda, Dyumptalej intrusives (n = 25)														
MD-48 Mikchangda intrusive														
48-18 1,1	0.34	4230	4899	1.20	1.0	26.0	±1.9	216	±117	0.286	5.1	0.0412	0.7	0.143
48-18 1,2	0.63	2028	1943	0.99	1.0	229.8	±2.5	182	±245	0.249	10.6	0.0363	1.1	0.105
48-18 2,1	0.55	3042	3227	1.10	1.0	231.5	±2.3	153	±216	0.248	9.3	0.0366	1.0	0.107
48-18 2,2	0.58	2491	3403	1.41	1.0	254.3	±2.4	145	±229	0.272	9.8	0.0402	1.0	0.098
48-18 21,1	1.13	3597	4131	1.19	1.0	247.6	±3.0	–	–	0.240	13.7	0.0392	1.2	0.088
48-18 21,2	2.54	2366	4068	1.78	1.0	259.3	±4.3	263	±479	0.292	20.9	0.0410	1.7	0.082
48-18 5,1	0.64	1566	1162	0.77	1.0	293.3	±5.2	230	±271	0.326	11.9	0.0466	1.8	0.152
48-18 5,2	0.71	1192	919	0.80	1.0	224.2	±5.8	261	±236	0.251	10.6	0.0354	2.7	0.250
48-18 5,3	0.69	2492	3104	1.29	1.0	254.1	±2.7	95	±204	0.266	8.7	0.0402	1.1	0.124
48-25 6.1	0.01	1087	1949	1.85	1.0	242.4	±3.0	426	±138	0.292	6.3	0.0383	1.3	0.202
48-25 6.2	2.20	795	1101	1.43	1.0	236.5	±3.9	–	–	0.236	19.3	0.0374	1.7	0.088
48-30 1.1	0.88	2174	3366	1.60	1.0	263.2	±2.6	76	±237	0.273	10.0	0.0417	1.0	0.102

(continued)

Table 6 (continued)

Crater	$^{206}Pb_c$, %	U, ppm	Th, ppm	$^{232}Th/^{238}U$	$^{206}Pb^*$, ppm	$^{206}Pb^*/^{238}U$ Age		$^{207}Pb^*/^{206}Pb^*$ Age		$^{207}Pb^*/^{235}U$	±%	$^{206}Pb^*/^{238}U$	±%	KK
48-30 10.1	0.57	8282	10,038	1.25	1.0	244.7	±1.5	255	±86	0.274	3.8	0.0387	0.6	0.169
48-30 5.1	0.85	4412	7272	1.70	1.0	255.9	±1.7	119	±117	0.270	5.0	0.0405	0.7	0.134
48-30 18.1	0.07	5840	10,549	1.87	1.0	275.7	±1.6	271	±50	0.311	2.2	0.0437	0.6	0.262
48-30 18.2	0.75	4352	5033	1.19	1.0	215.6	±2.6	104	±302	0.226	12.8	0.0340	1.2	0.095
48-30 10.2	0.84	2495	3843	1.59	1.0	222.0	±2.7	21	±219	0.224	9.2	0.0350	1.2	0.133
S-1 Binyuda														
S1-6 4	1.03	454	712	1.62	14.6	235.0	±5.6	95	±390	0.245	17.0	0.0371	2.4	0.146
S1-6-1.1	0.01	424	203	0.49	14.4	249.4	±5.7	474	±75	0.308	4.1	0.0395	2.3	0.561
S1-6-1.2	0.17	411	333	0.84	13.8	247.1	±5.4	278	±120	0.279	5.5	0.0391	2.2	0.400
S1-6-2.1	0.43	1248	1122	0.93	44.7	262.3	±5.4	165	±120	0.283	5.7	0.0415	2.1	0.367
TP-43 Dyumptalej														
43-27.28 1	0.01	2918	4320	1.53	96.6	244.0	±4.8	278	±30	0.276	2.4	0.0386	2.0	0.835
43-27.28 4	0.02	6223	8989	1.49	208.0	245.5	±4.8	261	±20	0.276	2.2	0.0388	2.0	0.912
43-27.28 5	0.01	2090	2341	1.16	68.7	242.1	±4.8	281	±43	0.274	2.7	0.0383	2.0	0.733
43-27.28 6	0.12	1671	1584	0.98	55.5	244.2	±4.9	203	±60	0.267	3.3	0.0386	2.0	0.616
NP-37, F-233, MP-26, PУ Lower Norilsk, Zelenaya Griva, Kruglogorsky, Morongo intrusives (n = 20)														
NP-37 Lower Norilsk														
37-12 1	0.45	771	637	0.85	25.4	241.9	±5.0	208.0	±110	0.265	5.2	0.0382	2.1	0.411
37-12 4	0.61	1192	869	0.75	41.4	253.8	±5.3	–	–	0.251	10.0	0.0402	2.1	0.211
37-12 7	0.09	733	581	0.82	24.6	247.2	±5.1	248.0	±67	0.276	3.6	0.0391	2.1	0.583
37-23.27-2(4)	0.56	783	1545	2.04	25.3	236.8	±5.0	64.0	±220	0.244	9.4	0.0374	2.2	0.231
37-23.27 1	1.18	356	2872	8.33	11.7	238.2	±5.9	–	–	0.236	21.0	0.0376	2.5	0.119
37-23.27 2	1.03	997	2329	2.41	32.3	236.2	±5.0	–	–	0.235	11.0	0.0373	2.2	0.194
37-23.27 3	0.11	926	6973	7.78	29.7	236.3	±4.8	258.0	±66	0.264	3.6	0.0373	2.1	0.584
F-233 Zelenaya Griva														
F233-2 1.1	0.99	1013	1843	1.88	–	279.0	±4.2	–	–	0.265	18.7	0.0442	1.5	0.082
F233-2 2.1	1.34	1225	2137	1.80	–	264.9	±4.5	–	–	0.250	25.1	0.0420	1.7	0.069
F233-4 1.1	0.45	1372	3779	2.85	–	259.9	±2.8	262.7	±141	0.292	6.2	0.0411	1.1	0.176

(continued)

Table 6 (continued)

Crater	$^{206}Pb_c$, %	U, ppm	Th, ppm	$^{232}Th/^{238}U$	$^{206}Pb^*$, ppm	$^{206}Pb^*/^{238}U$ Age		$^{207}Pb^*/^{206}Pb^*$ Age		$^{207}Pb^*/^{235}U$	±%	$^{206}Pb^*/^{238}U$	±%	KK
F233-4 17.2	0.85	3146	3627	1.19	–	266.8	±2.0	144.0	±145	0.285	6.2	0.0423	0.8	0.125
233-6 2	0.27	359	399	1.15	11.6	237.5	±5.2	288.0	±130	0.269	6.1	0.0375	2.3	0.368
233-6 3	0.49	1709	2510	1.52	56.2	241.0	±4.8	17.0	±120	0.244	5.3	0.0381	2.0	0.387
233-6 4	0.04	3414	8765	2.65	113.0	244.1	±4.8	252.0	±30	0.273	2.4	0.0386	2.0	0.836
F233-7 1.1	1.60	878	1173	1.38	–	268.1	±4.9	–	–	0.243	28.1	0.0425	1.9	0.067
F233-7 1.2	0.46	2242	5980	2.76	–	265.7	±2.4	122.2	±186	0.281	8.0	0.0421	0.9	0.115
MP-2bis Kruglogorsky														
K9 1.1	0.48	6622	8903	1.39	–	241.0	±1.6	244.0	±106	0.268	4.7	0.0381	0.7	0.144
K9 2.1	0.12	3879	10,876	2.90	–	262.5	±1.7	305.8	±56	0.301	2.5	0.0416	0.7	0.258
K9 2.2	0.20	4700	7484	1.65	–	227.0	±1.7	232.8	±92	0.251	4.0	0.0358	0.8	0.186
NP-3734, MD-48, NP-3752 Daldykan, Oganer, Ergalakh intrusives (n = 8)														
NP-3734 Daldykan														
Ru-2.5 1	0.97	365	304	0.86	12.2	243.0	±5.7	–	–	0.229	15.0	0.0384	2.4	0.160
37-34 _22	0.08	373	185	0.51	12.8	252.5	±5.6	260.0	±92	0.283	4.6	0.0399	2.3	0.492
MD-48 Oganer														
37-34 6	0.46	195	64	0.34	6.5	243.3	±6.1	103.0	±230	0.255	10.0	0.0385	2.6	0.253
48-7 1.1	2.15	1211	960	0.82	–	234.9	±3.7	–	–	0.206	24.4	0.0371	1.6	0.067
48-7 2.1	2.93	481	299	0.64	–	255.3	±10.4	–	–	0.235	58.8	0.0404	4.2	0.071
48-7 3.1	1.44	1053	907	0.89	–	235.3	±4.4	–	–	0.206	29.3	0.0372	1.9	0.065
48-7 3.2	2.11	1260	1310	1.07	–	242.0	±3.3	109.1	±317	0.254	13.5	0.0383	1.4	0.104
NP-3752 Ergalakh														
37-52 1.1	0.15	389	40	0.11	115.0	1898.0	±34.0	2404.0	±18	7.330	2.3	0.3425	2.0	0.893
37-52 1.2	0.02	1419	3	0.01	400.0	1828.0	±32.0	1951.8	±8.9	5.410	2.1	0.3279	2.0	0.970

Notes Age—In Ma, taking into account errors in decay constant determination. All errors are ±1 sigma. $^{206}Pb_c$ and $^{206}Pb^*$—common and radiogenic lead. Isotopic ratios and contents of non-radiogenic Pb are corrected by the measured ^{204}Pb. KK is correlation coefficient between $^{206}Pb/^{238}U$ and $^{207}Pb/^{235}U$ isotope ratio detection errors. Errors in calibration of the standard are 0.25–0.64%. Dash—The parameter was not determined

2 Rhenium–Osmium Isotope Systematics

Methods of Re–Os dating, samples. Dissolution of samples and chemical separation of Re and Os were performed in CIR VSEGEI laboratories according to the method [5]. Samples weighing 50–200 mg were digested in a mixture of reagents (1 ml of Br_2 + 2 ml of 7 N HNO_3 + 0.5 ml of 40% CrO_3 in 7 N HNO_3) in 5 ml Teflon Seville vials at 90 °C for 48 h. Then, osmium was separated by microdistillation, and rhenium, by liquid extraction with isoamyl alcohol. To determine Re and Os concentration and $^{87}Re/^{188}Os$ ratio, isotope dilution with mixed tracer ^{85}Re–^{190}Os was used. Isotopic composition of osmium was measured by solid phase multi-collector mass spectrometer TIMS Triton TI (Thermo Scientific) in a static mode in negatively charged ions on Faraday collectors. Os in the form of bromide was applied to platinum band of evaporator. To improve ionization efficiency, device puffing oxygen in the mass spectrometer chamber was used. For correction for mass fractionation, $^{192}Os^{16}O_3/^{188}Os^{16}O_3 = 3.092016$ ratio was used. Measurement of isotopic composition and determining Re concentration from solution were made on a single-collector inductively coupled plasma mass spectrometer ICP-MS ELEMENT2 (Thermo Scientific).

One of the most effective indicators of commercial-scale mineralization is the initial isotopic composition of osmium. This is primarily due to the fact that massive ore from commercial deposits, according to Walker et al. [18], is characterized by values corresponding to a mixture of crustal and mantle material with the largest share of mantle component. In the disseminated ores of the same deposits mantle component is strongly "diluted" with crustal one, as disseminated ore is more "sensitive" to contamination by crustal material in comparison with massive ores. It was also found that the proportion of mantle component is minimal in weakly mineralized Lower Talnakh and Lower Norilsk intrusions [18], in which the low mass of sulphide material was, probably, least protected from crustal contamination. We have characterized not only ore from commercially mineralized (Talnakh and Kharayelakh) and weakly mineralized (Lower Talnakh, Lower Norilsk, and Zelenaya Griva) intrusions, but also disseminated ore from reserve deposits of mineralized intrusions (Chernogorsky, Zub-Marksheidersky, South Pyasino, and Vologochan) in the Norilsk Province and occurrences of potentially mineralized Binyuda intrusion in the Taimyr Province.

A total of 31 sulphide ore samples from commercially mineralized Talnakh and Kharaelakh intrusions have been studied; 12 samples from mineralized Chernogorsky, Zub-Marksheidersky, South Pyasino, and Vologochan intrusions; two samples from potentially mineralized Binyuda intrusion, and 12 samples from weakly mineralized Lower Talnakh, Lower Norilsk, and Zelenaya Griva intrusions.

Results of Re–Os investigations. ***Isotope-geochemical characteristics of commercially mineralized intrusions***. Talnakh intrusion, disseminated and massive sulphide ores. Re–Os method was used to analyse a series of disseminated and massive ore samples from OUG-2 drill core. Re–Os isochrone with age of 247 ± 18 Ma and initial osmium ratio of $^{187}Os/^{188}Os = 0.1366 \pm 0.0037$ was

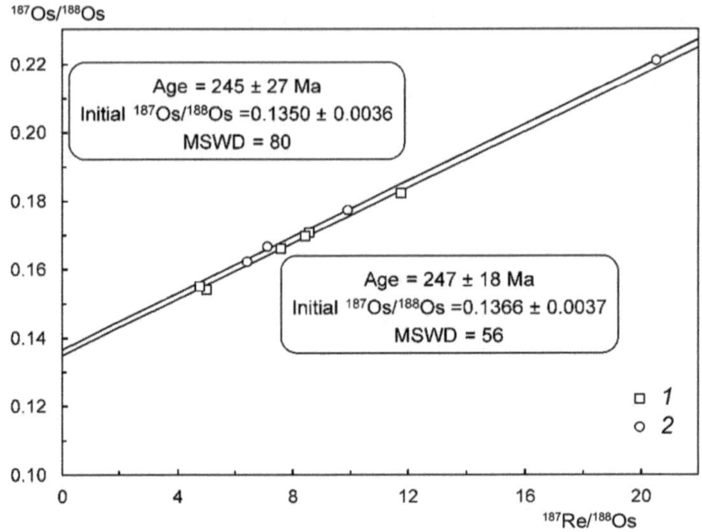

Fig. 4 Re–Os isochrons after disseminated (1) and massive (2) sulphide ores of Talnakh intrusive

obtained from massive ore samples; isochrone corresponding to the age of 245 ± 27 Ma and initial isotopic composition of $^{187}Os/^{188}Os = 0.1350 \pm 0.0036$ was constructed from disseminated ore samples (Fig. 4). Since age and initial Os/Os ratio in both ore types are similar, their analytical data can be combined into a single isochrone corresponding to the age of 251 ± 13 Ma ($^{187}Os/^{188}Os$ = 0.1348 ± 0.0021, mean square weighted deviation (MSWD) = 125).

Significant isochron MSWD are due to variations in initial isotopic composition of sulphides from various horizons in sequence. Age data coincide with the results by Walker et al. obtained earlier [18], which cite a combined Re–Os isochron with age of 245.7 ± 0.6 Ma and initial ratio of $^{187}Os/^{188}Os = 0.1326 \pm 0.0025$, received from sulphides of Norilsk-1 and Talnakh intrusions.

According to our data, disseminated ores have lower initial osmium ratio ($^{187}Os/^{188}Os = 0.1350$) compared to massive ores ($^{187}Os/^{188}Os = 0.1366$), that may be both due to a lower degree of osmium contamination from host rocks by disseminated ores and by slightly earlier crystallization of disseminated ores compared with massive ores. Model initial osmium ratio computed for the age of 250 Ma (osmium gamma, γOs) for disseminated ore is also lower than that of massive ore (5.7–7.4 vs. 7.5–8.8). Model ages for both ore types overlap: 270–320 Ma for massive and 292–333 Ma for disseminated. Re/Os ratio in massive ore samples (1.3–4.2) varies over a wider range than in disseminated ore samples (1.0–2.4). Rhenium and osmium concentrations, as well as the range of fluctuations in massive ore are larger than in disseminated ones.

Kharaelakh intrusion, disseminated and massive sulphide ores. Re–Os method was used to analyse a series of disseminated and massive ore samples from KZ-844

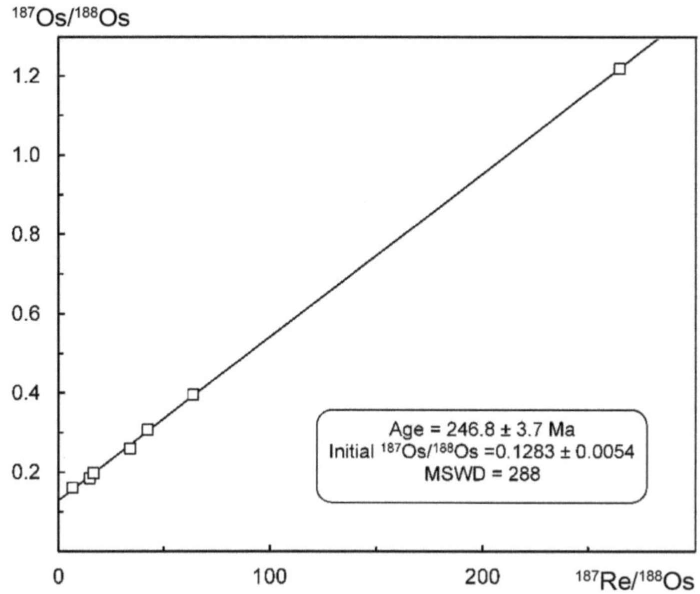

Fig. 5 Re–Os isochron after massive sulphide ores of Kharaelakh intrusive

and KZ-963 drill core. Re–Os isochron with age of 246.8 ± 3.7 Ma and initial osmium ratio of $^{187}Os/^{188}Os$ = 0.1283 ± 0.0054 was obtained from massive ore samples (Fig. 5).

At the same time, the analysed disseminated ore samples are characterized by a series of parallel isochrones with similar ages but different initial osmium ratios (Fig. 6). The main analysis sampling forms an isochron corresponding to the age of 247 ± 21 Ma and initial isotopic composition of $^{187}Os/^{188}Os$ = 0.1331 ± 0.0052. Furthermore, two of the remaining four analyses form an isochron with age of 248.7 ± 2.9 Ma and higher $^{187}Os/^{188}Os$ = 0.13642, while two analyses lie on the isochron with age of 249.6 ± 2.2 Ma and even higher $^{187}Os/^{188}Os$ = 0.14356 (Fig. 6).

These age data on the majority of disseminated ore samples are consistent with the Re–Os results by Walker et al. [18], giving a Re–Os isochron for the Kharayelakh intrusion sulphides with age of 247.0 ± 3.8 Ma and initial ratio of $^{187}Os/^{188}Os$ = 0.133 ± 0.021.

According to our data, massive sulphide ores have significantly lower initial osmium ratio ($^{187}Os/^{188}Os$ = 0.1283) in comparison with disseminated ores ($^{187}Os/^{188}Os$ = 0.1331), that may be both due to a lower degree of osmium contamination from host rocks by massive ore and a ***more primitive nature*** of massive sulphide ore source. Moreover, heterogeneity of the initial osmium isotope ratio $^{187}Os/^{188}Os$ (from 0.1331 to 0.1436) with coinciding within the limits of age error is observed in disseminated ores. Revealed heterogeneity may also be associated with mixing and contamination in the magma chamber. Model initial osmium ratio

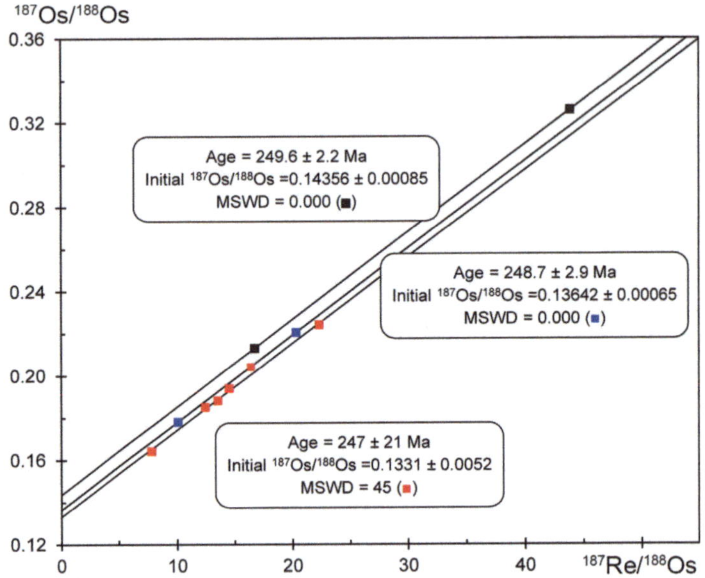

Fig. 6 Re–Os isochron after disseminated sulphide ores of Kharaelakh intrusive

computed for the age of 250 Ma for disseminated ores is from 4–6 to 13–14. Re/Os ratio in massive ore samples (1.4–47.8) varies in a much greater range than in disseminated ore samples (1.6–8.8) due to reduced osmium content in massive ores (1–15 vs. 30–55 ppm). Rhenium concentration in disseminated ores is also higher than in massive ores. Model age of massive ores varies in a considerably wider range than the model age of disseminated ores (227–322 and 256–303 Ma).

Re–Os isotope-geochemical characteristics of potentially mineralized intrusions. Chernogorsky intrusion, disseminated ores. Re–Os method was used to analyse two samples of disseminated sulphide ores from MP-2 bis drill core. Along with disseminated ore samples from Zub-Marksheidersky, South Pyasino, and Vologochan intrusions with off-balance ores, samples from the Chernogorsky intrusion form a Re–Os isochron with age of 248 ± 14 Ma and initial osmium ratio of $^{187}Os/^{188}Os$ = 0.1381 ± 0.0014 (Fig. 7).

Disseminated ore from the Chernogorsky intrusion is characterized by high Re (155–170 ppm) and Os (230–305 ppm), low Re/Os ratio (0.6–0.7), arelatively older model age (418–457 Ma) and moderate osmium gamma values (9.5–9.6).

Zub-Marksheidersky intrusion, disseminated ores. Re–Os method was used to analyse two disseminated sulphide ore samples from MP-27 drill core. One of Zub-Marksheidersky samples, along with samples from other mineralized intrusions (Chernogorsky, South Pyasino, and Vologochan), participates in the construction of Re–Os isochron with age of 248 ± 14 Ma and initial osmium ratio 0.1381 ± 0.0014 (Fig. 7).

Disseminated ore from the Zub-Marksheidersky intrusion is characterized by average Re (185–275 ppm) and Os (53–142 ppm) contents, average Re/Os ratio

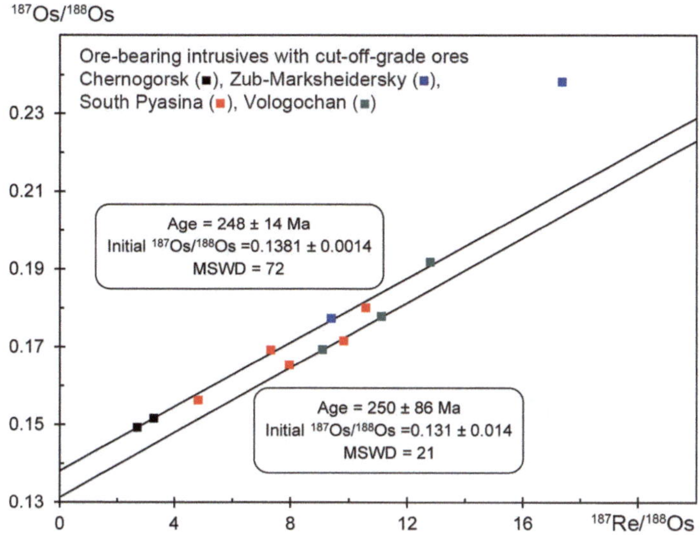

Fig. 7 Re–Os isochron after disseminated ores of ore-bearing intrusives in Norilsk Province

(1.9–3.5), relatively increased model age (310–377 Ma), and moderate to high osmium gamma values (9.7 and 31.6). One of the samples (corresponding to disseminated ores from an extremely altered olivine gabbro from the lower intrusion) shows signs of rhenium loss in the Re–Os isochron diagram, being much left of the isochron (Fig. 7).

South Pyasino intrusion, disseminated ores. Re–Os method was used to analyse five samples of disseminated sulphide ore from OM-25 drill core (Table 8 in Chapter "Strontium and Neodymium Isotopes"). Along with samples from other mineralized intrusions (Chernogorsky, Zub-Marksheidersky, and Vologochan), samples from the South Pyasino intrusion form Re–Os isochrons with age of 248 ± 14 Ma with ($^{187}Os/^{188}Os_i$) = 0.1381 ± 0.0014 and 250 ± 86 Ma with ($^{187}Os/^{188}Os_i$) = 0.131 ± 0.014 (Fig. 7).

Disseminated ore from the South Pyasino intrusion is characterized by average Re (49–207 ppm) and Os (24–168 ppm) contents, average Re/Os ratio (1.0–2.2), average model age (262–343 Ma), moderate osmium gamma values (3.7–10.1), and bimodal nature of the initial osmium ratio (0.131 and 0.138).

Vologochan intrusion, disseminated ores. Re–Os method was used to analyse three disseminated sulphide ore samples from OB-29 drill core. Along with samples from other mineralized intrusions, samples from Vologochan intrusion form Re–Os isochrons with age of 248 ± 14 Ma, where ($^{187}Os/^{188}Os_i$) = 0.1381 ± 0.0014 and 250 ± 86 Ma with ($^{187}Os/^{188}Os_i$) = 0.131 ± 0.014 (Fig. 7).

Disseminated ore is characterized by average Re (194–262 ppm) and Os (74–115 ppm) contents, average Re/Os ratio (1.9–2.6), average model ages (265–295 Ma), and moderate osmium gamma values (4.3–9.9). Like ores from the South Pyasino intrusion, disseminated ores from the Vologochan intrusion exhibit bimodal initial osmium ratio (0.131 and 0.138).

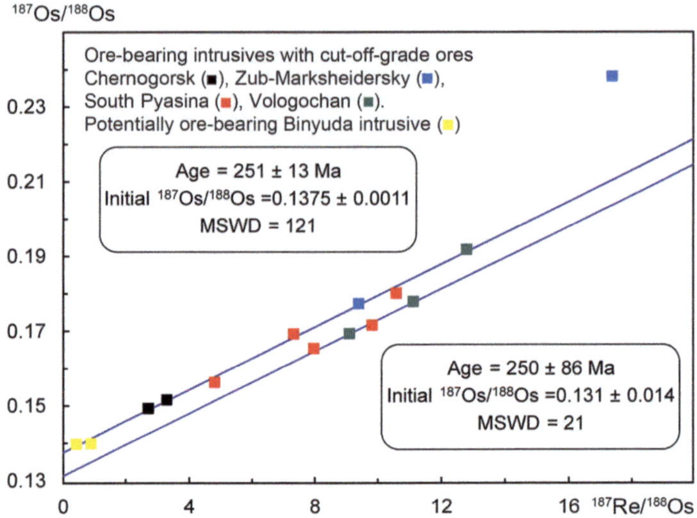

Fig. 8 Re–Os isochron after sulphide ores of ore-bearing intrusives in Norilsk Province and potentially ore-bearing Binyuda intrusive in Taimyr Province

Binyuda intrusion, vein ore. Re–Os method was used to analyse two vein sulphide ore samples from C-2 drill core (Table 8 in Chapter "Strontium and Neodymium Isotopes"). Along with mineralized intrusion samples (Chernogorsky, Zub-Marksheidersky, South Pyasino, and Vologochan) with off-balance ores, veinlet ore samples from the Binyuda intrusion correspond to previously given Re–Os isochron (Fig. 7), and are characterized by age of 251 ± 13 Ma and initial osmium ratio of $^{187}Os/^{188}Os = 0.1375 \pm 0.0011$ (Fig. 8). Similarity in age and initial osmium ratio, despite the fact that this intrusion is significantly distant from the remaining massifs, indicates osmium source similarity between these two regions.

Binyuda intrusion ore is characterized by average Re (22–264 ppm) and increased Os (256–1437 ppm) contents, very low Re/Os ratio (0.09–0.18), ancient model ages (750–1566 Ma), and moderate osmium gamma values (8.1–9.5).

Isotope-geochemical characteristics of weakly mineralized intrusions. Lower Talnakh intrusion, disseminated ores. Re–Os method was used to analyse five sulphide ore samples from TG-31 drill core. Disseminated ore from the Lower Talnakh intrusion forms Re–Os isochron with age of 247 ± 45 Ma and ($^{187}Os/^{188}Os_i$) = 0.197 ± 0.095 (Fig. 9).

Disseminated ore has increased Re (119–316 ppb) and low Os (4.4–20.6 ppb) content, high Re/Os ratio (13.7–71.6), average model ages (261–318 Ma), and high osmium gamma values (35.6–117.8), consistent with the data by Arndt et al. [3] (γOs = 46–71).

Lower Norilsk intrusion, disseminated ores. Re–Os method was used to analyse 6 sulphide ore samples from NP-37 drill core. Disseminated ore from the Lower Norilsk intrusion is characterized by Re–Os isochron with age of 251 ± 30 Ma and ($^{187}Os/^{188}Os_i$) = 0.196 ± 0.033 (Fig. 10).

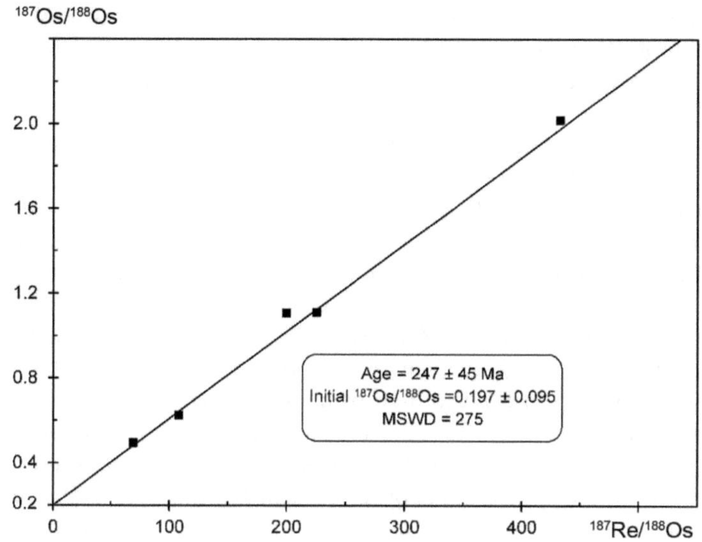

Fig. 9 Re–Os isochron after disseminated ores of Lower Talnakh intrusive

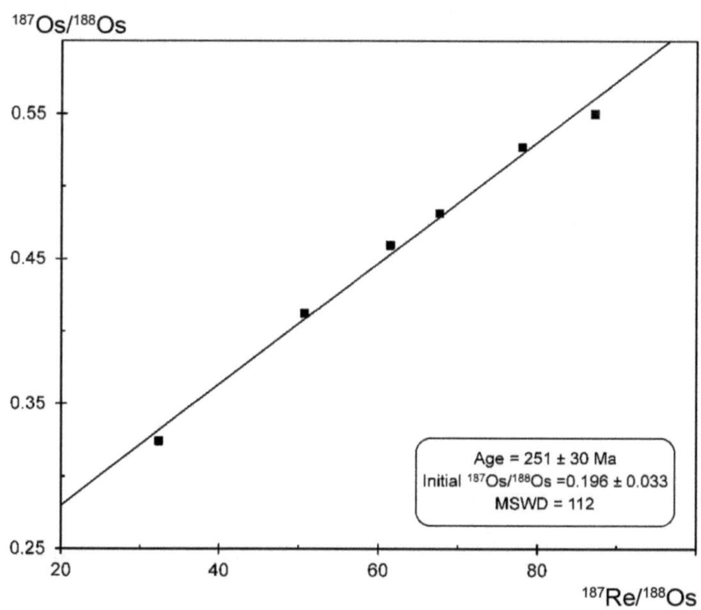

Fig. 10 Re–Os isochron after disseminated ores of Lower Norilsk intrusive

Disseminated ore has increased Re (119–203 ppm) and low Os (7.0–28.3 ppm) content, high Re/Os ratio (6.5–17.1), average model ages (289–361 Ma), and high osmium gamma values (47.7–61.2), which are close to the data by Arndt et al. [3] (γOs = 10.2–67.4).

If we exclude the extreme upper and lower points from isochrons in Fig. 10, we obtain the age of 251 ± 20 Ma.

Zelenaya Griva intrusion, disseminated ores. Re–Os method was used to analyse one disseminated sulphide ore sample from NP-37 drill core. Disseminated ore from the Lower Norilsk, Lower Talnakh, and Zelenaya Griva intrusions form a Re–Os isochron with age of 250 ± 14 Ma ($^{187}Os/^{188}Os_i$) = 0.195 ± 0.013 (Fig. 11).

Disseminated ore from the Zelenaya Griva intrusion is characterized by increased Re (255 ppm) and average Os (33 ppm) content, high Re/Os ratio (7.8), average model age (352 Ma), and high osmium gamma values (54.5).

We have obtained new Re–Os age and isotope-geochemical data for the Talnakh and Kharayelakh intrusions separately for massive and disseminated ores. Age data coincide within the age determination error with the data by Walker et al. [18]. Initial isotope ratios are somewhat different from those in the literature; differentiation of ore types has been revealed. Additional Re–Os results are also obtained for the Lower Talnakh and Lower Norilsk intrusions compared with data by Arndt et al. [3], which made it possible to calculate the Re–Os isochrone age for weakly mineralised intrusions. Re–Os data for the Chernogorsky, Zub-Marksheidersky, Vologochan, South Pyasino, and Zelenaya Griva intrusions have been first obtained.

Initial Re–Os isotopic data for rocks and ores are given in Tables 7, 8, 9 and 10.

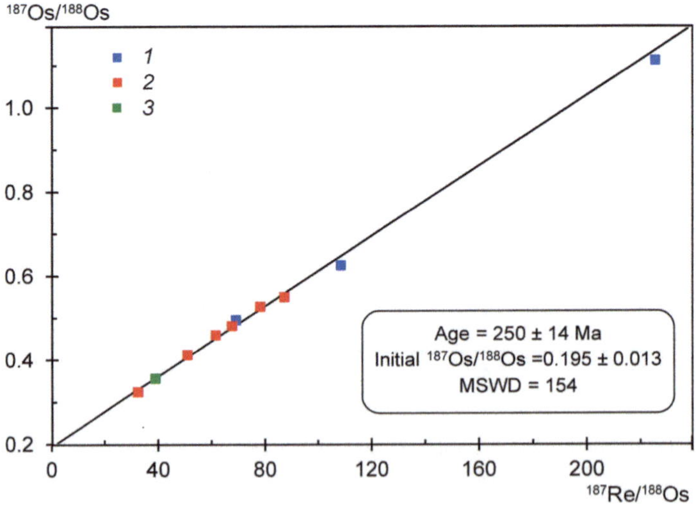

Fig. 11 Re–Os isochron after disseminated ores of Lower Talnakh (1), Lower Karelia (2) and Zelenaya Griva (3) weakly ore-bearing intrusions

Table 7 Re-Os isotopic characteristic of disseminated and massive sulphide ores of Talnakh intrusive (Bh. OUG-2), n = 10

Sample number	Test portion, g	Re, ppm	Os, ppm	Re/Os	$^{187}Re/^{188}Os$	±%, 2σ	$^{187}Os/^{188}Os$	±, 2σ	Model age	γOs
Disseminated ores										
T-13	0.04273	192.2	185.4	1.0	5.01	0.5	0.15412	0.00016	304.4	5.7
T-19 (duplicate of T-14)	0.05720	139.0	141.3	1.0	4.76	0.5	0.15510	0.00019	333.1	7.4
T-15	0.01514	157.1	89.3	1.8	8.57	0.5	0.17086	0.00009	295.1	7.3
T-16	0.04658	171.3	109.3	1.6	7.59	0.5	0.16607	0.00075	295.4	6.7
T-17	0.05481	188.2	108.2	1.7	8.43	0.5	0.16975	0.00057	292.1	6.8
T-18	0.04797	213.3	88.1	2.4	11.75	0.5	0.18237	0.00017	273.8	5.8
Massive ores										
87sr (chalcopyrite)	0.06923	188.9	45.1	4.2	20.53	0.5	0.22113	0.00022	269.9	7.5
87s	0.07799	221.3	109.1	2.0	9.89	0.5	0.17742	0.00007	295.4	8.1
79s	0.07414	369.3	252.9	1.5	7.11	0.5	0.16667	0.00013	320.3	8.8
96s	0.09219	366.9	278.5	1.3	6.41	0.5	0.16225	0.00018	313.8	7.6

Note γOs value is calculated for the age of 250 Ma. Model age in Ma

Table 8 Re–Os isotopic characteristics of disseminated and massive sulphide ores of Talnakh intrusive (Bh. OUG-2), n = 10

Sample number	Test portion, g	Re, ppm	Os, ppm	Re/Os	$^{187}Re/^{188}Os$	±%, 2σ	$^{187}Os/^{188}Os$	±, 2σ	Model age	γOs
Disseminated ores										
T-13	0.04273	192.2	185.4	1.0	5.01	0.5	0.15412	0.00016	304.4	5.7
T-19 (duplicate of T-14)	0.05720	139.0	141.3	1.0	4.76	0.5	0.15510	0.00019	333.1	7.4
T-15	0.01514	157.1	89.3	1.8	8.57	0.5	0.17086	0.00009	295.1	7.3
T-16	0.04658	171.3	109.3	1.6	7.59	0.5	0.16607	0.00075	295.4	6.7
T-17	0.05481	188.2	108.2	1.7	8.43	0.5	0.16975	0.00057	292.1	6.8
T-18	0.04797	213.3	88.1	2.4	11.75	0.5	0.18237	0.00017	273.8	5.8
Massive ores										
87 sr (chalcopyrite)	0.06923	188.9	45.1	4.2	20.53	0.5	0.22113	0.00022	269.9	7.5
87s	0.07799	221.3	109.1	2.0	9.89	0.5	0.17742	0.00007	295.4	8.1
79 s	0.07414	369.3	252.9	1.5	7.11	0.5	0.16667	0.00013	320.3	8.8
96 s	0.09219	366.9	278.5	1.3	6.41	0.5	0.16225	0.00018	313.8	7.6

Note γOs was calculated for the age 250 Ma. Model age in Ma

Table 9 Re–Os isotopic characteristics of disseminated sulphide ores of Chernogorsk (boreholes MP-2bis, Ch-11, Ch-13), Zub-Marksheidersky (Bh. MP-27), South Pyasina (Bh. OV-25), Vologochan (Bh. OV-29) and Binyuda (Bh. C-2) intrusives, n = 14

Intrusive, sample number	Test portion, g	Re, ppm	Os, ppm	Re/Os	$^{187}Re/^{188}Os$	±%, 2σ	$^{187}Os/^{188}Os$	±, 2σ	Model age	γOs
Chernogorsk, Bh. MP-2bis										
CH-11	0.05991	169.7	305.5	0.6	2.70	0.5	0.14927	0.00008	457.4	9.6
CH-13	0.01485	155.3	230.1	0.7	3.28	0.5	0.15160	0.00071	418.8	9.5
Zub-Marksheidersky, Bh. MP-27										
27-13, 87.2 m	0.05126	275.3	143.0	1.9	9.39	0.5	0.17737	0.00013	310.8	9.7
MP-27/96.9 m	0.13771	185.8	52.6	3.5	17.36	0.5	0.23823	0.00036	377.8	31.6
South Pyasina, Bh. OV-25										
25-20	0.07222	166.9	168.4	1.0	4.82	0.5	0.15631	0.00035	343.8	8.1
25-31	0.09783	150.1	99.9	1.5	7.32	0.5	0.16926	0.00028	332.0	10.1
25-35	0.10404	97.4	59.5	1.6	7.97	0.5	0.16540	0.00055	276.4	4.9
25-36	0.07782	207.5	95.6	2.2	10.59	0.5	0.18013	0.00011	291.3	7.9
25-41	0.0559	49.1	24.4	2.0	9.82	0.5	0.17160	0.00011	262.2	3.7
Vologochan, Bh. OV-29										
OV-29/852.3	0.08436	194.8	74.4	2.6	12.79	0.5	0.19182	0.00014	295.8	9.9
29-24 (854–855)	0.09490	207.6	111.2	1.9	9.10	0.5	0.16935	0.00015	268.1	4.3
OV-29/867.5	0.12113	262.4	115.1	2.3	11.12	0.5	0.17791	0.00011	265.4	4.4
Binyuda, Bh. S-2										
S-2/1	0.08512	22.2	256.7	0.1	0.42	0.5	0.13973	0.00014	1566.9	9.5
S-2/2	0.11355	264.4	1437.2	0.2	0.89	0.5	0.13986	0.00014	750.5	8.1

See note to Table 7

Table 10 Re–Os isotopic characteristics of disseminated sulphide ores of Lower Talnakh (Bh. TG-31), Lower Norilsk (Bh. NP-37) and Zelenaya Griva (Bh. F-233) intrusives, n = 12

Intrusive, sample number	Test portion, g	Re, ppm	Os, ppm	Re/Os	$^{187}Re/^{188}Os$	±%, 2σ	$^{187}Os/^{188}Os$	±, 2σ	Model age	γOs
Lower Talnakh, Bh. TG-31										
31 3	0.05422	281.2	20.6	13.7	68.94	0.5	0.49512	0.00149	318.2	64.6
31-_3 (duplicate)	0.09788	316.2	4.4	71.6	432.61	0.5	2.01893	0.00102	261.7	69.4
31-10	0.08567	196.5	5.4	36.6	199.80	0.5	1.10827	0.00115	293.6	117.8
31-11	0.00888	119.4	5.7	21.0	108.20	0.5	0.62474	0.00044	274.6	37.5
31-11 (duplicate)	0.04216	250.9	6.1	41.2	225.39	0.5	1.11154	0.00121	261.2	35.6
Lower Norilsk, Bh. NP-37										
37-9g	0.06144	184.4	28.3	6.5	32.42	0.5	0.32423	0.00027	361.1	50.0
37-11a	0.06969	166.0	12.5	13.3	67.54	0.5	0.48133	0.00086	312.6	58.3
37-12a	0.14567	191.8	19.0	10.1	50.72	0.5	0.41226	0.00029	334.7	59.2
37-12b	0.09187	119.3	7.0	17.1	87.17	0.5	0.54982	0.00061	289.3	47.7
37-12v	0.06896	203.4	13.3	15.3	78.00	0.5	0.52682	0.00058	305.7	59.8
37-12g	0.16664	191.3	15.7	12.2	61.43	0.5	0.45941	0.00108	322.3	61.2
Zelenaya Griva, Bh. F-233										
F-233-10	0.05475	255.1	32.9	7.8	38.73	0.5	0.35628	0.00039	351.7	54.5

See note to Table 7

3 Results and Discussion

Injection time of three main mineralized intrusions in the Norilsk district is practically the same. Natural reference points limiting the injection event are reliable ages of felsic volcanics (rhyodacites)—270 ± 3 Ma, U–Pb, zircon, SIMS SHRIMP-RG (lower limit) and the age of conjugate mafic trap lava flows—249 ± 2 Ma, Ar–Ar (upper limit). Formation of magmatic zircon generations in each of the studied intrusions in the area reflects the crystallization time of silicate portion of the geological objects.

Two consecutive episodes (pulses) of injection and crystallization of silicate melt can be distinguished on the basis of the data obtained: early, 254 ± 4 Ma, and later, 244 ± 4 Ma. The total duration of magmatic activity of this type is about 10 million years.

Sulphide matrix (mineralization process) is almost synchronous to the silicate one; its age is 245–250 Ma according to Re–Os ID-TIMS method.

Powerful secondary alteration processes with preferential removal of Th and active migration of U and REE developed at about 220–230 (225 ± 5) Ma. Against the background of active recrystallization (both partial and complete) of the matrix of primary magmatic generations, proper metasomatic generation of zircons was formed, confined to the expressed zones of metasomatic reworking of rocks within intrusions.

Age of the secondary (alteration) processes manifested in all intrusions is the same (220–230 Ma) and exactly matches the time of plagiogranite massif injection into the Norilsk district rocks during tectonic and magmatic activity, for example, Bolgotokh, 229.0 ± 0.4 Ma (U–Pb, zircon) and universal biotite development with age of 225–230 Ma ($^{40}Ar/^{39}Ar$).

Age values in the range of 260–270 Ma are overstated and reflect redistribution and migration of excess uranium. U–Pb system of such zircons is often characterized by reverse discordance. The presence of inherited zircon microdomains playing a role of seed cores during crystallization is also possible.

There is a clear trend towards an increase in the number and variety of inherited zircons from the Kharaelakh (maximum) to the Talnakh (minimum) intrusion. The inherited Variscan zircons (P-C) are clearly distinguished by a significantly lower uranium content.

Large number of xenogenic Permian and Carboniferous (290, 300, 330, and 350 Ma) zircons in mafic rocks of the Norilsk district intrusions have been revealed. No Devonian (360–420 Ma) zircons have been found. These zircons reflect the age and composition of crustal rocks hosting intrusions, formations, from which they are trapped during injection as well as indicate their degree of assimilation. Individual Proterozoic and Archean zircon grains aged at 1.9 and 2.7 Ga (in the Norilsk and Pyasino-Vologochan intrusions) have been detected; this indicates the presence of ancient basement in the Norilsk district.

Mafic igneous rocks formed at the turn of the Jurassic and Cretaceous, 145–150 Ma, are first identified. Their genetic relationship with the complex of the Norilsk mineralized intrusions is not determined.

4 Geochronology Isotope Mineralization Criterion and Metallogenic Consequences

Silicate matrix of intrusions in both pulses (especially the early one), along with other lithophylic elements is enriched in uranium, whose significant amounts may be localized in intrusion areals, which was reflected in extremely high concentrations of this element in primary magmatic zircons (thousand ppm at a standard of hundred ppm) and high (>2) Th–U ratio. This indicates a high probability of the presence of uranium mineralization in the spatial and genetic association with mafic intrusions.

Search and study of xenogenic zircons is no less important than dating of igneous populations. Also of interest is the study of detrital zircons from the host sedimentary rocks and directly from sulphide ores.

Xenogenic zircons reflect the age and composition of crustal rocks, from which they are trapped during intrusion injection, and their number also indicates the degree of rock assimilation.

In our opinion, it requires a special comparative study of xenogenic zircon population in various intrusions and ores with a view to identifying the direct connection between PGE-sulphide mineralization and a specific horizon of assimilated Paleozoic and/or Precambrian host rocks. Such rocks can claim as an additional source of mobilized and redeposited useful component (lateritic crust, paleo-placers etc.).

Productive intrusions belong to the earlygroup; their injection initiated migration and accumulation of components (PGE, sulphur, uranium).

The most promising intrusions in the Norilsk district:

- magmatic generation of accessory zircons is no younger than 250–255 Ma, i.e. when crystallization of silicate part slightly preceded the formation of sulphide (ore) part (Re–Os isochron method);
- maximum component of xenogenic (captured during injection) Paleozoic zircons, indicating a substantial assimilation of host rock substance of this age, is revealed.

Secondary Late Triassic (225–230 Ma) alterations do not have a negative impact on commercially valuable sulphide ore stocks.

The absence of large amounts of sulphide ores in the late group intrusions (240–247 Ma), even in the presence of similar to the early group intrusions evidence of the Paleozoic rock assimilation (zircon xenocryst), points to the injection of the late group intrusions into enclosing rocks already depleted in ore-forming components.

It is surprising, that the relationship between hafnium isotopic composition and ore content is preserved for zircons, which are 20 million years younger than the ore material. Obviously, the isotopic characteristics of silicate material that is not related to mineralization may not reflect the extent of ore content.

Direct age determination of sulphide ores by Re–Os isochron method allows us to understand the geological interpretation of U–Pb data in zircons. Re–Os data with a surprising constancy match age of about 250 Ma for all the studied intrusions. This correspondence reflects a high degree of almost synchronous sulphide Cu–Ni ore formation with the time of magmatism (injection and crystallization of silicate intrusion melt), leaving no grounds for assumptions about the ore system development for tens or hundreds of millions years prior to this stage.

The secondstage dated by zircon at 230 Ma is not associated with sulphide ore formation and reflects the age of the imposed process. Its geological essence requires further study, but is most likely to be identified with Co–Ni–Sb–As vein mineralization of essentially another type manifested in the same massifs.

Thus, commercially mineralized intrusions of Norilsk-1, Talnakh, and Kharaelakh formed in the Norilsk ore district at the early magmatism stage at 254 $\pm$ 4 Ma, likely due to a deep mantle plume source. Abundance of fluid phase and a high mole fraction of water in the fluid, on the one hand, led to the formation of horizons with low-sulphide Pt–Pd mineralization, on the other hand, a high capacity for assimilation of crustal material, which contributed to the formation of massive sulphide ores. Multiple isotope systematics of radiogenic Sr, Nd, Hf, and Pb isotopes indicates that the crustal material was intensely drawn into rock and ore formation in the Norilsk ore district.

Later stage of deep magmatism at 244 $\pm$ 4 Ma in the traditional isotopic and geochemical classification of mantle material was accompanied by active participation of depleted mantle source, resulting both in a smaller number of exceptionally disseminated Cu-Ni sulphide ores and, probably, in the lack of economically valuable platinum group metal deposits.

References

1. Lyakhnitskaya IV, Tuganova EV (1977) Regional and local patterns of copper-nickel sulphide deposits distribution. L.: Nedra, 77 p
2. Malich NS, Masaitis VL, Staritsky Yu G (1974) Geologic formations of the pre-cenozoic cover of the Siberian platform and its mineralization. M.: Nedra, 280 p
3. Arndt NT, Czamanske G, Walker RJ et al (2003) Geochemistry and origin of the intrusive hosts of the Noril'sk-Talnakh Cu-Ni-PGE sulphide deposits. Econ Geol 98:495–515
4. Black LP, Kamo SL, Allen CM et al (2003) TEMORA 1: a new zircon standard for U–Pb geochronology. Chem Geol 200:155–170
5. Birck JL, Barman MR, Campas F (1997) Re–Os isotopic measurements at the femtomole level in natural samples. Geostandards Lett 20(1):19–27
6. Campbell IH, Czamanske GK, Fedorenko VA et al (1992) Synchronism of the Siberian traps and the Permian-Triassic boundary. Science 258:1760–1763

7. Czamanske GK, Gurevitch AB, Fedorenko V, Simonov O (1998) Demise of the Siberian plume: Paleogeographic and paleotectonic reconstruction from the prevolcanic and volcanic record, north-central Siberia. Int Geol Rev 41:95–115
8. Czamanske GK, Wooden JL, Walker RJ et al (2000) Geochemical, isotopic, and SHRIMP age data for Precambrian basement rocks, Permian volcanic rocks, and sedimentary host rocks to the ore-bearing intrusions, Noril'sk-Talnakh district, Siberian Russia. Int Geol Rev 42: 895–927
9. Dalrymple GB, Czamanske GK, Fedorenko A et al (1995) Areconnaissance $^{40}Ar/^{39}Ar$ geochronological study of ore-bearing and related rocks, Siberian Russia. Geochim Cosmochim Acta 59:2071–2083
10. Dalrymple GB, Czamanske GK, Lanphere MA et al (1991) $^{40}Ar/^{39}Ar$ ages from samples from the Noril'sk-Talnakh ore bearing intrusions and the Siberian flood basalts, Siberia. EOS 72:570
11. Kamo SL, Czamanske GK, Krogh TE (1996) Aminimum U–Pb age for Siberian flood-basalt volcanism. Geochim et Cosmochim Acta 60:3505–3511
12. Kamo SL, Czamanske GK, Amelin Y et al (2003) Rapid eruption of Siberian flood-volcanic rocks and evidence for coincidence with the Permian-Triassic boundary. Earth Planet Sci Lett 214:75–91
13. Ludwig KR (2001) SQUID 1.02, A User Manual, A geochronological toolkit for microsoft excel. Berkeley Geochronology Center Special Publication, Berkeley, USA
14. Ludwig KR (2003) User'sManual for Isoplot/Ex, Version 3.00, AGeochronological Toolkit for Microsoft Excel. Berkeley Geochron. Center Spec. Publ. Berkeley, USA
15. Renne PR (1995) Excess 40Ar in biotite and hornblende from the Norilsk I intrusion, Siberia: implications for the age of the Siberian Traps. Earth Planet Sci Lett 131:165–176
16. Schuth S, Gornyy VI, Berndt J et al (2012) Early proterozoic U–Pb zircon ages from Basement Gneiss at the Solovetsky Archipelago, White Sea, Russia. Int J Geosci 3(2): 289–296
17. Stacey S, Kramers JD (1975) Approximation of terrestrial lead isotope evolution by a two-stage model. Earth and Planet Sci Lett 26:207–221
18. Walker RJ, Morgan JW, Horan MF et al (1994) Re–Os isotopic evidence for an enriched mantle source for the Noril'sk-type ore-bearing intrusions, Siberia. Geochim et Cosmochim Acta 58:4179–4197
19. Wetherill GW (1956) Discordant uranium-lead ages. Trans Amer Geophys Union 37:320–326
20. Wiedenbeck M, Allé P, Corfu F et al (1995) Three natural zircon standards for U–Th–Pb, Lu–Hf, trace element and REE analyses. Geostand Newslett 19:1–23
21. Williams IS (1998) U–Th–Pb geochronology by ion microprobe. In: McKibben MA, Shanks III WC, Ridley WI (eds) Applications of microanalytical techniques to understanding mineralizing processes. Rev Econ Geol 7:1–35
22. Wooden JL, Czamanske GK, Fedorenko VA et al (1993) Isotopic and trace-element constraints on mantle and crustal contributions to characterization of the Siberian continental flood basalts, Norilsk area, Siberia. Geochim Cosmochim Acta 57:3677–3704

Conclusion

Sources of rock and ore substance. Complex isotopic criterion for mineralization of mafic intrusions in the Norilsk District. Integrated isotopic studies of the Norilsk-Taimyr district intrusions held at CIR VSEGEI allowed us to clarify and specify a number of questions on the origin of rock and ore substance, formation of Cu-Ni-PGE-deposits. Particular attention was paid to the diagnosis of substance sources by isotopic genetic markers, as well as rock and ore age determination, duration of geological processes using isotopic techniques. In addition, the isotopic criteria for mineralization scale of ultramafic intrusions were sought.

Almost all isotopic indicators pointed at an extended participation of crustal material in the ore genesis. For example, the composition of helium, an effective indicator of fluid "mantle nature", incorporates a very high proportion of the Earth's crust helium. Paleofluids from rocks and ores are dominated by helium with a low $^{3}He/^{4}He$ isotope ratio, formed only in the Earth's crust rocks. In commercially mineralized (rich) intrusions, the share of mantle helium does not exceed 4%. Low-ore intrusions contain much more mantle helium, to 22%.

Ore mineral sulphur in rich intrusions is only crustal. Its isotopic composition ($\delta^{34}S$) is very different from that in most cases and similar to that observed in anhydrite of enclosing sedimentary rocks. In accordance with the pronounced correlation links in the isotopic composition of sulphur, helium, and argon, in rich intrusions one can assume the presence of two crustal sources. The temperature of ore mineral formation is estimated by sulphur isotopic geothermometers at a few hundred degrees.

Set of data on the isotopic composition of strontium and neodymium clearly indicates a significant influence of the crustal component on the injected mantle fluid. Computation of crust and mantle contribution in this two-component system, unfortunately, is impossible due to an uncertain knowledge of crustal isotopic coordinates.

Features of the isotopic composition of lead, lutetium–hafnium system elements also indicate an intense crust–mantle interaction. Apparently, zircon source rocks with high Lu/Hf and Yb/Hf also had high values of these ratios. It is assumed that

O. Petrov (ed.), *Isotope Geology of the Norilsk Deposits*, Springer Geology,
https://doi.org/10.1007/978-3-030-05216-4

these formations were phosphate-bearing, clayey and evaporite rocks providing fluids with halogens. These rocks are adequately presented in the sequence of enclosing sedimentary beds.

It can be reasonably assumed that the contribution of crustal material is great, especially in ores, and that this substance is not only a passive content, but may provoke ore genesis.

Not less important information has been revealed about the other isotopes of another noble gas, argon, an indicator of underground fluid communication with the atmosphere. Mafic rocks of Norilsk, especially ores, are dominated by argon of atmospheric origin, which indicates an extremely high contribution of subsurface water from rocks enclosing intrusions into mineralizing fluid formation. A particularly high proportion of atmospheric argon (88–100%) is in rich intrusions. It is typical of gas from shallow sedimentary strata (1–2 km).

Apparently, ore deposition was accompanied/caused by intense water circulation in sedimentary strata, penetrating into the zone of intrusion crystallization. Circulation was obviously initiated by magma injection into the sedimentary layers. Water streams could carry ore components and deposit them at the postmagmatic stage. In case of sufficient mineral wealth and water volume involved in this process, one may be able to solve the problem of copper balance in deposits and source rocks.

Clear correlation of helium and argon isotope ratios in three richest intrusions has been revealed; the presence of, at least, two different fluid sources is admitted.

Copper and lead isotopes of ore minerals from these intrusions also exhibit a correlation with the composition of noble gases and sulphur. Apparently, these metals from sulphides also owe their origin to the sources mentioned above.

Nickel isotopic composition variations are not related to those in noble gases, copper, lead, and sulphur. The immediate source of nickel is, probably, mantle protoliths of mafic rocks.

Isotope systematics indicates that ore formation is directly related to the assimilation by mantle magmas of enclosing sedimentary rock substance (including fluids). This follows from the isotopic characteristics of noble gases, sulphur and copper, neodymium, strontium, and hafnium. It is hoped that the "isotopic" arguments will be used in the construction and testing of deposit formation models for the Norilsk district. Academician A.P. Vinogradov, an enthusiast and organizer of isotope geochemical studies in our country, called for developing this very approach to grading of existing models and choice of the most appropriate ones.

Age limits for formation of intrusions differing in mineralization scale, as well as of sulphide ores themselves have been established. We received a large amount of data about the time of geological processes on the basis of uranium-lead local dating of accessory zircons from a wide variety of rocks. Magmatic generation of accessory zircons is aged at 250–255 Ma. This means that crystallization of the silicate part of material slightly preceded the formation of the sulphide (ore) part (245–250 Ma). The presence in rocks of productive intrusions of xenogenic (captured during injection) Paleozoic, Proterozoic, and Archean zircons is also important. This, in turn, suggests a substantial assimilation of host rock matter of these ages. Secondary Late Triassic (225–230 Ma) alterations did not have a negative impact on commercially valuable sulphide ore stocks.

Absence of large amounts of sulphide ores in intrusions of the second (late) group (240–247 Ma), even given assimilation evidence of the same Paleozoic and older rocks (xenocrysts zircon) similar to intrusions of the early group, points to the injection of the late group intrusions into enclosing rocks already depleted in ore-forming components.

Direct age determination of sulphide ores by Re–Os isochron method allows to understand the geological interpretation of U-Pb data in zircons. Re–Os data with a surprising constancy match age of about 250 Ma for all studied intrusions. This correspondence reflects a high degree of almost synchronous sulphide Cu–Ni ore formation with the time of magmatism (injection and crystallization of silicate intrusion melt), leaving no grounds for assumptions about the ore system development for tens or hundreds of millions years prior to this stage.

Stage dated by zircon at 230 Ma is not associated with sulphide ore formation and reflects the age of the imposed process. Its geological essence requires further study, but is most likely to be identified with Co–Ni–Sb–As vein mineralization of essentially another type manifested in the same massifs.

In the course of studies, we have identified significant differences in the isotopic characteristics of intrusions with varying degrees of ore content. These differences can be seen in the summary table; they can be involved as isotopic criteria for the scale of intrusion mineralization. Isotopic characteristics are ranked according to the degree of reliability. Of the three groups, the greatest predictive capabilities provide data on the isotopes of noble gases, helium and argon.

We offer a comprehensive isotopic criterion for mineralization of the Norilsk type intrusions:

Degree of reliability	Good	Good	Average	Weak	Weak
Element	He	Ar	S	Cu	Pb
Isotope ratio	**m, %**	**a**, %	$\delta^{34}S$, ‰	$\delta^{65}Cu$, ‰	$^{206}Pb/^{204}Pb$ etc.
Rich	1–4	90–100	9–13 **f** (He, Ar)	-2…0 **f** (S, He)	17.90–18.35 **f** (S)
Satellites	0.3–0.7	84–87	5–7	–0.6…–0.3	17.95–18.3
Average	0.5–4	50–80	0–8	–0.8…–0.4	18–15–18.70
Poor	5–22	70–90	0–8	–0.8…–0.4	18.15–18.40

m—share of mantle helium, **a**—share of atmospheric argon, **f**—correlation of isotopic parameters

The most promising are intrusions aged at 250–255 Ma with a predominance of atmospheric-crustal argon and helium, with the presence of excess radiogenic hafnium δHf(T) > 5, isotopically heavy sulphur, and fractional copper from sedimentary rocks. The presence of the pre-Triassic zircon xenocrysts is also important. The research results have important theoretical and practical value and can be effectively applied during forecasting and assessment work in the Norilsk district.

Improvement of measurement techniques and interpretation of results are required, including the development of a rational complex of isotopic measurements as well as carrying out practical assessment of mafic intrusion mineralization in the Norilsk-Taimyr district.

Completeness of the collected and prepared material allows in-depth treatment of the following specific tasks:

- identify association between the obtained isotopic ages and specific geological events observed in a geological site;
- binding dated growth zone of mineral-geochronometer to a complex development stage for local dating methods (ion probe or laser ablation);
- computations of the duration and direction of rock evolution, indicative minerals, ore material, duration and thermodynamic regime of ore stage, cooling or rock uplifting rate (geochronology, Cu, Ni);
- quantitative assessment of the magnitude and time of crust-mantle interaction as the major ore-controlling factor (isotopic composition of lead, hafnium, sulphur);
- assessment of ore-controlling zones depth (by He and Ar systematics);
- interpretation and comparison of isotopic dating results of ore and ore-bearing matrix material (Re–Os, Rb–Sr, Sm–Nd systematics);
- assessment of the extent and magnitude of fluid—rock interaction as the major ore-controlling factor (isotopic composition of oxygen, carbon, sulphur).

The authors would like to thank E.V. Tuganova, K.N. Malich, and I. Yu. Badanina for a great help in getting the unique materials for investigations.

MIX
Papier aus verantwortungsvollen Quellen
Paper from responsible sources
FSC® C105338

If you have any concerns about our products,
you can contact us on
ProductSafety@springernature.com

In case Publisher is established outside the EU,
the EU authorized representative is:
Springer Nature Customer Service Center GmbH
Europaplatz 3, 69115 Heidelberg, Germany

Printed by Libri Plureos GmbH
in Hamburg, Germany